Making Sense of Microsoft Access

Making Sense of Microsoft Access

First Edition

John P. Herzog

Wittenberg University

Bassim Hamadeh, CEO and Publisher
Jennifer Codner, Senior Field Acquisitions Editor
Michelle Piehl, Senior Project Editor
Abbey Hastings, Associate Production Editor
Jess Estrella, Senior Graphic Designer
Stephanie Kohl, Licensing Associate
Natalie Piccotti, Director of Marketing
Kassie Graves, Vice President of Editorial
Jamie Giganti, Director of Academic Publishing

2970 Sorrento Valley Blvd., Ste. 600, San Diego, CA 92121

Contents

Chapter 4   Making Easier Forms to Enter Data   181

Chapter 5 Making a Main Menu for Transaction Data Entry and Reports 253

Chapter 6 Viewing and Printing Data of a Given Record 293

Chapter 7   Creating Show-All Queries, IIF Expressions, Pop-Ups, Option Groups, and Conditional Macros   317

Chapter 8   Document Management   349

Epilogue   369

Index   371

Introduction

WHAT IS A DATABASE?

A **database** is a computer program that allows you to keep track of people or things; people such as customers, donors, students, patients, and employees; things such as purchases, donations, classes, emergency room visits, labor hours, inventories, and much, much more. When you want to, you can then run **queries** and **reports** regarding all of those people and things in order to be able to view important information about them. **Queries** and **reports** make it much easier to accomplish routine tasks, such as tracking sales, sending invoices, sending thank-you letters to customers and donors, creating payroll checks, and making lists of all of these things when and where needed. **Databases** also give you forms in order to make it easier to enter such data into the database in a way that is user friendly.

WHY DO WE NEED A DATABASE?

Many people believe the myth that if they use word processors and spreadsheets that they are fully automated. Not so!

Some may say that lists of things can be entered into a spreadsheet and that databases are nothing more than lists and thus are not needed. While spreadsheets are extremely valuable, there are several reasons that databases can be far superior to spreadsheets.

First, in a database, one row of data can be converted into a letter-sized document without having to retype it. For example, if you are making a list of your customers' purchases in a database, you can take each row of that list and use it for sending out statements to each of those who made the purchases. The statements can include very attractive and easy-to-read information. However, if you enter all of that data into a spreadsheet, you would have to retype all of the information into a Word document or something that is similar, even though you had already typed into the spreadsheet. Typing the same information twice or more is called **data redundancy**. It is time consuming, inefficient, and costly. In many cases it is also less accurate. When a person types the same information twice, there will almost always be differences and inconsistencies.

Some argue that spreadsheet information can be used in what is called a **mail merge**, where the names from a list are put into letters. That is true, but mail merges take several steps. They are not always a quick, easy, or accurate way to create such documents, especially if those documents must be sent out regularly. A database will merge data routinely by the click of the mouse and in one step.

Second, still using the example of the recording of purchases, if you enter customers' purchases into spreadsheets, along with customers' addresses and telephone numbers, each time they come in to make another purchase, you would have to retype that same information. Once again, you would have *data redundancy*, a lack of consistency and a lack of accuracy. Some argue that copying and pasting can be done to solve this problem in a spreadsheet, but copying and pasting is often done incorrectly and can still take a great deal of time, especially if your list holds thousands of lines.

Third, as mentioned earlier, screens for entering data into a database can be created in a way that is easier to use and that makes data entry much faster. Such screens will be shown to you in this book as well as the procedure for creating them.

Fourth, databases have the ability to make sure that there are fewer data-entry errors in your organization. They do so with things such as **data validation** and **referential integrity**. **Data validation** is also available in spreadsheets. For example, it makes sure that no one will enter 12/88/2018 into a database or spreadsheet as a date, because there is no such date.

Referential integrity, on the other hand, prevents people from entering bogus data in another sense. Suppose, for example, that you are entering sales transactions into a table and you enter the purchaser's *ID* number. If you enter the wrong *ID* number of someone who does not exist in your purchasers' files, referential integrity will stop you. **Referential integrity**, in this example, would permit your computer to check to see if an entered purchaser is actually a bona fide customer. **Referential integrity** (often called **one-to-many relationships**) will be discussed in much greater depth later in this textbook. It is vital, and it cannot be done in a spreadsheet.

Sixth, sometimes more than one person needs to use your data for a different purpose. In manufacturing, for example, when you enter an order, it must also be shared with employees who work in accounting, production, and shipping in order to create production schedules, bills of lading, packing lists, invoices, and much more. Without a database, all of the data for these things would have to be entered each time they are needed. More time would be wasted and there would very likely be many inconsistencies. Figure 0.1a gives an example of how inefficient a spreadsheet would be in comparison to using a database in manufacturing. Figure 0.1b shows how much more efficient a database would be. Another example would be the legal profession.

The data about a client is sometimes needed for many different documents such as a complaint, a motion, correspondence letters, docket information, court data, and much more. Figures 0.1c and 0.1d give an example of how inefficient a word document would be in comparison to using a database in the legal profession. This is often called **document management** (which will be discussed in chapter 8).

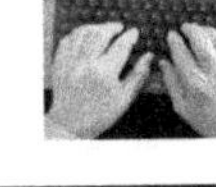

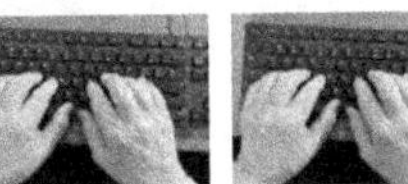

FIGURE 0.1a   Manufacturing data entry: The slow way example

FIGURE 0.1b   Manufacturing data entry: The database way example

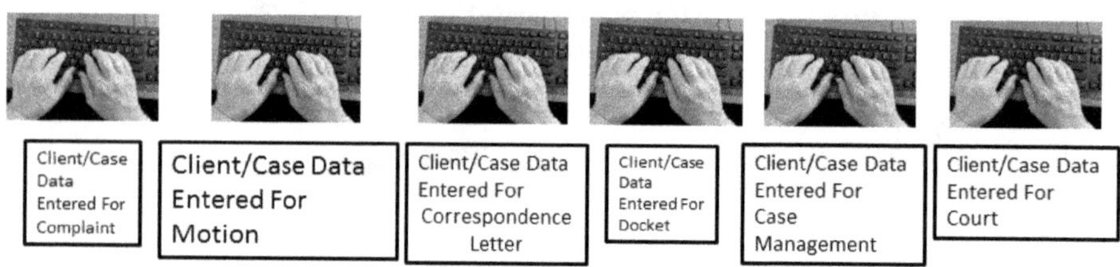

FIGURE 0.1c   Legal firm data entry: The slow way example

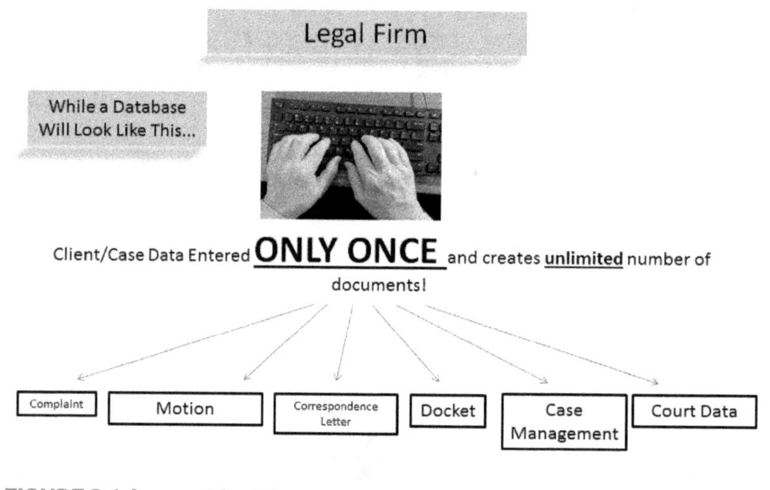

FIGURE 0.1d   Legal firm data entry: The database way example

A database would make it so that data would only need to be entered once. **Reports** would then be created. When you run **Reports**, it will make all of these documents read that data within seconds, by a simple click of the mouse.

TYPES OF DATABASES

There are two major types of databases. The first is called a **flat database**. Such a database is almost the same as what you can create in a spreadsheet. An example might be if you are just making a list of those who attended a given event and there is very little chance that the list will be used more than once or twice. The other is called a **relational database**. That is the kind we will be creating in this book. A **relational database** removes the need for **data redundancy**. The process for this will be explained in much depth later in this textbook.

ACCESS DATABASE OBJECTS AND THEIR CAPACITIES

Microsoft Access databases are what we will be using in this book. Decades ago, when **Access** was first introduced, tables could not hold more than 20,000 to 30,000 **records** (or lines/rows of data) and the database could accommodate only one user at a time. Today, Access can have an unlimited number of records and an unlimited number of users. While more records and users can slow down the speed of the database, the capabilities are still far greater than years ago.

Each **record**, such as a customer, an item of inventory, or whatever you are tracking, is comprised of **fields**. **Fields** are different attributes of a record. For example, a field could be a person's first name, last name, address, date of birth, and much more. In the case of tracking inventory, a field could be an identification number for each of the stock items, their cost, the number of them on hand, and more. **Fields** are like and are often called **column headings** as are used in spreadsheets.

Records are stored in what is called a **table**. **Tables** are the most basic part of a database that hold its information. Without **tables**, you have no database. In Access, you can have no more than 255 **fields** in a table.

Queries are used to extract given information from a **table**. They can do so from two perspectives: removing unnecessary fields and removing unnecessary records. For example, suppose you have a table that contains customers' information (such as their addresses, telephone numbers, and more). Also suppose you only need to see their names and telephone numbers. **Queries** will let you choose only the fields that you need.

Also, in some cases, you may only want to see customers of a certain area code. **Queries** will allow you to extract only the records of the customers with such area codes. Hence, you can hide unneeded data from **queries** by reducing the number of **fields** and **records** that are not necessary to view at a given time. You can have no more than 255 **fields** in a **query**.

Forms are created to read the tables you create and make the data entry easier to enter (or more user friendly).

Reports allow you to present the information you need in a way that is easier to read and, in some cases, in a way that is required by certain users. For example, if you are tracking payroll information, you can create a **report** that will print payroll checks for you. There are some government agencies that require a report to have a certain look and format. **Reports** can be tailored to whatever format you need in showing the information stored in a database.

Macros, among other things, allows you to open and close these objects. **Macros** can be attached to **command buttons** that will make it easier to run them in order to carry out these tasks.

Tables, **forms**, **queries**, and **reports** are examples of what are referred to as database objects. There is no limit to how many objects you can put into your database, unless of course your computer simply runs out of main memory or storage space.

There is no limit to how many users can use an Access database **SIMULTANE-OUSLY**. However, if the speed and main memory of your network and computers used to run the database are limited, the database may become slower, and can in fact become corrupt (or unusable). In cases such as this, it is best to create copies of the database with the **forms**, **reports**, **queries**, and **macros**. They would be called **front-end databases**. The copied databases should be placed on the computer of each user and then linked to the tables of the database in one central file. The central file would be called the **back-end database**. **Linking** is also discussed in a later chapter.

WHAT IS NEEDED TO MAKE AN ACCESS DATABASE?

Before you even begin to open Access and create a database, there is a lot of preliminary work to be done. Building a database, just like building a house, needs a plan.

The first thing you must do is ask the users or potential users of the database what kind of data they need to capture. In other words, **THEY MUST TELL YOU WHAT FIELDS ARE NEEDED IN THE TABLES**. For example, suppose you are building a database for a hospital emergency room. You would need to sit down and discuss with the emergency room staff what kind of data they might need. Fortunately, Access, as well as most other databases, has the flexibility to add fields and other database attributes as an organization's needs change. For example, years ago, databases didn't track information such as mobile phone numbers or email addresses, because many people didn't have them. Such fields have since been added to many databases. As society changes and as we begin to need to track more things, fields can quickly and easily be added.

Second, you must figure out what parts of your tracked data will be **static** and what parts will be **dynamic**. **Static data** changes very seldom, while **dynamic data** changes often. Let's return to the example of the emergency room. For example, if your emergency room database is tracking visits of patients who come there, the patients' names, addresses, allergies, and family physicians will not necessarily change each time they visit. On the other hand, the dates and reasons for the visits and actions taken would each change. Therefore, the patients' names, addresses, allergies, and family physicians' data would be considered **static**, while the visit information would be considered **dynamic**. Figure 0.2 is an example of fields that your table may require in order to capture needed information as patients enter the ER.

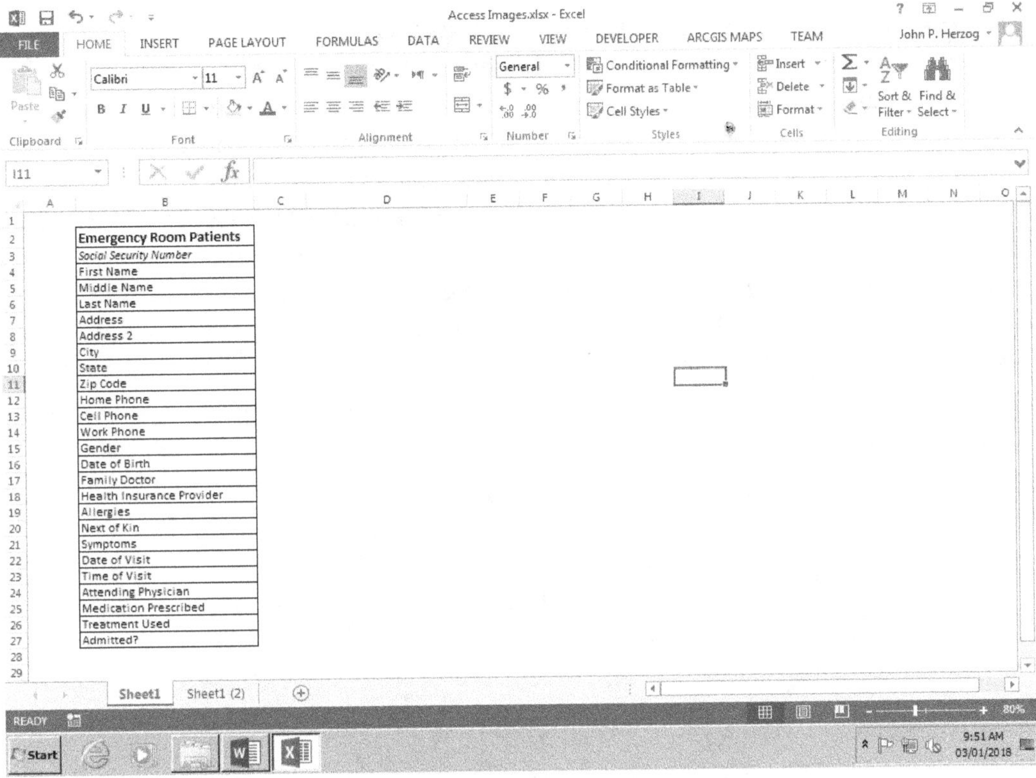

FIGURE 0.2 Emergency room visit example

The problem with putting this data into a spreadsheet or using a flat database is that you would have to type the static data each time the same person came to the ER. Once again, that would cause you to have data redundancy. That is what separates databases from spreadsheets. To save a great deal of time, the table you see in figure 0.2 should be split into two related or **relational tables**: one of them holding the *Emergency Room* data, and the other holding the data regarding each *Emergency Room visit*. Then each *patient* should be given an ID number, perhaps their *Social Security number*. The tables would then be designed as you see in figure 0.3. The ID number (or in this case, the *Social Security number*) would be called the **primary key** field. In almost every case a table needs a **primary key** field. If you have two *Jane Smiths* as *patients*, the **primary key** field will determine which one is which. You cannot have two records in a table with the same **primary key** data. **YOU CANNOT LEAVE THE PRIMARY KEY BLANK IN A RECORD!**

Notice that the *Social Security number* is then added to the table with the *Emergency Room visits* data. It is the only thing that you would need to enter into the *Emergency Room visits* table regarding the *patient*, rather than having to reenter all of their other data. It would then link up with the *patients'* static data. That field then is the way we can link to the *Emergency Room patients* table to know the information related to the *Social Security number* that is entered. That field is then called the **foreign key** (figure 0.3). There could still be changes to some of the *Emergency Room patients'* data, such as addresses and telephone numbers, but those things don't necessarily change between each ER visit. This relational table system stops the need for typing in patient data at every visit. **THAT IS A HUGE DECREASE IN REQUIRED DATA ENTRY.**

One thing to note is that in recent years there are many who are reluctant to give out their Social Security numbers; therefore, you can set the system up to have it automatically assign the ID numbers without using the same number twice. That is the type of field that would be used as the *visit ID* field in the *Emergency Room visits* table. Many organizations have changed from using Social Security numbers to using telephone numbers.

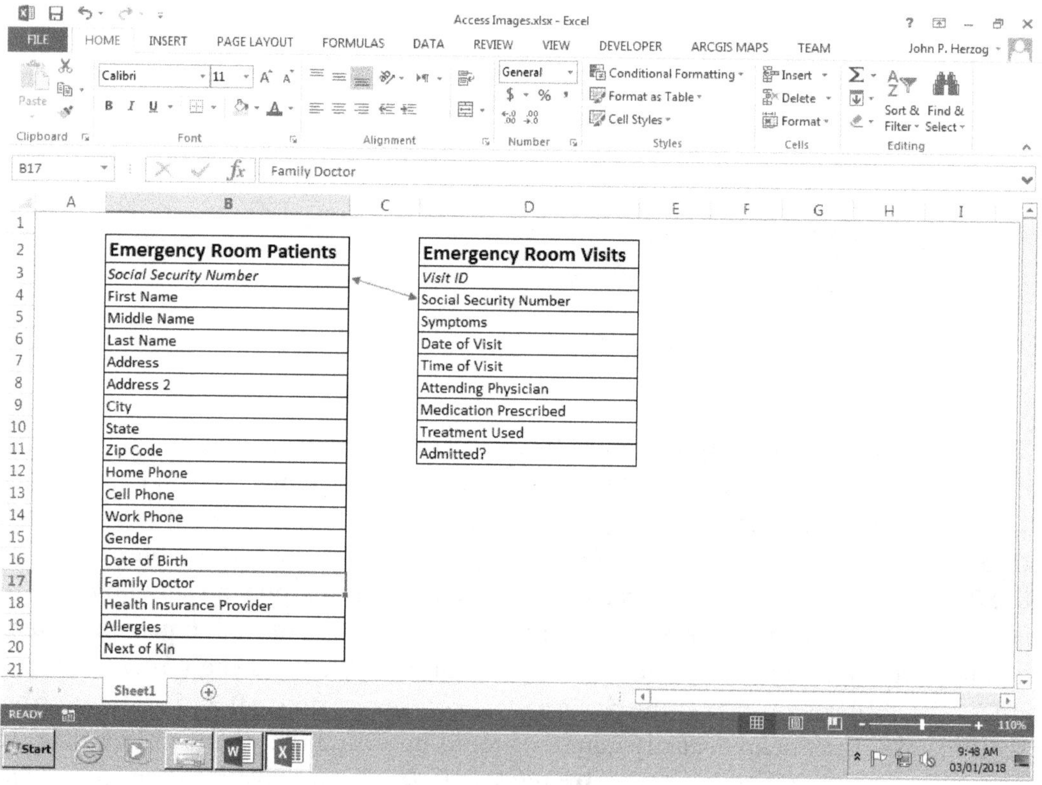

FIGURE 0.3   Emergency Room visit example with split tables

Is it possible for one patient to have more than one emergency room visit? Absolutely. But no hospital visit would ever be attributed to more than one patient. At the same time, you could ask yourself, can a mother have more than one child? Yes, but no child can have more than one biological mother. That is why in this example, the

Emergency Room patients table would be called the **parent table**, because one *patient* can have many *Emergency Room visits.* The *Emergency Room visits* table would be called the **child table**, because one *Emergency Room visit* could only be attributed to one *patient.* Similarly, if you are relating two tables in an order-processing database, you might have a table called "Customer Information" holding all of the *customer* data that is related to a table called *Orders.* The *Orders* table would hold all of the *order* data. Is it possible for one *customer* to have more than one *order*? Absolutely. But no *order* would be attributed to more than one *customer.* Therefore, in this example, the *customer information* table would be called the **parent table** and the *orders* table would be called the **child table**.

There will be two databases we will use in this textbook. The first will be tracking donations for a homeless shelter and soup kitchen. We will build it together on a step-by-step basis. It will be used for showing you how to build a database and how to apply these concepts. It will track *donors* (the **parent table**) and their *donations* (the **child table**). It will also be the database you will build for your assignments.

The other database will be used to give you a second perspective for creating a database. It will be for a bookstore, tracking *vendors* (the **parent table**) and tracking *inventory orders* (the **child table**) placed to those *vendors.* It will also be the database you will build for an alternate assignment. We will not be using the bookstore database in the chapter discussions.

Creating Tables, Forms, and Reports

IN THIS CHAPTER you will learn how to do the following:

1. Create **tables** from scratch
2. Create **tables** from other sources
3. Enter **data**
4. Create **forms**
5. Create **reports**

CHAPTER 1 TASK: CREATING A NEW DATABASE TRACKING DONORS FOR A COMMUNITY SHELTER AND SOUP KITCHEN

You will be building a database in Access that will track the *donations* given to a *community homeless shelter and soup kitchen*. In order to begin building your database, there are a few things that you need to know.

At this point it is assumed that all of the interrogation of the *community homeless shelter and soup kitchen* staff has been done and the database fields needed are clearly known. It has been found that they receive *donations* to make food available to the poor and to make a shelter available for the homeless. All *donors* can give a monetary *donation* outright or they can make installment payments on *donations*. They can also donate food or other non-monetary items needed.

TO CREATE A NEW DATABASE

Step 1: Open **Microsoft Access** in whatever **Microsoft Office** arrangement you have on your computer. When you do, you have the option of starting a new database from scratch (called a **blank database**), or you can use one that they have already created in the form of a **template** (figure 1.1). **Template databases** are good to get you started in tracking a specific type of data such as event management, school students, asset tracking, task management, inventory, and more. Once you create such a database from a template, you can then modify it to your specific needs. In our case, we will be using strictly new databases made from scratch. Therefore, you will need to create a **Blank Database**.

Step 2: After opening Access, and getting the menu seen in figure 1.1, click **Blank Database** (figure 1.1, arrow).

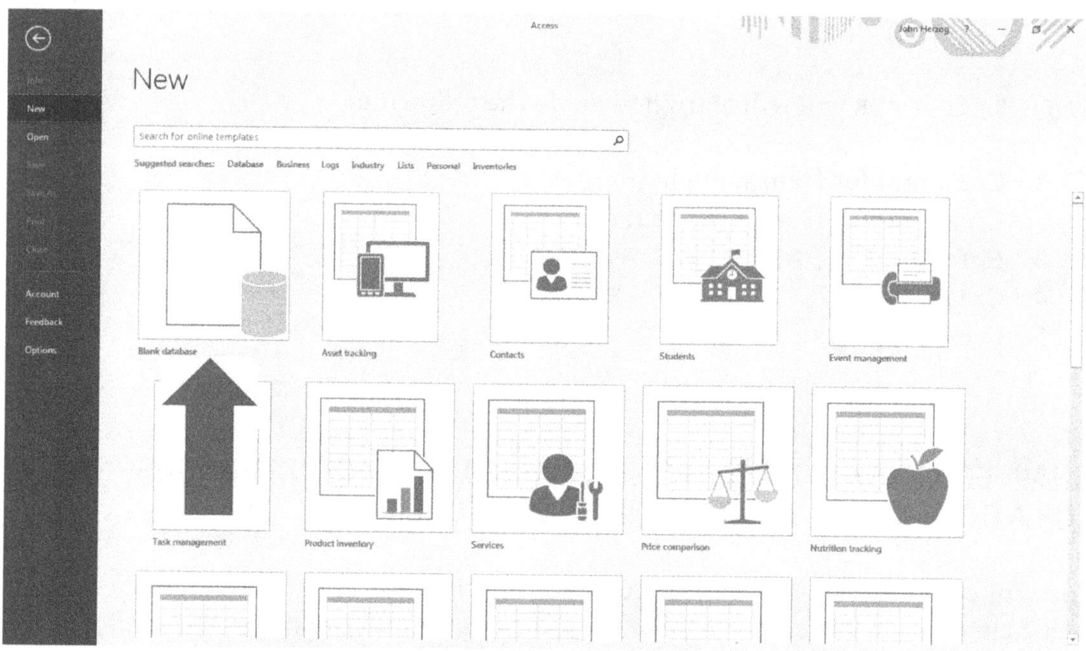

FIGURE 1.1   Access opening menu templates

Step 3: That will give you the menu you see in figure 1.2. When you do, click on the file folder to the right of where it says **File Name** (figure 1.2, arrow).

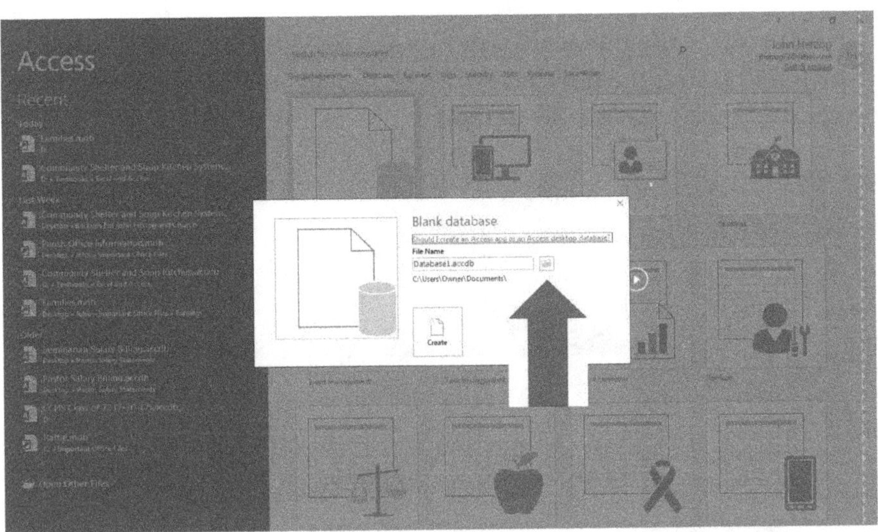

FIGURE 1.2   Folder for saving your database

Step 4: That will give you the menu seen in figure 1.3a that allows you to save the database wherever you wish. Make sure you have a computer drive or a folder location in mind where you can save your database. In our example we will use the **desktop**. Please type the name of the database to the right of the **text box** labeled **File Name:** (figure 1.3a, arrow A). Name it "Community Shelter and Soup Kitchen System" (figure 1.3a, arrow A).

Step 5: Click **OK** (figure 1.3a, arrow B).

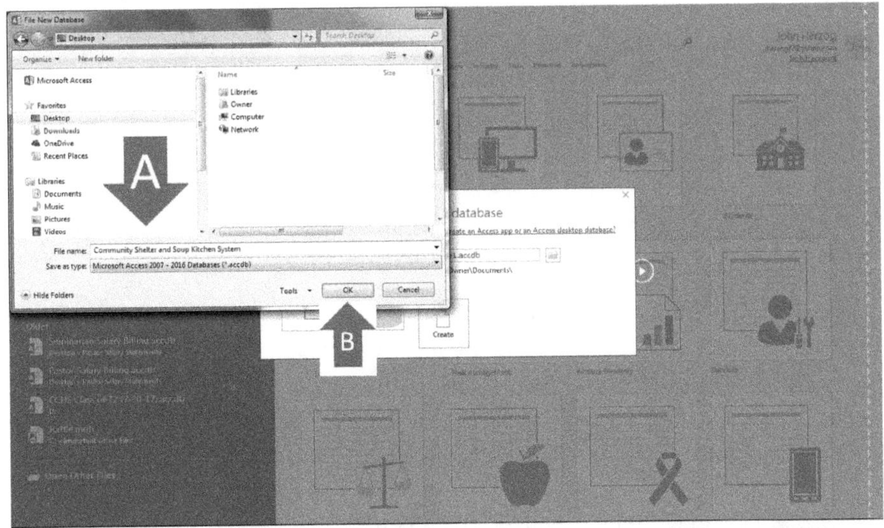

FIGURE 1.3a   Naming your database

That will take you to the menu seen in figure 1.3b. The menu will want you to confirm that you are creating the database with the name and location you have chosen. Step 6: Click **Create** (figure 1.3b, arrow).

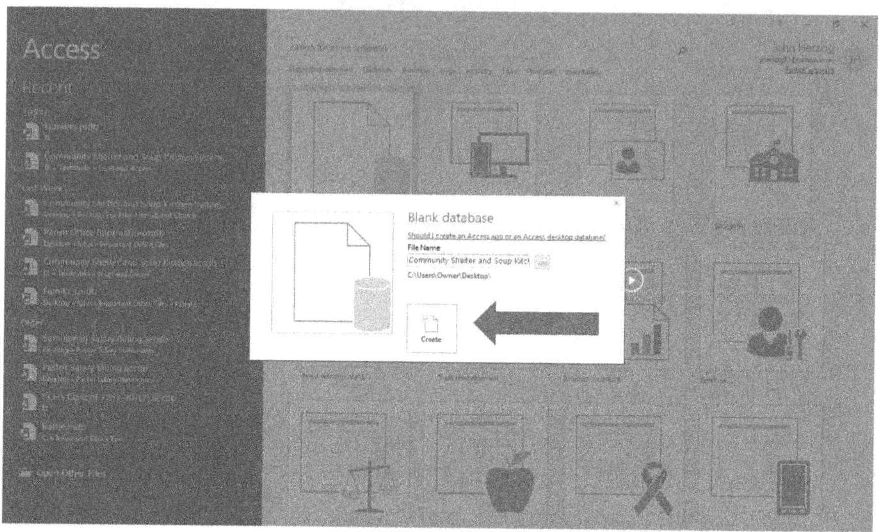

FIGURE 1.3b   Creating your database

That will take you to the menu seen in figure 1.4a.

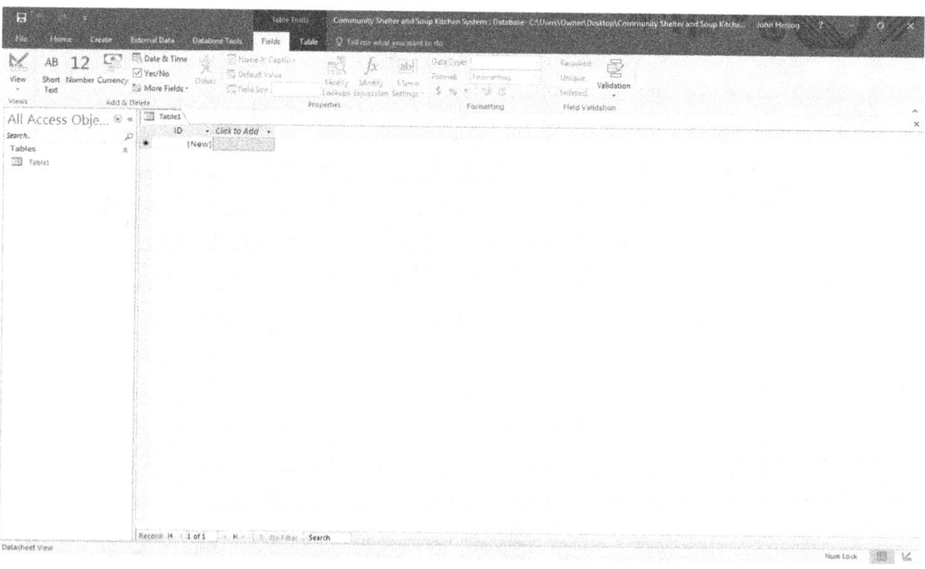

FIGURE 1.4a   First view after creating your database

NOTE: When you create other files in **Microsoft Office**, you can name them after you begin to work on them. Access is different. You must name the database **BEFORE** creating any objects or entering data. This is because the database holds all of the database objects (**tables**, **queries** etc.) in one file.

THE LOCKING FILE AND DESIGN LIMITATIONS WHILE OTHERS ARE USING THE DATABASE

When you create an Access Database, it will automatically add an extension to the file name of **.accdb** (figure 1.4b), which stands for **Access database**.

Another thing that happens whenever you create and/or open an Access database is that it will automatically create what is called a **locking file**. Its extension is always going to be **.laccdb** and it will have a small icon attached to it that looks like a lock (figure 1.4b).

The **locking file** will store the names of all of the computers that are currently using your database. Initially you will be the only user. If your database is on a network and another person opens it while you are still working on it, it will **NOT** allow you to save changes to the **DESIGN** of **forms**, **reports**, or **macros**. It will also prohibit you from changing the **DESIGN** of a **table** if the other current users are viewing that table or if they are viewing a **form**, **report**, or **query** that is reading that **table**. If you attempt to make **DESIGN** changes under those circumstances, an error message will pop up on your screen that will inform you of your inability to do so. When that happens, you will need to locate the other people using the file and ask them to close it in order for you to be able to make **DESIGN** changes. Data can be changed while others are simultaneously using the database. Only **DESIGN** changes are prohibited. The locking file is helpful toward finding those other users, when you open it by double clicking it. It will ask you what program you would like to use to view it. The most typical choice is **Microsoft Word**. At that point it will usually tell you the serial numbers or names of the computers of the other current user or users who have opened the database.

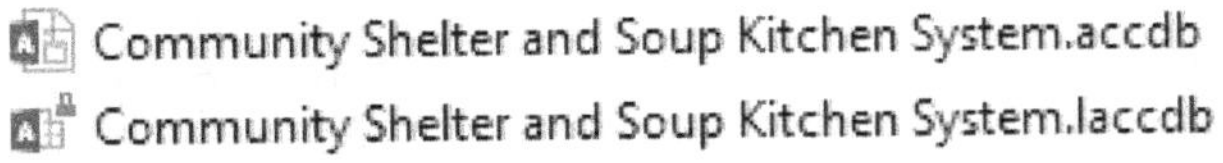

Community Shelter and Soup Kitchen System.accdb

Community Shelter and Soup Kitchen System.laccdb

FIGURE 1.4b　The locking file

TO CREATE A TABLE

Remember, a database is no use if it has no tables. When you build a database, you can create tables from scratch, while other tables can be imported from other sources. We will be doing both of those methods of creating a table in the *Community Shelter and Soup Kitchen System* database. Access gives you the ability to import data from your old sources when you begin to build your new database. This prevents you from having to retype all of the old data into your new database when it has been designed and created. This happens quite often, because people often use spreadsheets and other databases that are inefficient. When they decide to use Access or a database-building software to improve their means of tracking information, they will want to use the old data they have already captured in the past.

TO MAKE A TABLE FROM SCRATCH

Let's now make our first table for the *Community Shelter and Soup Kitchen System*. The table will store *donation payments* made by *donors* who wish to make installments on a *donation*.

The first field we will enter is the *payment ID* field, but it is already created for us with the name of *ID*. When you build tables using the **datasheet view**, Access will always create the first field for you and it will be the **primary key** field. They will name it *ID*, but we will change its name to *Payment ID* in a moment. The second field will store the *donation ID* of each donation toward which the payments will be applied. The *donation ID* field will be a **number** field type with a width of **long integer**.

NOTE: Field names cannot contain periods or exclamation points. While spaces and slashes are permitted, it is not a good idea to use them if you believe that a day will come when another, larger database will need to interact with yours. If you want to see spaces and slashes in the field names of your **tables**, **queries**, **forms**, or **reports** you can do so in the fields' **captions**. Later, when you make forms and **reports** from your **tables**, they will always use your **tables' captions** as they display field names. If you do not add **captions**, Access will always use your original field names as **captions** instead.

NOTE: Certain field names are not recommended in an Access **table**, because they may be **reserved words**. That means that they have certain meanings already defined in Access. If you use reserved words, such as *value*, *date*, *name*, *text*, or *year* as field names, you may receive a message stating that they are **reserved words** and that your field could cause error messages later as you use your database. Therefore, instead of naming a field *Date*, it is best to pair it with a name that is relevant to your database. For example, in this database, if we want to name the field *Date*, which would indicate the date of the *donation*, it would be better if we would use the field name of *Date* *of Donation.* Instead of naming a field *Value*, which would indicate the value of the donation, it may be best to name it something like *Donation Value.*

Now let's make the table.

Step 1: At the menu seen in figure 1.5, click the tiny arrow that is to the right of the words **Click to Add** (figure 1.5, arrow A). That will give you the menu that drops down. In that menu, choose **Number** (figure 1.5, arrow B).

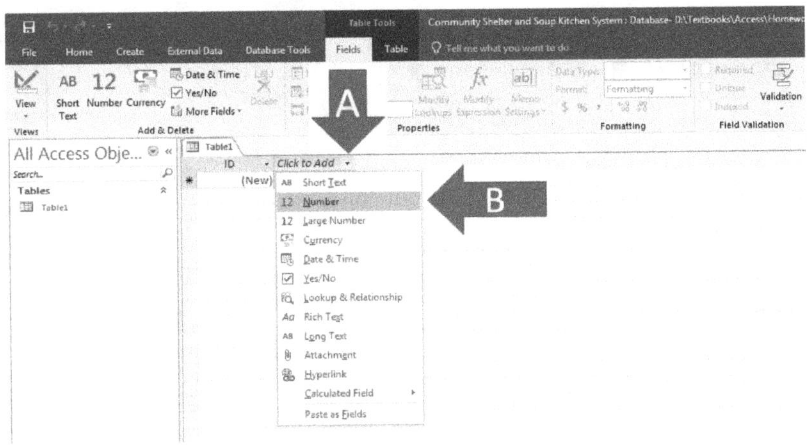

FIGURE 1.5 Adding a field in the datasheet view

Step 2: You will then get a screen that looks like figure 1.6a.

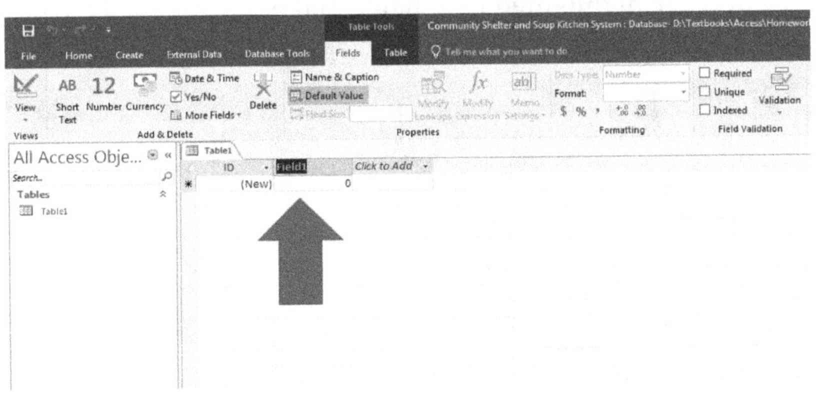

FIGURE 1.6a Naming a field in the datasheet view

Notice that Access will automatically give the field the name of *Field1.* **NEVER** name a field *Field1* or *Field2* or anything of that nature. A field should be named in a way that describes the data it holds. At this point it is ready for you to enter a field name. Since we are going to store the *donation ID of* each donation in the field, it is best to type *Donation ID* in the box and then press **Enter**. Doing so will overwrite the word *Field1* that is now there.

After doing so, the **Field List Type** menu will pop up again, so you will be ready to add the next field as seen in figure 1.6b. When it does, choose **Currency** (figure 1.6b, arrow).

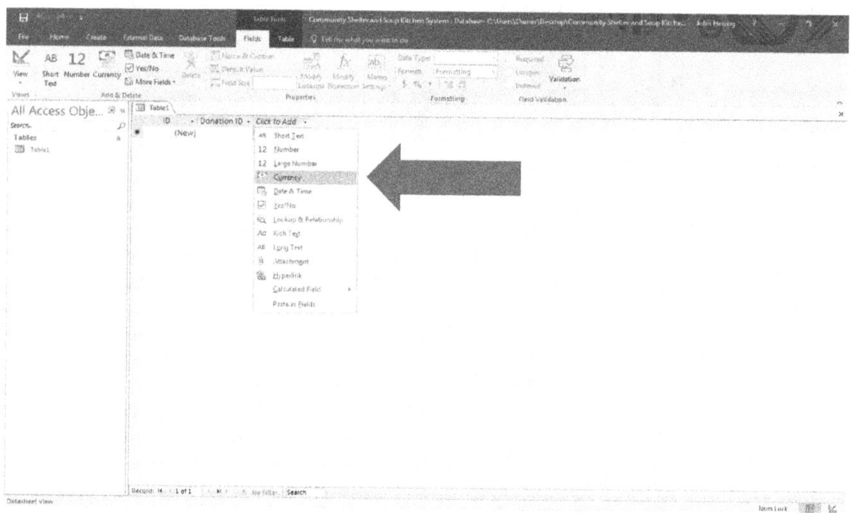

FIGURE 1.6b   Declaring a field type in the datasheet view

Step 3: Once again, when the field labeled *Field1* is highlighted as shown in figure 1.7, type over the highlighted field name and name the field *Payment Amount*.
Step 4: Press **Enter**.

FIGURE 1.7   Naming a field in the datasheet view

Continue this process until the remaining fields have been entered. The remaining field names with their associated field types are as follows:

FIELD NAME	FIELD TYPE
Check number	Short text
Payment date	Date/time
Comments	Long text

The following is a summary of the **field types**, their characteristics, and the options for controlling their widths. There are others that are not listed as they are not relevant in the discussion of this book, but they may be covered in more advanced books.

FIELD TYPE	CHARACTERISTICS	FIELD WIDTH OPTIONS
Autonumber	Automatically assigns a new, unused number to each record, usually as a primary key. By default, they will increase the number value of each newly added record by increments of 1, unless you change the setting.	Width cannot be controlled and is the equivalent to a number field with the column width of **long integer**.
Short text	Maximum data entered is 255 characters, which is the default width. Permits any kind of data to be entered.	Width can be controlled and thus changed to a width of 1 to 255.
Long text	Formerly called the memo field in earlier versions of Access. Permits any kind of data to be entered.	Width cannot be controlled. There is no limit to the characters that can be entered but permits no more than 1gb of data.
Number	Only raw numbers can be entered.	**Byte** width: Range is 0 through 255. **Integer** width: Range is –32,768 through 32,767. No decimals are permitted and if entered will be rounded. **Long integer** width: Range is –2,147,483,648 to 2,147,483,648. **Single** width: Permits decimals only –3.402823E38 to –1.401298E-45 for negative values and from 1.401298E-45 to 3.402823E38 for positive values. **Double** width: Permits large numbers with decimals with up to 15 places to the right of the decimal. Uses 8 bytes of disk space.

(continued)

FIELD TYPE	CHARACTERISTICS	FIELD WIDTH OPTIONS
Currency	Designed to accept currency data and will add currency format of your choice (the default currency in America is dollars). Permits no more than 4 digits on the right of the decimal point and up to 15 on the left and uses 8 bytes of disk space.	Width cannot be controlled.
Date/time	Will only allow legitimate dates to be entered such as 5/5/1955, but not 5/55/1955. Will accept entry of a date or a time or a combination of both.	Width cannot be controlled.
OLE object	Allows you to store files such as Word documents, PDFs, jpeg, sound files, and so on. Takes up storage space equal to at least the size of the file stored.	Limit to the file size is determined by your computer's memory. Width cannot be controlled.
Hyperlink	Allows the data entered to be clicked, which takes you to another object in the database, another file on your computer network, or a website. When an email address is entered and clicked, it will begin an email.	Width cannot be controlled.
Yes/no	In default it is usually displayed as a check box, containing a value of either 0 for no or –1 for yes. Can use yes/no, on/off, or true/false.	Width cannot be controlled.
Calculated	Allows you to do math within the table fields and/or fixed numbers.	Width is controlled by the format you choose for the calculated number.

TO SAVE A TABLE OR ANY OBJECT

At this point your screen should look like figure 1.8. It is now a good idea to save the table.

Step 1: Click the **Save** icon in the upper left of your menu in the **quick access toolbar** (figure 1.8, arrow).

NOTE: Sometimes the quick access toolbar is found in different places, but it is standard for it to be in the upper left of your screen.

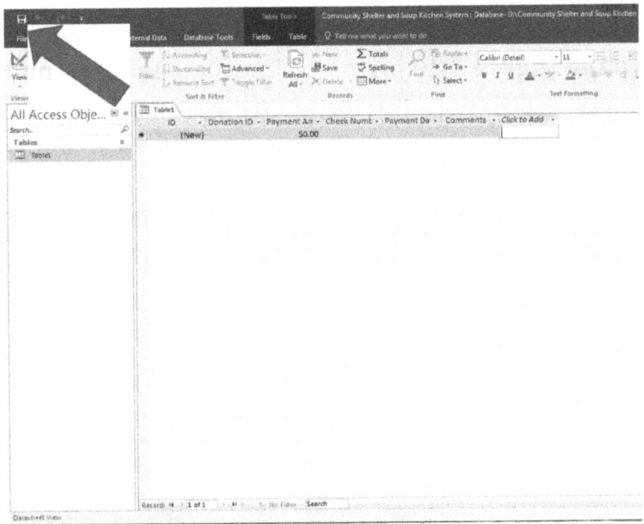

FIGURE 1.8   Completely designed table

Step 2: When you click the **Save** icon, you will get a **pop-up menu** prompting you to name the **object**. **NEVER** name a **table** *Table1* or *Table2*! Name it in a way that describes the data it holds, as mentioned earlier when you name fields. Enter the name in the **Table Name** box (figure 1.9, arrow). In this example, name it *Donation Payments* as you can see has been done in figure 1.9.

Step 3: Click **OK**.

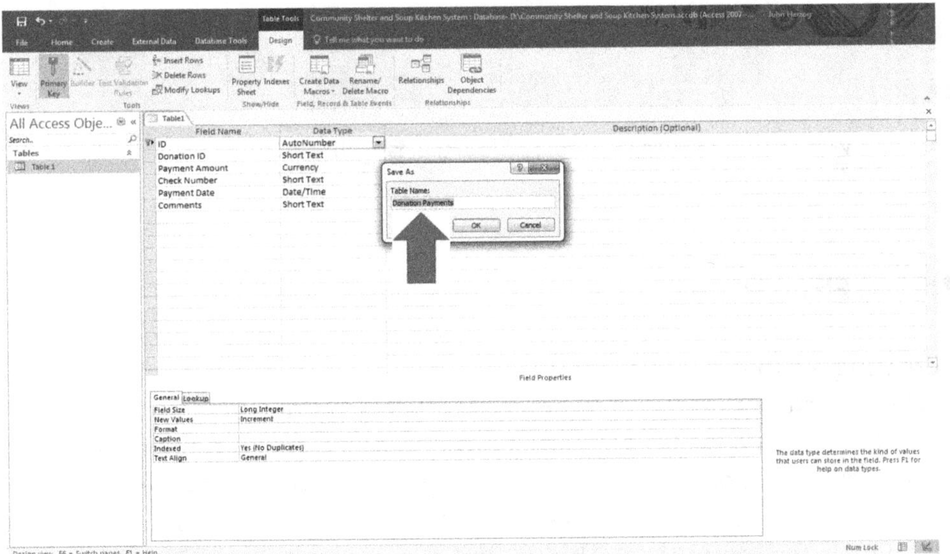

FIGURE 1.9   Saving the donation payments table

Please notice that after you save the **table**, its name will be in the list of tables in the **navigation pane** (figure 1.10, arrow). That is where the names of all **objects** (**tables**, **queries**, **forms**, **reports**, and **macros**) will be listed once they are created.

Congratulations! You have just created your first **object** in an Access database.

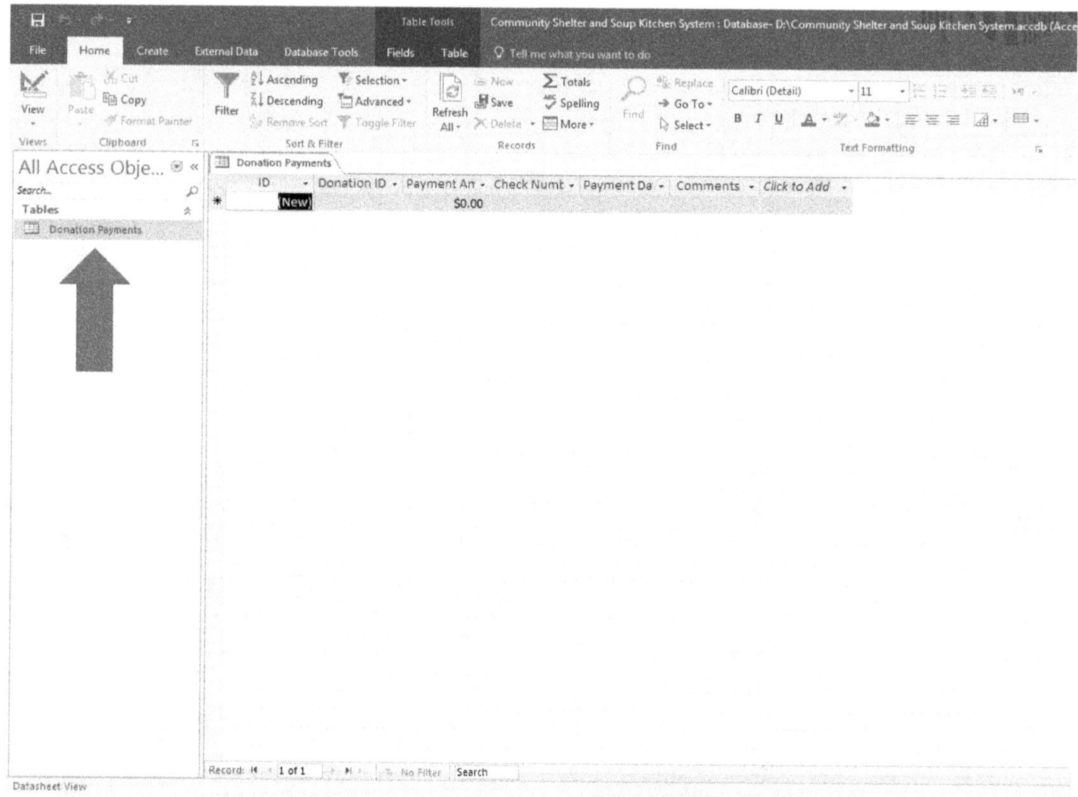

FIGURE 1.10 Viewing the donation payments table in the navigation pane

CHANGING FIELD NAMES

After creating a **table**, there are times that its field names need to be changed for better clarity or they may simply be misspelled. It is not recommended that you change field names in **tables** very often after creating **queries**, **forms**, **reports**, and **macros** referring to those **tables** and their original field names. Sometimes changing a field name will make those objects stop working. In many cases, Access will change the field names for you in those objects, but sometimes they do not. For clarity purposes, let's change the *ID* field in the *donation payments* **table** to *Payment ID*. You can do so by taking the following steps:

Step 1: Double-click the *ID* **Field Heading** (figure 1.11, arrow)

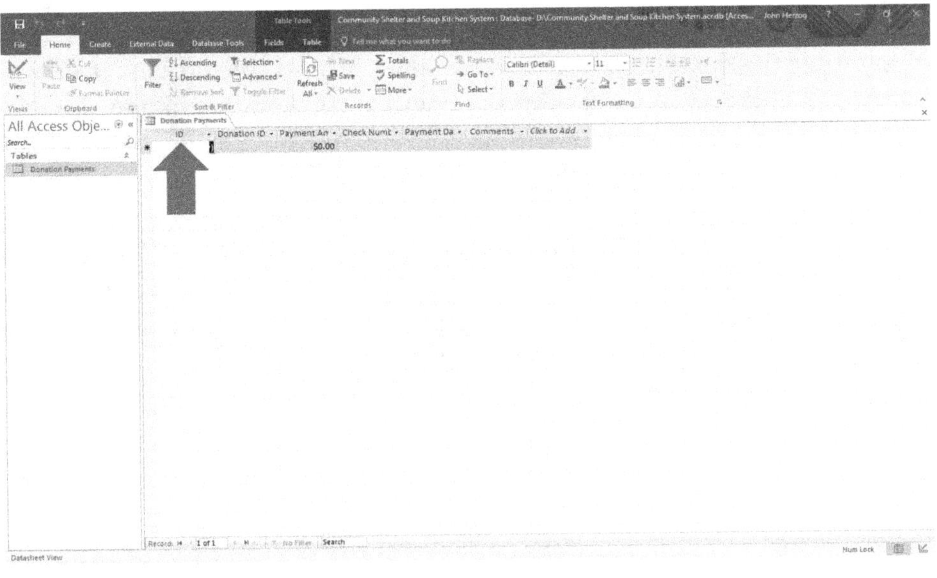

FIGURE 1.11   Changing field names (step 1)

Step 2: When you do, it will be selected. When you type selected data, what you type will overwrite the data. Type over (or overwrite) the selected *ID* field name and type *Payment ID* in its place (figure 1.12, arrow)

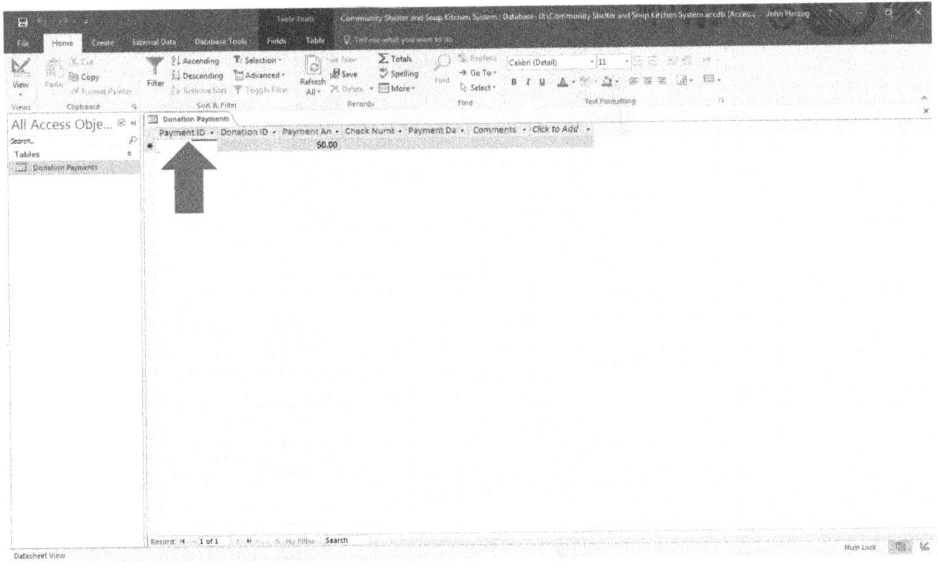

FIGURE 1.12   Changing field names (step 2)

CHANGING A FIELD TYPE OR FIELD SIZE

You can also change the **settings** for fields if you so choose. The settings can be changed in either the **design view** (discussed shortly) and the **datasheet view**. What kind of settings? There are several things you will often want to change. For example, you may want to change the **field type**. At arrow A in figure 1.13a, you will see that the *donation ID* field is selected, and in the **fields tab** of the **Table Tools** ribbon it is already designated as a **number field** (figure 1.13a, arrow C). If you wanted to change the field to any of the other field type, you would do so by choosing the field types listed in the **Data Type** box.

Notice that the **Field Size** box (figure 1.13a, arrow B), is ghosted now. That is because Access will only permit you to set the field width of a **number** field in the **table design view** (which will be discussed shortly). If you had designated the field as a **short text** field, the **field size** box would have allowed you to change the field size by selecting it and typing in the width you desired in the **field size** box (Figure 1.13a, arrow B).

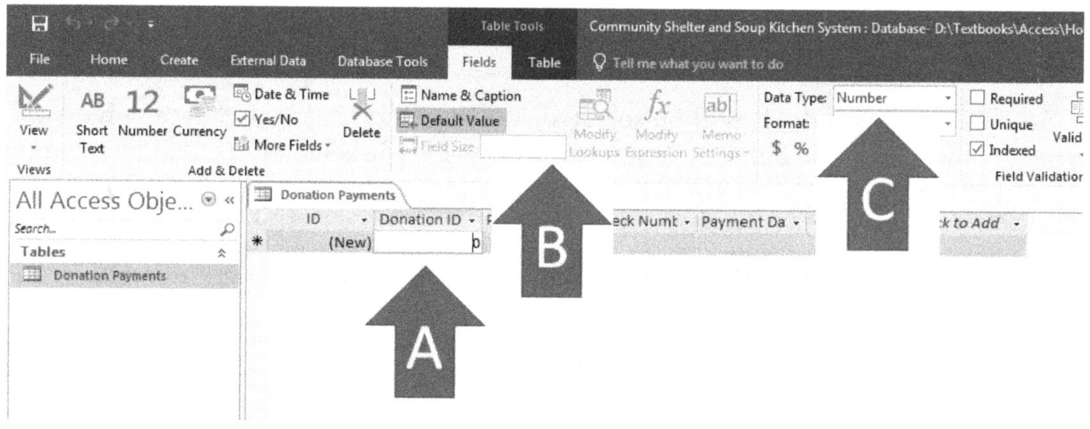

FIGURE 1.13a   Changing a field's default

NOTE: The **short text** fields are the only fields where you can usually enter a number for the field size. The **default column width** for **short text** fields is 255 characters. Seldom do you want to use that width, because each character requires the storage of one byte. If you have one million records in a table, and if you only need 50 of the **short text** characters, by using 255 characters, you are reserving room for, and thus wasting storage space for, 205 characters per record. That would be more than 200 megabytes that you are not using. That is using up storage space on your hard drive and in your computer's main memory. That wasted, reserved space would make your database slower.

CHANGING, ADDING, OR REMOVING A FIELD'S DEFAULT VALUE

Suppose you would like a field to always have a zero in it before a user would enter any data into any record. In some cases, users would also like to have the state entered of a customer if they only do business in one state. Still others would like to have the current date automatically entered into a field of each record when records are entered. These are all examples of setting a **default value** for a field, and they remove the need for users to enter repetitive data. This saves a lot of time. In order to add, change, or remove a field's **default value**, you can take the following steps:

Step 1: In the **Fields** tab of the **Table Tools** ribbon, you will see that the **default value** (figure 1.13b, arrow A) option (directly above the Field Size box). Click it. When you do, you will see that it gives you the **Expression Builder** menu seen in Figure 1.13b.

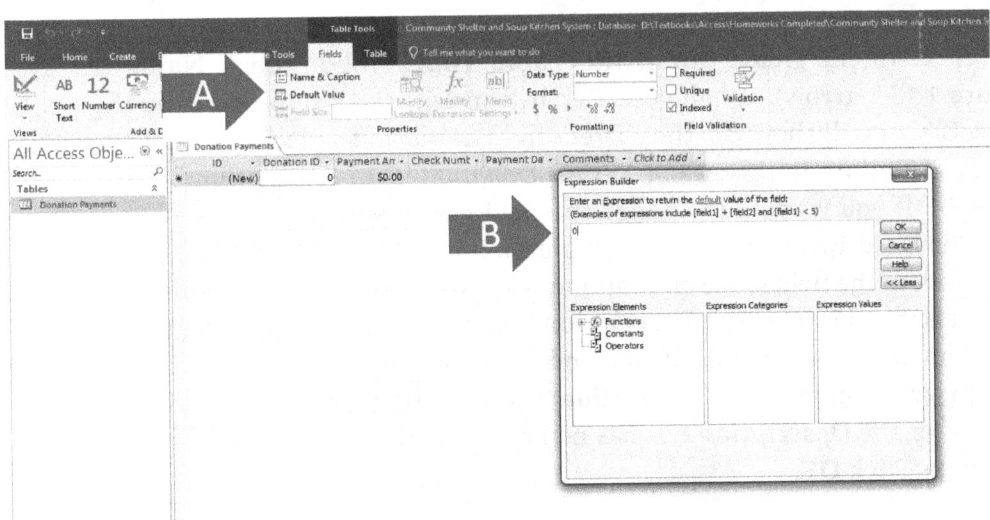

FIGURE 1.13b Changing a field's default

Step 2: If you type in the box (figure 1.13b, arrow B), you can make sure that the default will be something that will save you time.

In this example, we will delete the zero, because the *donor ID* field is not being used for calculations, therefore it is not appropriate to have a default value. Therefore, click the zero (0) and press the **Delete key**.

NOTE: It is **HIGHLY RECOMMENDED** that every table that is constructed in a database should **ALWAYS** have a field named "**Date and Time Entered**." The field should have a **default value** using what is called the **NOW function**. The reason for this is that on many, many occasions it will be very important to know when a record was entered. When a record is entered, the =**NOW()** **function** will freeze the exact time and date in that field that the record was created. **THERE WILL BE NO FIELD**

BY THAT NAME IN THIS DATABASE IN ORDER TO KEEP THE TABLES AS SIMPLE AS POSSIBLE TO AVOID CONFUSION.

CHANGING A FIELD CAPTION

Many institutions will have larger databases that may draw data from your databases, but they can't if there are spaces or slashes in the field names. The problem is that when you remove the spaces in the field names, it isn't as easy for some of the common users to read or understand the field name, thus they may not know what data to enter into the fields. That's why you have the option to change the **caption** to make it more user friendly if in fact you name the field with no spaces. In this example, if we had named the **donation ID** field with no space, you could have changed it by doing the following:

Step 1: Click the *Donation ID* field.

Step 2: In the **fields tab** of the **Table Tools** ribbon, click **Name & Caption** (Figure 1.13c, arrow). When you click it, you will see a **pop-up menu** (figure 1.13c, arrow B) that will show you the current **field name** and a place that will permit you to type in a **caption**. In this example, type *Donation ID* in the **caption** box (Figure 1.13c, arrow B). If you wanted to change the field name, you could have done so there as well. The **Description** box can be left blank. It is only useful if there is something very unique about the field name and caption that would not make it obvious as to what data must be entered into it. The **Description** box is also useful for writing notes to yourself about the field. For example, if you modify a table that someone else created, you may want to write a note in the **Description** box indicating who added or changed the field. Otherwise, the **Description** box has no impact on the operation of the database.

Step 3: Click **OK**.

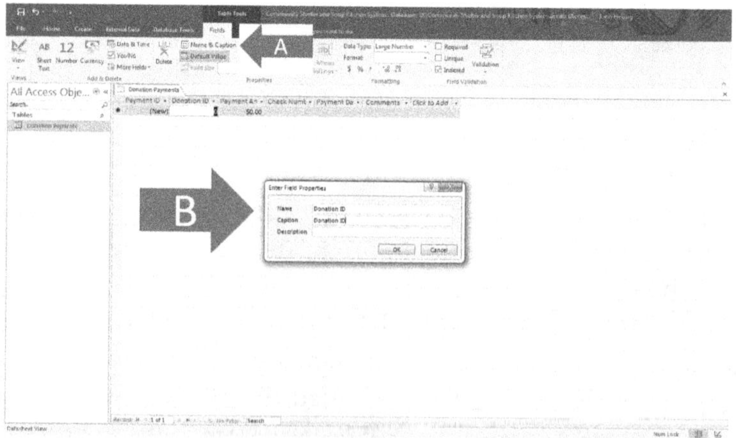

FIGURE 1.13c   Changing a field's caption

CHANGING A FIELD'S FORMAT

There will be times that you will want to change a field's format, particularly if they are **number** or **date/time** fields. For example, you may want **number** field data to be displayed as percentages (when they are decimals) or you may want them displayed with commas (when their values exceed 1,000). You may want to change a **date/time** field to show dates as *Tuesday, January 1, 2019* instead of *1/1/2019*. In a **table**, you can set the format for fields, **ONLY** if they are **number** or **date fields**, but not **text fields**. When you click on a **text field**, the **format box** becomes ghosted or unusable (figure 1.13d, arrow A). To change a field's format, select the field and make the choice of format you desire.

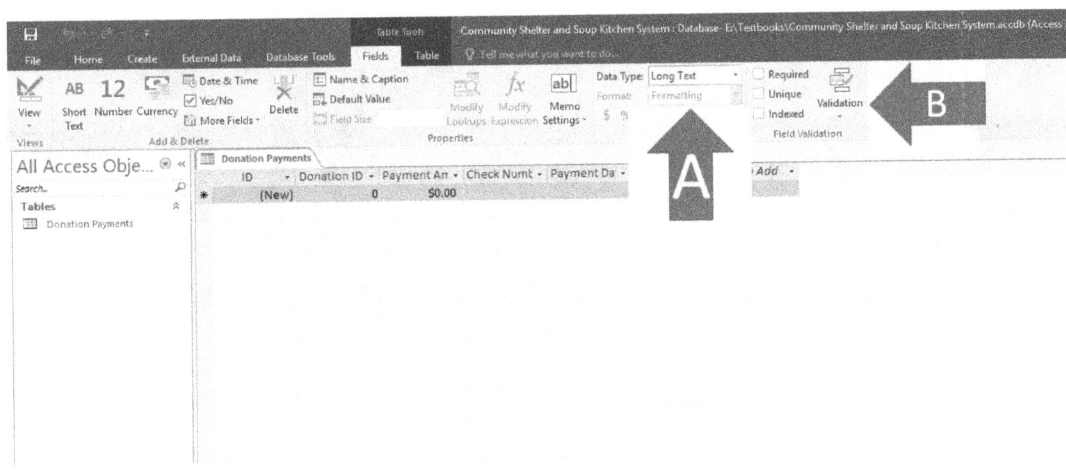

FIGURE 1.13d   Changing a field's format and validation rule

SETTING VALIDATION RULES FOR FIELDS

Validation rules can also be added to fields. This allows you to keep bad data from being entered. For example, if it is not possible to have a *payment amount* that is less than zero, you will want to prevent the user from entering a negative number into that field by giving it a **validation rule**. You can do so by taking the following steps:

Step 1: Select the *Payment Amount* field.

Step 2: In the **Fields** tab of the **Table Tools** ribbon, click Validation Rule (figure 1.13d, arrow B). Doing so will give you what you see in figure 1.13e.

Step 3: Click **Validation Rule** (figure 1.13e, arrow A). That will take you to the **expression builder** that you see in figure 1.13f.

Step 4: In the **Expression Builder** box, type in *[Payment Amount]>=0* (figure 1.13f, Arrow). By taking these measures, you will not be able to enter negative numbers into the *payment amount* field.

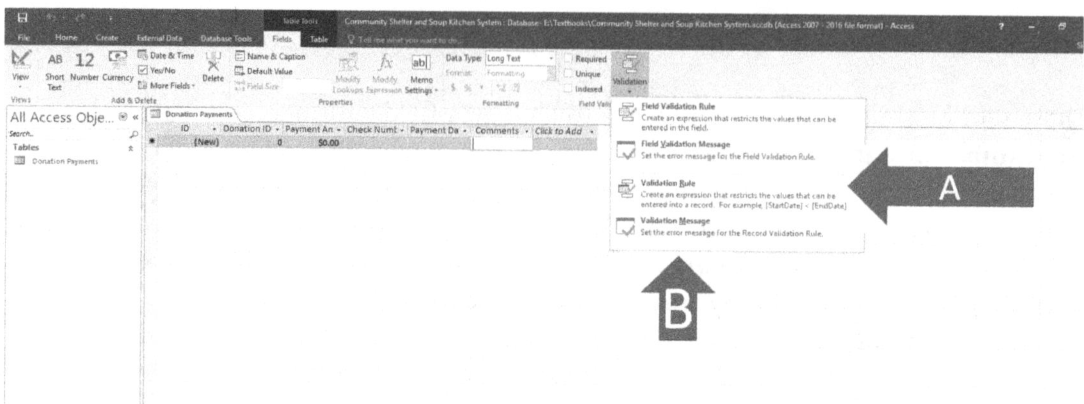

FIGURE 1.13e   Changing a field's format and validation rule

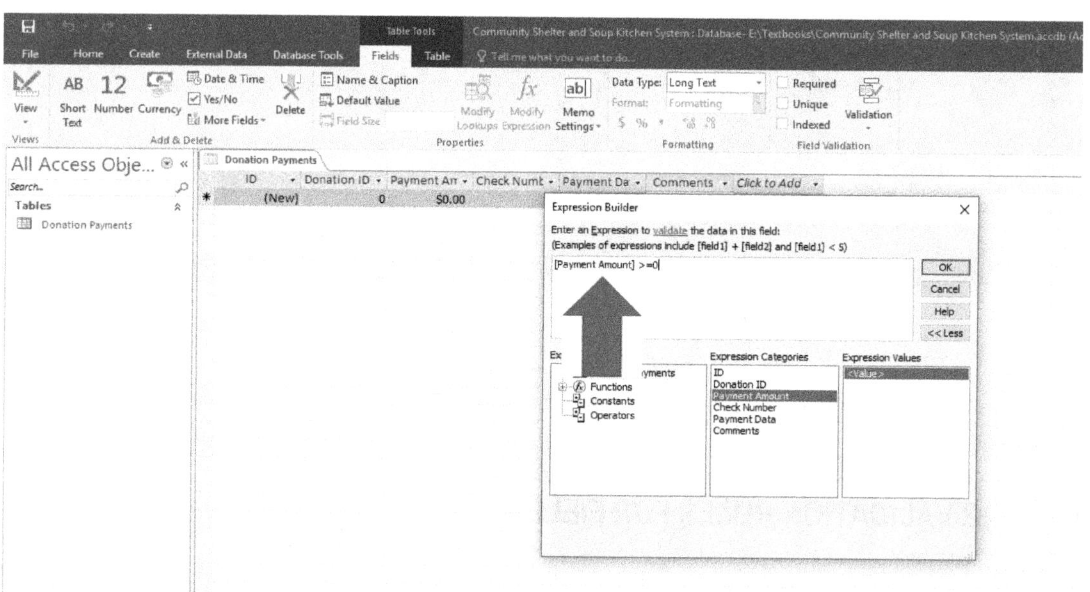

FIGURE 1.13f   Changing a field's validation rule and validation message

NOTE: If you return to the menu shown in figure 1.13e and and click **Validation Message** (figure 1.13e, arrow B) you will get the screen in figure 1.13g. In that box, type the error message that you want displayed if a user does in fact enter a negative number in that field. In this example, type in the **Validation Message.** *You must enter a value greater than or equal to zero!*

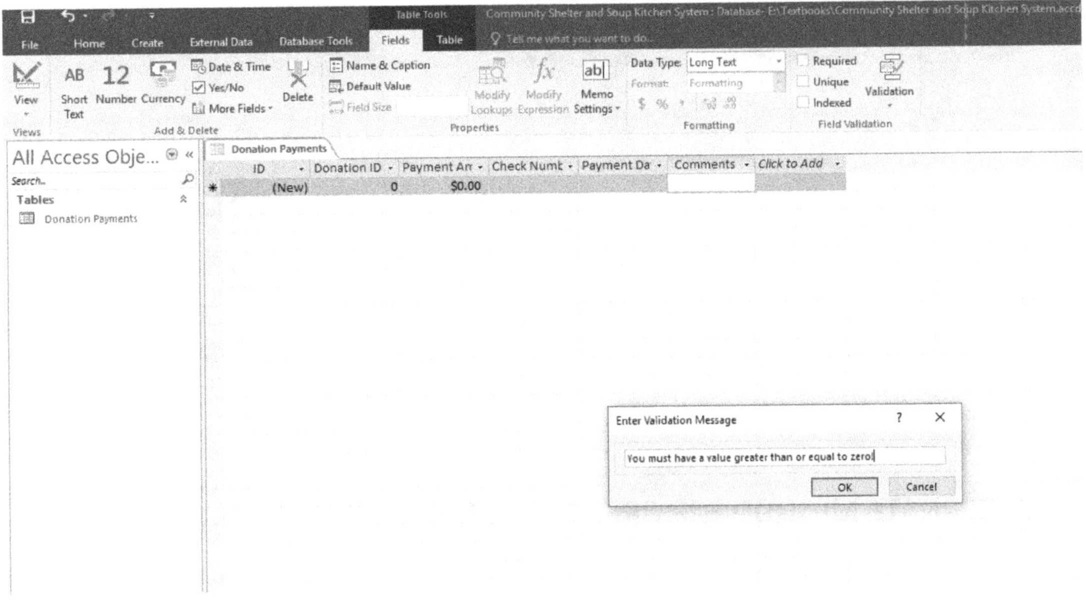

FIGURE 1.13g Changing a field's validation message

TO DESIGNATE FIELDS AS REQUIRED, UNIQUE, OR INDEXED

There are some fields that you don't want users to skip. You can designate them as **required** fields by clicking the **Required** checkbox in the **Fields** ribbon within the **Table Tools** ribbon (figure 1.13h, arrow). You can also make sure that some fields have no duplicates. You can designate them as **unique** fields (figure 1.13h, arrow). You can also designate a field as **indexed**. If you do, queries (which will be discussed later) will run faster with the indexed fields and it will also speed things up for sorting and grouping operations (which will also be discussed later). For example, if you search for a specific *donor* name in the *donor contact last name* field, you can create an index for that field to speed up the search for a specific *donor*. You can designate a field as indexed by clicking the **indexed** checkbox (figure 1.13h, arrow). When you set a field as a **primary key** field, it will be automatically designated as **required**, **unique**, and **indexed**. These settings are only used if there are fields in your **table** in addition to the **primary key** that **MUST** have these attributes.

FIGURE 1.13h   Changing a field's validation message

THE ALTERNATE METHOD FOR CREATING A TABLE FROM SCRATCH

The method for creating a table you did in the **datasheet view** is quicker and easier for many, because it resembles that of a spreadsheet, and many people are more familiar with spreadsheets. However, there are things you cannot do in the **datasheet view** that you can do in the **design view**. To create a table using the **design view**, do the following steps:

Step 1: Click the **Create** ribbon (figure 1.14a, arrow A).

Step 2: Click **Table Design** (figure 1.14a, arrow B). That will give you the menu you see in figure 1.14b.

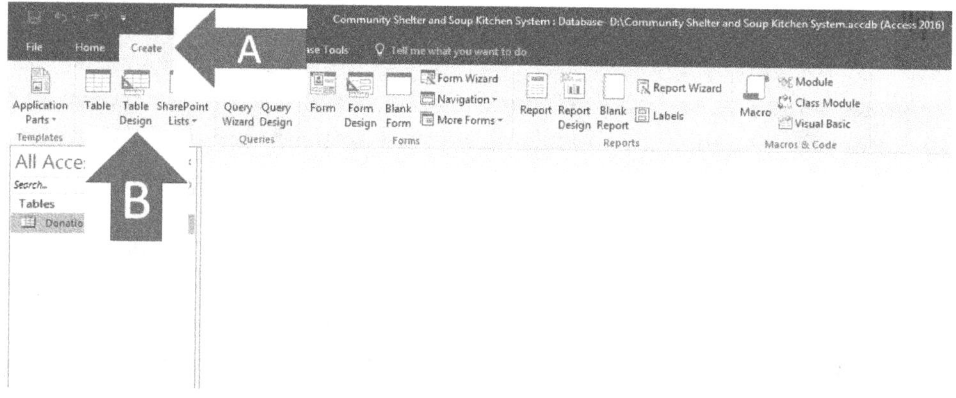

FIGURE 1.14a   Creating a table from scratch (step 1)

Step 3: You can add the field names where arrow A is pointing in figure 1.14b, and you can choose the data types in the **Data Type** menu seen here where arrow B is pointing in figure 1.14B. The **description (optional)** area at arrow C in figure 1.14b has no bearing on the operation of the table as mentioned earlier. Figure 1.14c, arrow A shows what the table would look like in the **design view** after the same fields have been added that were created earlier in the **datasheet view**.

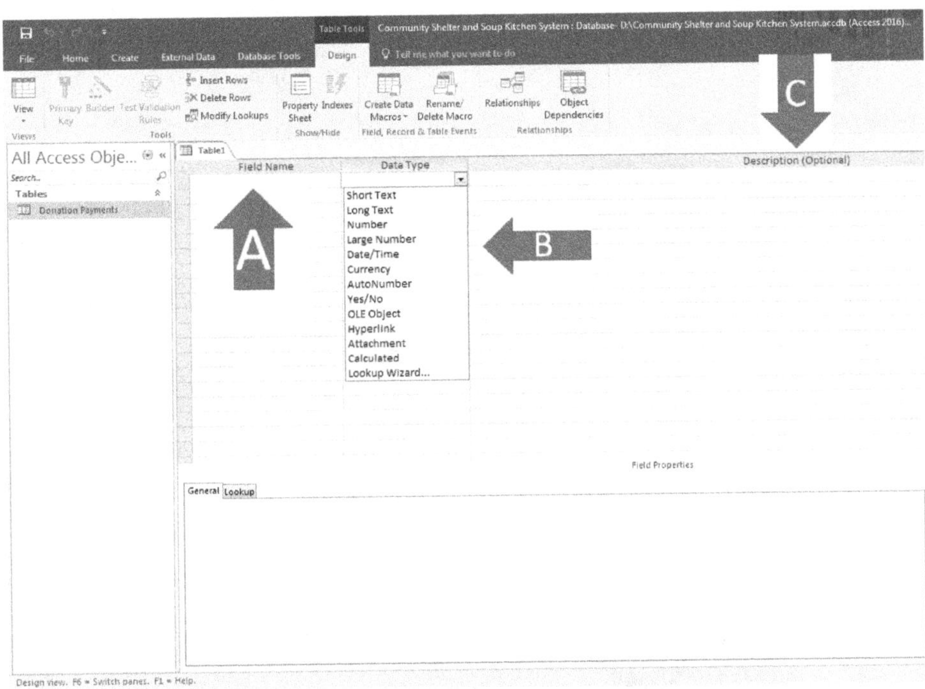

FIGURE 1.14b   Creating a table from scratch (step 2)

Step 4: To change **field properties** (which are the attributes of fields discussed earlier, such as field captions, formats, default values, and more), select a field and then click in the **field properties** area at the bottom of the screen (figure 1.14c, arrow B). In this case, the *payment ID* field is selected, thus it is showing in the **field properties** area that it has a **field size** of **long integer**.

Step 5: Designate a field (in this case, the *payment ID* field) as the **primary key** by clicking the field and then clicking the **Key** icon labeled as **primary key** (figure 1.14c, arrow C).

NOTE: The **primary key** is the field by which Access will sort your records unless you tell it to sort by a different field.

NOTE: The results of creating a table in the **table design view** or the **datasheet view** are absolutely the same. It's just a matter of doing it the way that you feel is easier. As already stated, there are some things you simply cannot do in the datasheet view that you can do in the **table design view**. For example, in the **datasheet view**, you cannot add a field that must be an **OLE object** field type. As also stated earlier, setting the **field width** of a **number** field can only be performed in the **table design view**.

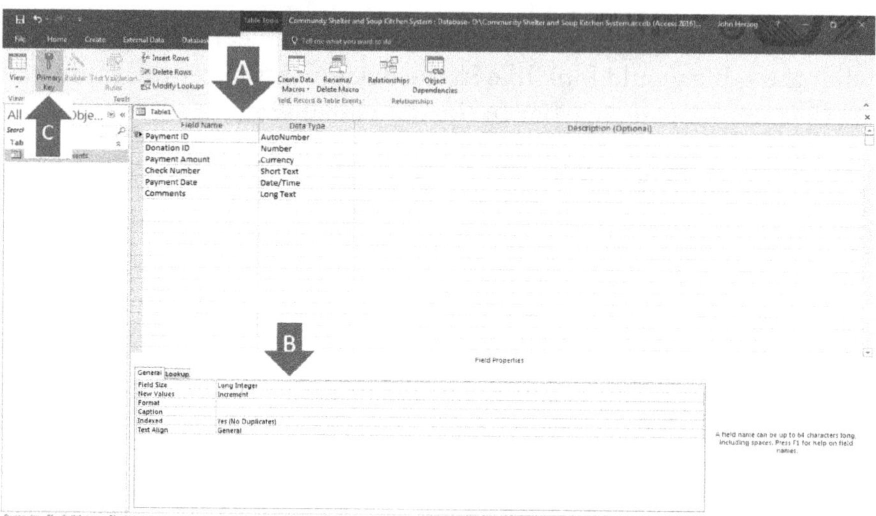

FIGURE 1.14c   Creating a table from scratch (step 2)

Step 6: To return to the **datasheet view**, click the datasheet view icon in the **Table Design** ribbon (figure 1.14d, arrow).

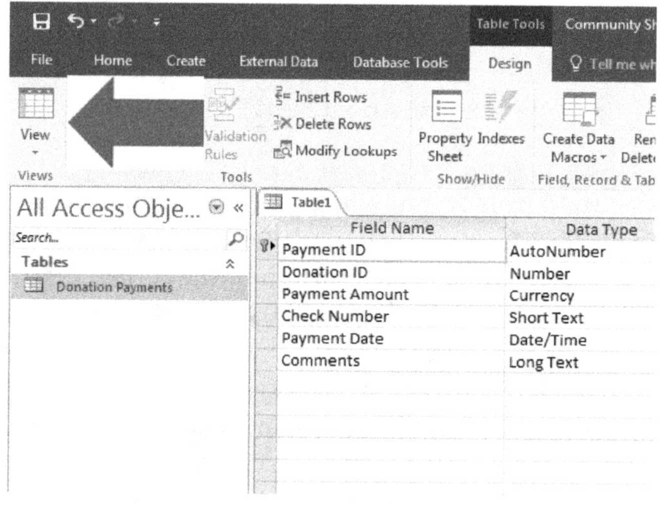

FIGURE 1.14d   To Return to datasheet view

ENTERING DATA

If you click the **Datasheet View** button in the **design view** or if you are already in the **datasheet view**, your screen will look like figure 1.15a. You can begin to add data. If you have ever used a spreadsheet, the process of entering data is the same in a **table** in Access. Simply type whatever data you wish to enter in the **cell** or **data box** (assuming that you are not trying to put text into a currency, number, or date field, or trying to put more characters in the field that you have permitted). Another similarity between entering data into Access **tables** and Excel spreadsheets is that you can change the column (or field) widths by placing your mouse on the tiny line that separates the field names (figure 1.15a, arrow A) and then clicking and dragging that line to the right or to the left until the field width is what you want. You can also double-click the line for a **best fit** setting. The **best-fit** setting will allow Access to set the column (or field) width to make sure it is just wide enough to display all of the field's data.

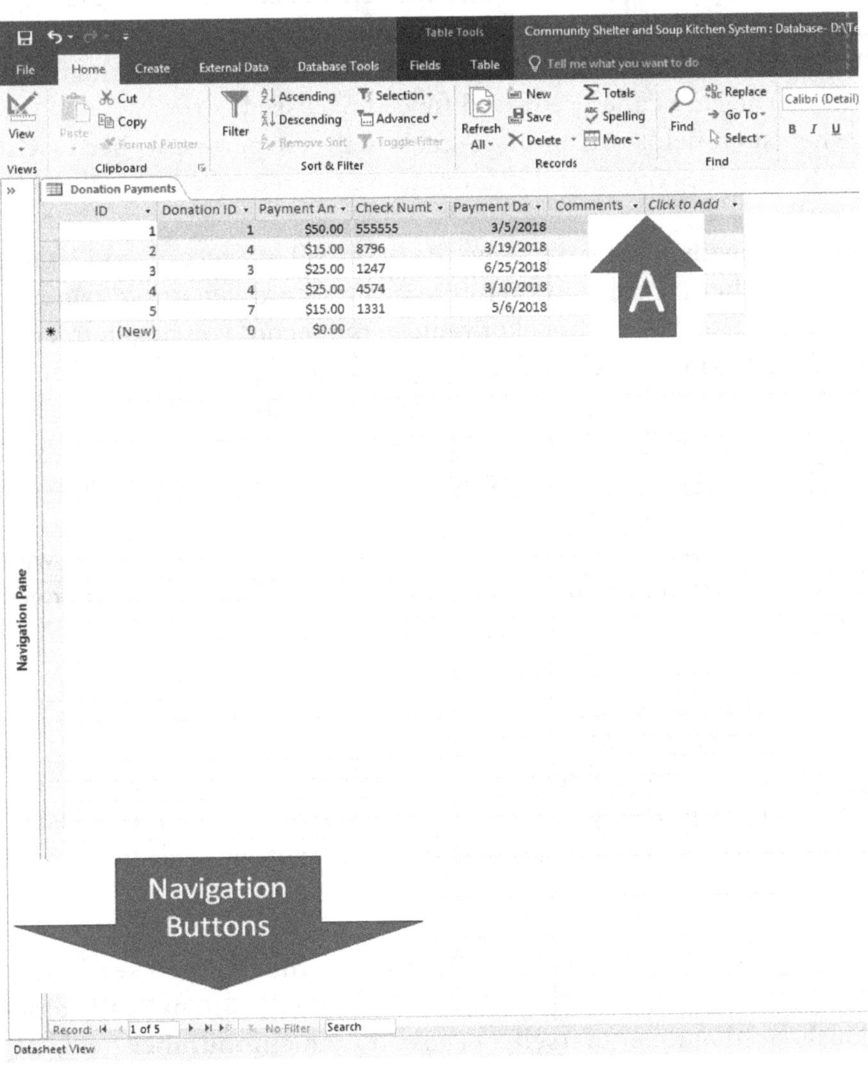

FIGURE 1.15a   Entering records

Another similarity between database **tables** and spreadsheets is that the intersection of a column (or field) and a row (or record) is considered a **cell**.

After typing the data, simply press **Enter**, **Tab**, **PgUp**, **PgDn**, or any one of the **cursor control arrow keys** ↑↓→← on your keyboard that would take you to another cell.

NOTE: If you type over data that is already in a cell the original data is lost forever. Be careful not to type over data unless it is your intent to remove it.

NOTE: WHEN YOU LEAVE A RECORD (ASSUMING THAT YOU HAVE NOT PUT IN DATA THAT VIOLATES VALIDATION RULES), THE RECORD IS AUTOMATICALLY SAVED! IT IS NOT POSSIBLE TO CLOSE A TABLE OR FORM WITHOUT SAVING A RECORD! If you try to enter invalid data into a cell, Access will stop you before you leave the cell, because it will not let you save the record until the invalid data is removed or corrected.

NOTE: There are some fields that must have data entered into them (such as the **primary key**). They are called **required** fields (discussed earlier). If you try to leave a record when such fields are empty, Access will not permit it, because it will be attempting to save the record, and it cannot save the record when a **required** field is empty (or **null**).

NOTE: If you begin to enter data into the first field of a record, and if you have not yet moved to another record (and therefore if the record you are in is not yet saved) press the **Escape (Esc) key** to cancel or remove the record you began to create.

NOTE: If you begin to enter data into any field other than the first one, and if you have not yet moved to another record (and therefore if the record you are in is not yet saved), press the **Escape (Esc)** key to cancel or remove the data in the field you are in currently. If you press the **Escape (Esc)** key twice, it will cancel or remove the entire record you began to create.

NOTE: Whenever you close a **table**, the only time it will ask you if you want to **save** is if you change the structure of the table, not the data. Structure changes include the following:

a. Changing the sort order of records
b. Widening or narrowing a field
c. Adding or removing a field in the **datasheet view** or the **design view**
d. Changing any property of a field in the **datasheet view** or the **design view**.
e. Adding or removing a **total row** at the bottom of the table (to be discussed in chapter 3)

By using these procedures, please enter the records that you see in the table in figure 1.15a. Please notice that the *payment ID* field will automatically assign a new, not-previously-used number to itself, because it is an **autonumber** field.

NAVIGATING IN AN ACCESS TABLE

There are a few things you need to know when it comes to moving around in an Access **table**:

1. You can move to the end of a record by pressing the **End** key on your keyboard.
2. You can move to the beginning of a record by pressing the **Home** key.
3. You can move to the last field and last record of a table (or the lower right of a table) by pressing **Ctrl End**.
4. You can move to the first field and first record of a table (or the upper left of a table) by pressing **Ctrl Home**.
5. You can edit data of a given cell within a table by pressing **F2**.

These things hold true if you are not in the middle of editing a cell. If you are in the middle of editing a cell, these keystrokes will take you to the beginning or the end of the data in the cell, not the table.

Navigation buttons are found at the bottom any **table**, **query**, **form**, or **report** as you are shown in figure 1.15a. Figure 1.15b shows a closer view of the **navigation buttons**. They can be used as follows:

1. Clicking where arrow A, figure 1.15b is pointing will take you to the **first record** of a table.
2. Clicking where arrow B, figure 1.15b is pointing will take you to the previous record. In this example it is ghosted (or unusable), because you are in the first record of the **table**, which does not have a previous record.
3. Arrow C, figure 1.15b is pointing to where you can see what record you are in **AND** how many records exist in this table. In this example, you are now in *record 1* of 5 records.
4. Clicking where arrow D, figure 1.15b is pointing will take you to the **next record**. If it is ever ghosted, it would be because you are already on the last record, which does not have a next record.
5. Clicking where arrow E, figure 1.15b is pointing will take you to the **last record**.
6. Clicking where arrow F, figure 1.15b is pointing will start a **new, blank record**.

NOTE: In **reports**, the clicking on these arrows will take you to the next, previous, first, or last **PAGE** in the **report**. The same area to which arrow C, figure 1.15b is pointing will also tell you what **PAGE** you are on and how many **PAGES** are in the **report**.

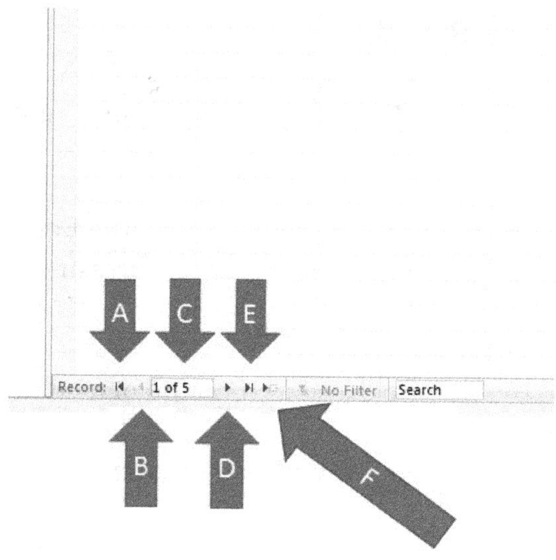

FIGURE 1.15b   The Navigation button arrows

CLOSING A TABLE OR ANY MICROSOFT ACCESS OBJECT

When you close an object, you can do so without closing the entire database file. You can do so by clicking the **lower X** in the upper right of your screen (figure 1.15c). Clicking the **upper X** in the upper right of the screen is used to close the entire database and Access.

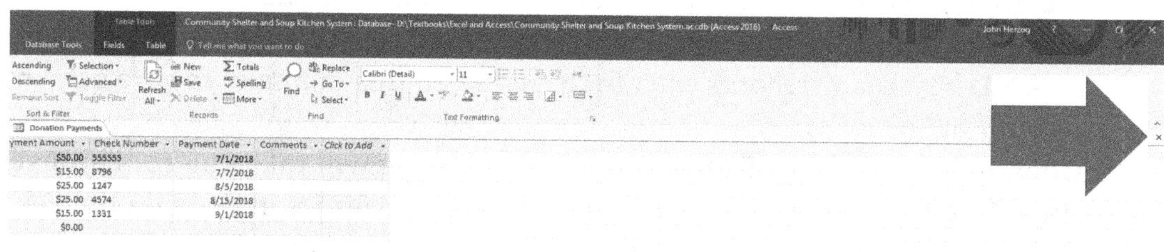

FIGURE 1.15c   Closing a table's datasheet view

CREATING (IMPORTING) A TABLE FROM AN EXISTING SPREADSHEET

Another very common practice, as discussed earlier, is to create a table using data that has already been entered into spreadsheets or other database types in past years. As discussed previously, people often enter things into spreadsheets, but when they see that databases are so much better in their situation and that they should no longer use spreadsheets, they fear that they will have to retype what they have created so far in those spreadsheets in years past. However, it is not necessary to retype the data, because the spreadsheets can be transformed into database tables.

Suppose that those working on the *community shelter and soup kitchen system* had been entering their *donors* into a spreadsheet that they would now like to import into their new database. It can be accomplished by the following steps:

Step 1: Any spreadsheet that you plan to import must **NOT** have labels that are merged and centered above their columns holding the data. The first row of the spreadsheet should only be column headings (which will shortly become the field names of the table once we import it). We have a spreadsheet for you that you can use to import that meets the standards needed. Copy the **Excel spreadsheet** file named *Donors* that is associated with this textbook to your flash drive or desktop. See figure 1.16. In our example, the file was copied to a particular folder in a flash drive.

NOTE: The reason that it is best to copy the file to be downloaded to your computer or flash drive is that if you attempt to download a data source from a network, you will often receive an error message if there is already someone else trying to download that same file.

NOTE: All names and addresses used in this textbook are fictional. If there are real people or addresses matching, or similar to, those in these examples (with exception to John Herzog) such instances are purely coincidental. You may also be wondering why the customers have no cities or states. The answer is that we will be using a *zip code* table that will relate to these *donors'* zip codes. The table will show their associated cities and states. Using this concept saves a great deal of data entry as it removes the need for having to enter cities and states in each record. It also creates a great deal of consistency. Zip codes for the entire country can be downloaded from the internet.

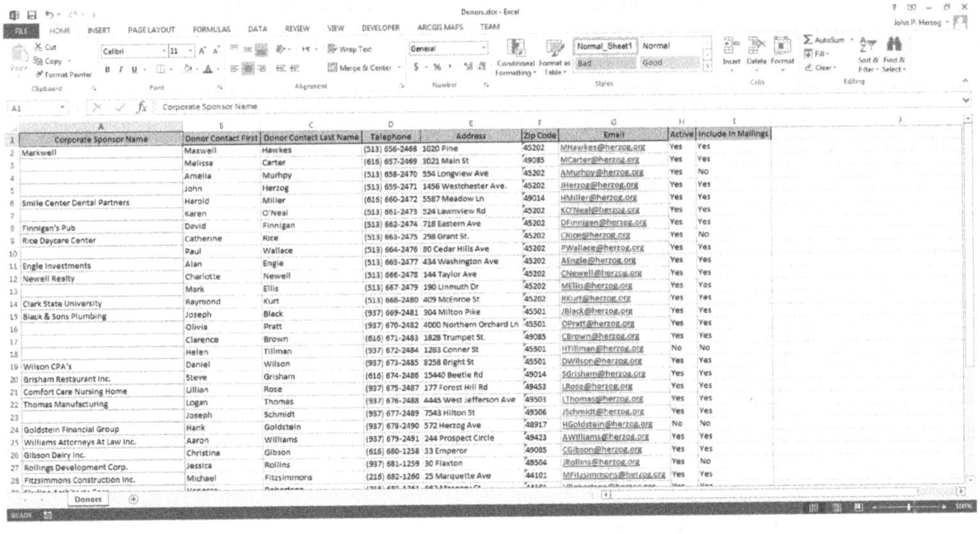

FIGURE 1.16 Viewing a spreadsheet before importing it (step 1)

Step 2: Click the **External Data** ribbon (figure 1.17, arrow A).

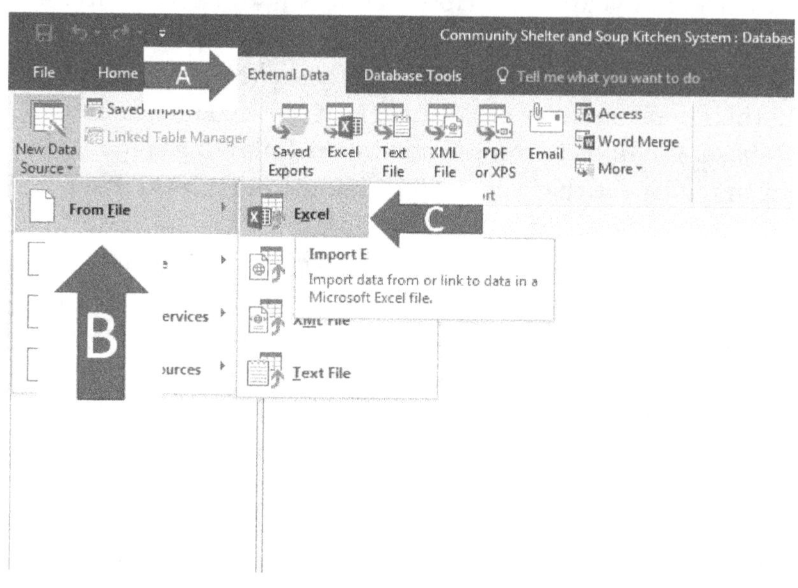

FIGURE 1.17   Importing a spreadsheet (step 2)

When you do, you will get the menu that you see in figure 1.17.

Step 3: Click **New Data Source** and then click From File (figure 1.17, arrow B).

Step 4: Click **Excel** (figure 1.17, arrow C). That will give you the menu you see in figure 1.18.

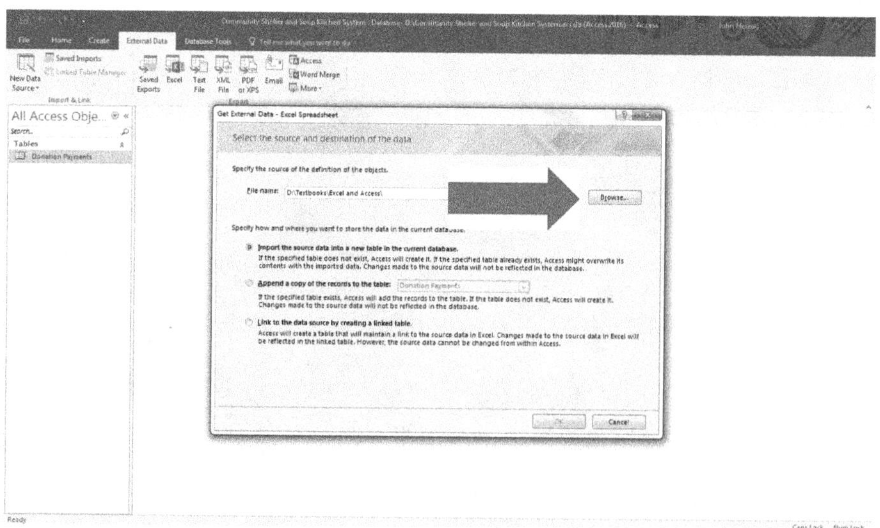

FIGURE 1.18   Importing a spreadsheet (step 3)

Step 5: Click **Browse** (figure 1.18, arrow) in order to locate the **Excel spreadsheet** named *Donors* that we observed earlier. It will give you the menu you see in figure 1.19. Since we chose to import an **Excel** file, only **Excel** files will be displayed. Click on the *Donors.xlsx* file and then click **Open** (figure 1.19, arrow).

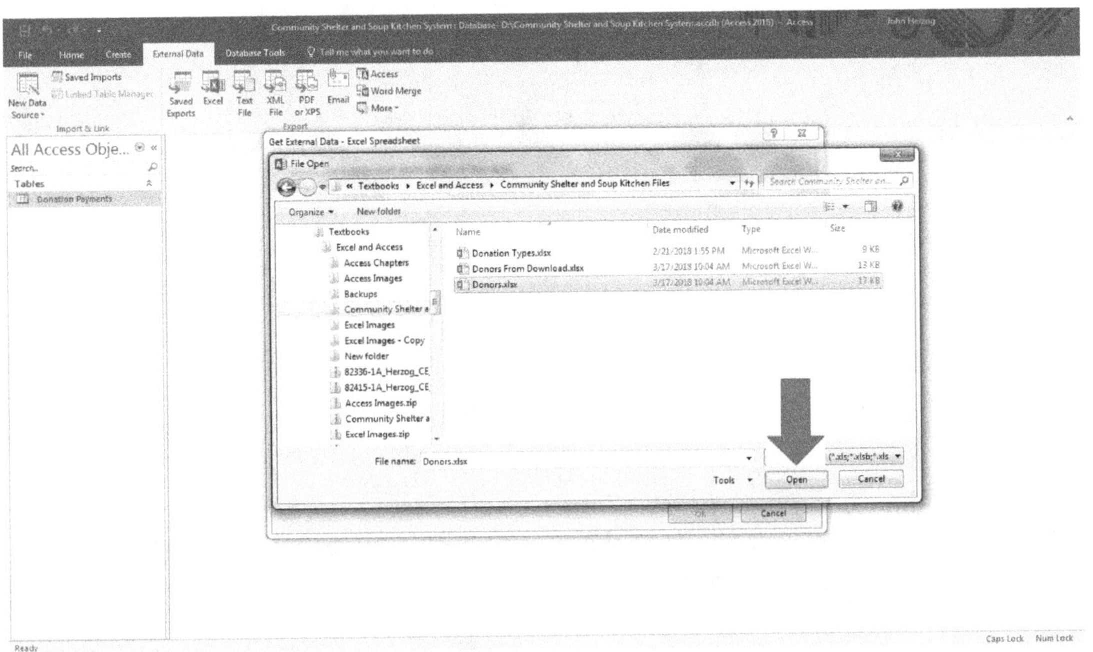

FIGURE 1.19 Importing a spreadsheet (step 4)

Step 6: That will take you to the menu shown in figure 1.20. Make sure that the **Import the source data into a new table in the current database** option button (figure 1.20, arrow A) is checked and then click **OK**.

NOTE: We don't want to link the spreadsheet in this example, but if you did, you would click the option button at arrow B in figure 1.20 that reads **Link to data source by creating a linked table**. Linking to a spreadsheet makes it possible to change the **table** data while simultaneously changing the spreadsheet that is its source and vice versa. Many people often like to use spreadsheets while still reaping the benefits of using database advantages, hence they often use both of them.

NOTE: Appending the data (or adding the data) to an existing **table** is not going to be used in this example or in this textbook, hence we will not be clicking on the **Append a copy of the records to the table** option button.

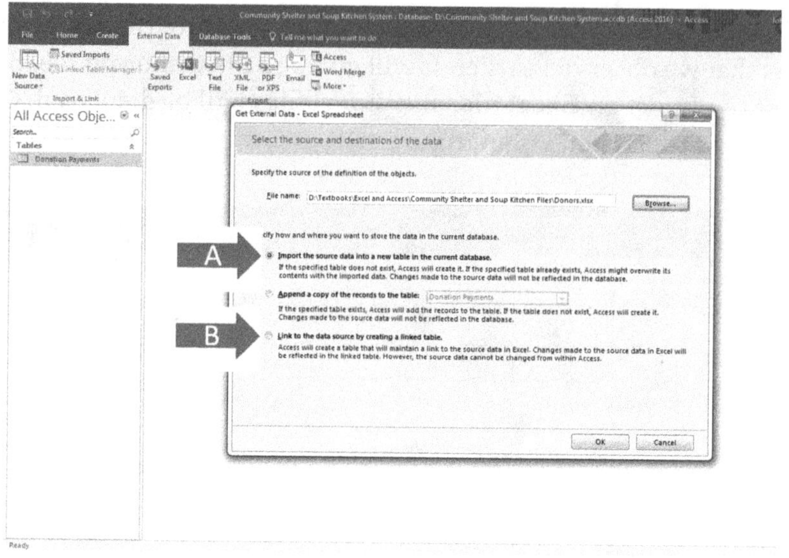

FIGURE 1.20   Importing a spreadsheet (step 5)

That will take you to figure 1.21.

Step 7: At this screen, make sure that the **Show Worksheets** option (figure 1.21, arrow A) is selected and that the worksheet that you want to import is selected. In this case, the Excel workbook has only one sheet, the *donors* worksheet, and it is already selected. Once you select the sheet to be imported, click **Next** (figure 1.21, arrow B).

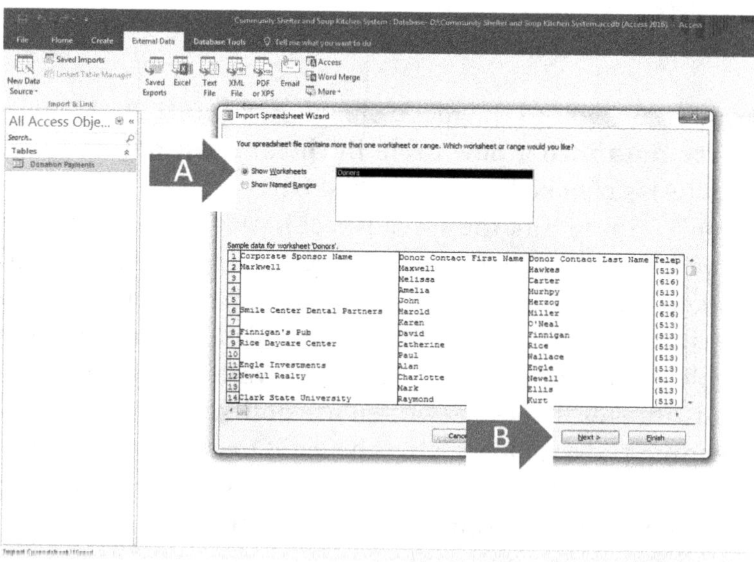

FIGURE 1.21   Importing a spreadsheet (step 6)

Step 8: That will take you to the screen shown in figure 1.22. At this screen, make sure that the **First Row Contains Column Headings** check box is checked (figure 1.22, arrow A). This will make sure that the column headings in the spreadsheet will become the field names in this new table when it is imported.

NOTE: You must remember that field names cannot contain periods or exclamation points. If a column heading in your spreadsheet has any of those things, it will rename the field to *Field1* for you.

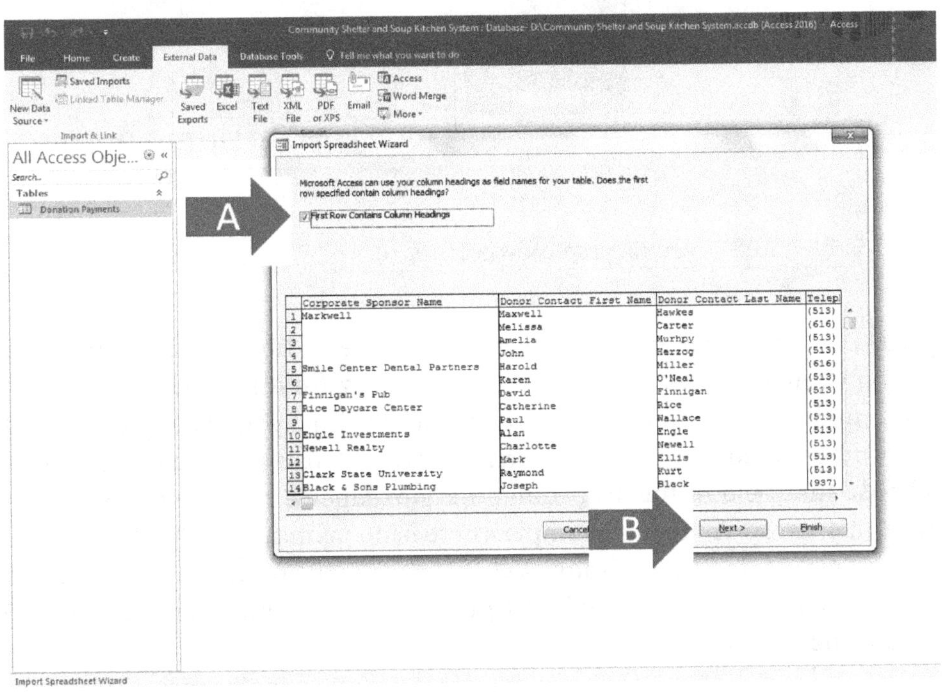

FIGURE 1.22   Importing a spreadsheet (step 7)

Step 9: Click **Next** (figure 1.22, arrow B).

Step 10: That will take you to figure 1.23. At that point you can click on the name of each field (figure 1.23, arrow B). When you do, the column of the field will go dark. When that happens, you can then click in the **Field Options** box to rename the field or to change the field type to something different, such as short text, date/time, or number and so on (figure 1.23, arrow A). You may also choose to skip the field if you decide you don't need it in the table. Be sure to look at all of the fields from the spreadsheet. The ones to the right will be hidden unless you use the **Scroll Bar** (figure 1.23, arrow C) in order to see them.

Step 11: In our example, we do not want to change any of the field names or the field types, and we do not want to skip any fields. Once all needed changes are made, click **Next** (figure 1.23, arrow D).

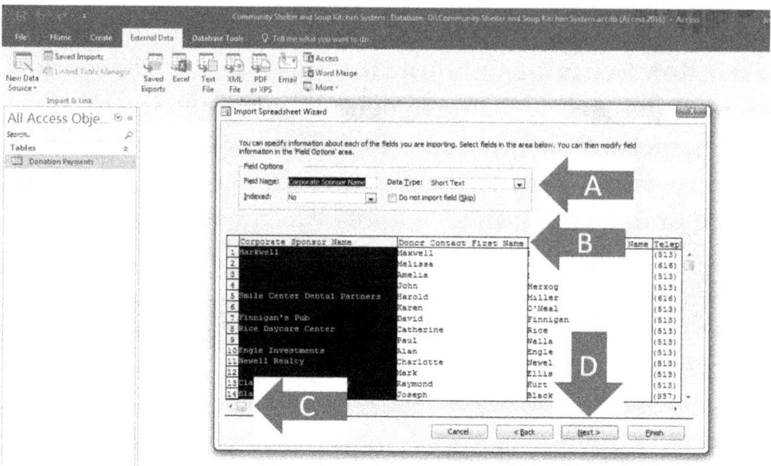

FIGURE 1.23   Importing a spreadsheet (step 9)

Step 12: That will take you to the screen you see in figure 1.24. It will ask you whether you want Access to create the **primary key field** for you or if you already had one in mind. If your spreadsheet already has a primary field key, such as a Social Security number, or some other numbering system that is required by the user, you would choose that field in the drop-down list box labeled **Choose my own primary key** (figure 1.24, arrow A). In our example, there is no **primary key** field and we indeed would like to let Access add a **primary key** by keeping the option button selected that is labeled **Let Access add primary key** (figure 1.24, arrow B). The field they add will be an **autonumber** field.

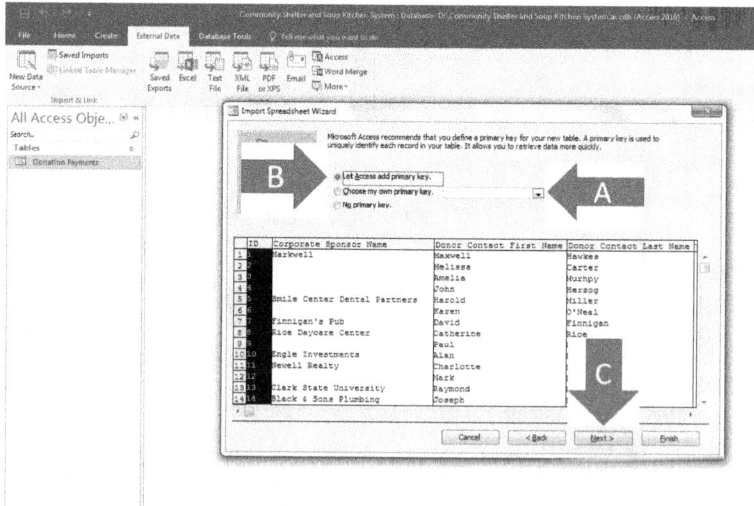

FIGURE 1.24   Importing a spreadsheet (step 11)

Step 13: Click **Next** (figure 1.24, arrow C).

Step 14: That will take you to the screen you see in figure 1.25. Notice that Access assumes that you want to name the table the same as the sheet tab that is in the workbook from which you downloaded this table. If you want to change it, simply type the name you want over what is there. In our example, we want to keep the name for the table as *Donors*. Click **Finish** (figure 1.25, arrow).

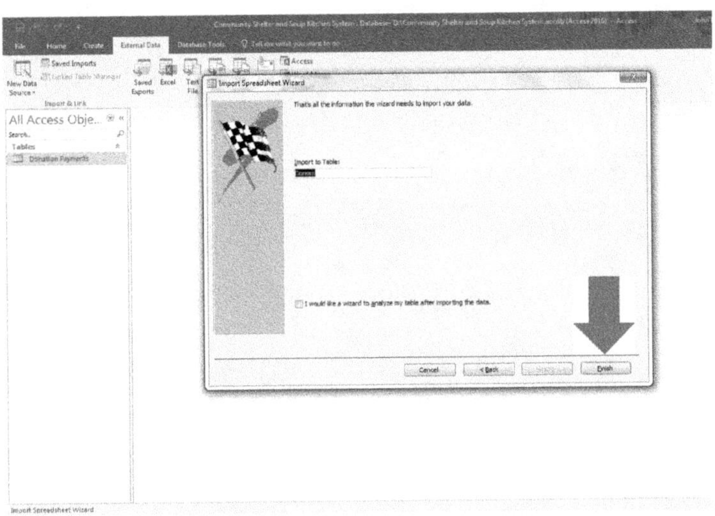

FIGURE 1.25 Importing a spreadsheet (finish)

Step 15: That will take you to the screen you see in figure 1.26. You do **NOT** need to save the steps to an import unless you plan to import this spreadsheet on a regular basis (which is not the case here). Therefore, just click **Close** (figure 1.26, arrow).

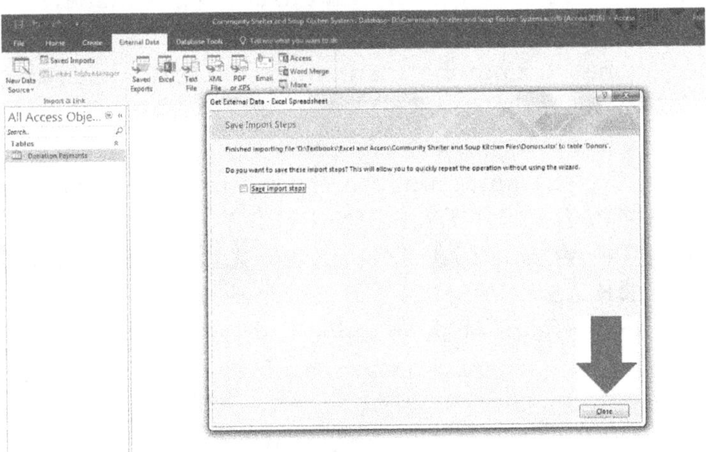

FIGURE 1.26 Closing the save import steps menu

Notice that the new *donors* table is now displayed in the **navigation pan**e as you can see at the arrow in figure 1.27.

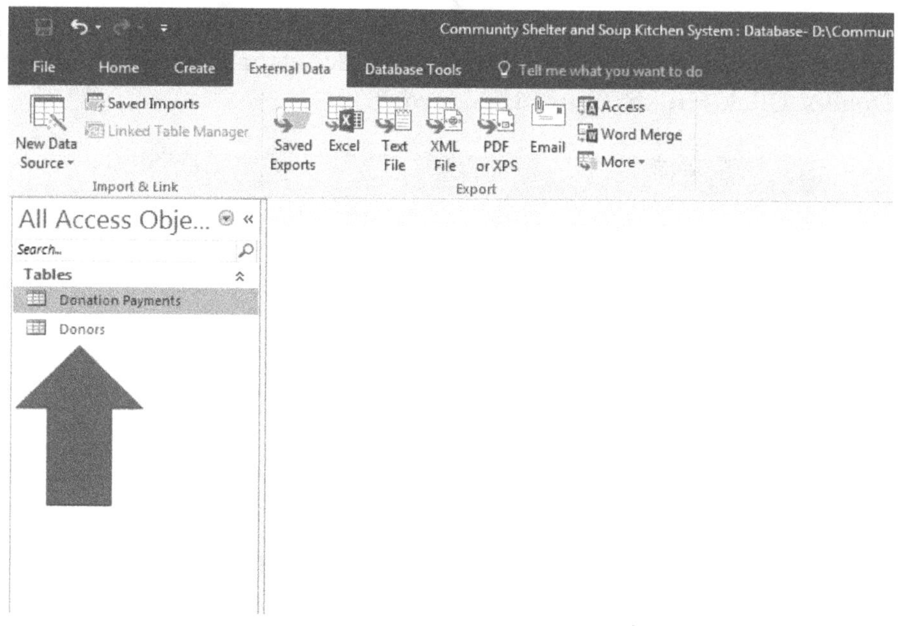

FIGURE 1.27   The donors table in the navigation pane

ADDING FIELDS TO IMPORTED OR EXISTING TABLES IN A QUICK WAY

Whether you are adding fields to a table that you have imported or to a table you have created from scratch, sometimes you will want to add fields that are commonly used. For example, there are often many telephone fields needed, such as home phones, cell phones, office phones, fax numbers, and more. Address fields are also time consuming to enter as you need the street address, the city, the state, the zip code, and more. To save time, you can use the **Add Fields** feature of Access to speed things up as you add fields. For example, suppose you are making a table with *customers*. **NOTE: THIS EXAMPLE IS ONLY BEING USED TO DEMONSTRATE THE ALTERNATE WAY TO ADD FIELDS. THIS TABLE IS NOT TO BE INCLUDED IN THE HOMEWORK ASSIGNMENT.** If you make such a table and you want to add the telephone number fields of them quickly, do the following:

Step 1: Get to the position where you want to add the fields (it will often be the last field) (figure 1.27a, arrow A).

Step 2: Click the **Fields** ribbon of the **table tools** (figure 1.27a, arrow B).

Step 3: Click **More Fields** (figure 1.27a, arrow C).

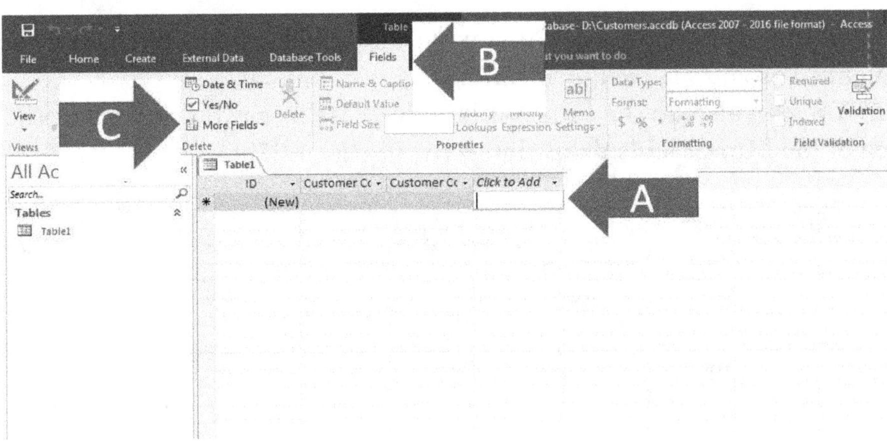

FIGURE 1.27a   Alternate way to add fields to a table in datasheet view (step 1)

Step 4: That will give you what you see in figure 1.27b. Scroll down in the list until you find *phone* (figure 1.27b, arrow).

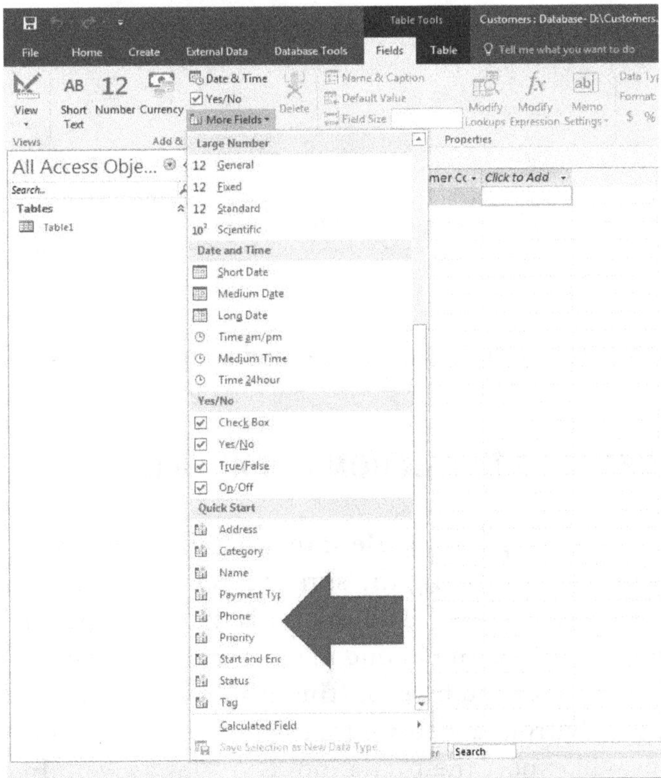

FIGURE 1.27b   Alternate way to add fields to a table in datasheet view (step 3)

It will add the telephone fields for you (*business phone, home phone, mobile phone, and fax number*) as you see in figure 1.27c.

NOTE: You can also do the same with address fields as you can see in the **Quick Start** area of the **More Fields drop-down list** (figure 1.27b).

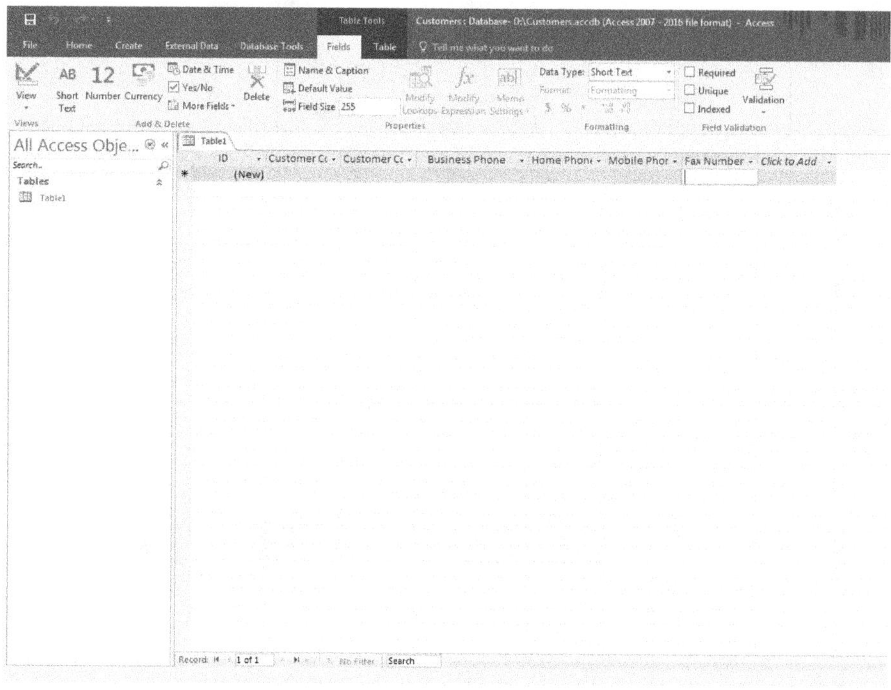

FIGURE 1.27c   The newly added phone fields

Let's now return to the community shelter and soup kitchen system.

CREATING (IMPORTING) A TABLE FROM A TEXT FILE

The procedure for importing a **text file** into a database as a new table is similar to what you just did with importing the spreadsheet. The questions is, why would someone want to import a **text file**? The answer is that when people build databases in Access it is often to replace an old one in a different program, and that program may only be able get data to you by exporting it into a text file. In some cases data is downloaded regularly from a website that is needed to be sent to your database, and many websites will export their data into text files as well. In our case, we are assuming that it is from a website and that the **Notepad** text file we are importing is **comma delimited** as shown in figure 1.28a. As you can see, **comma delimited** text files separate field names with a comma. The data in such files will be separated by quotation marks. In this example, we have a list downloaded from the web with *zip codes*. Notice that the first row of data does **NOT** have column headings or field names. Instead, the first row contains the first record.

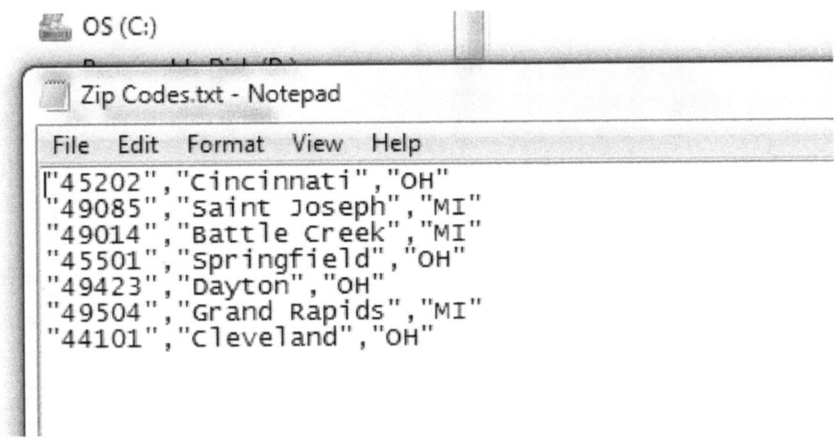

FIGURE 1.28a   Previewing the text file before importing it

Now let us begin to import the file named *Zip Codes* into a new table. You can do so using the following steps:

Step 1: Copy the text file named Zip Codes *that is associated with this book to your desktop or flash drive. The file can be found at* http://www.herzogsystems.com/resources

It is the same file that you see in figure 1.28a.

Step 2: Click the **External Data** ribbon (figure 1.28b, arrow A).

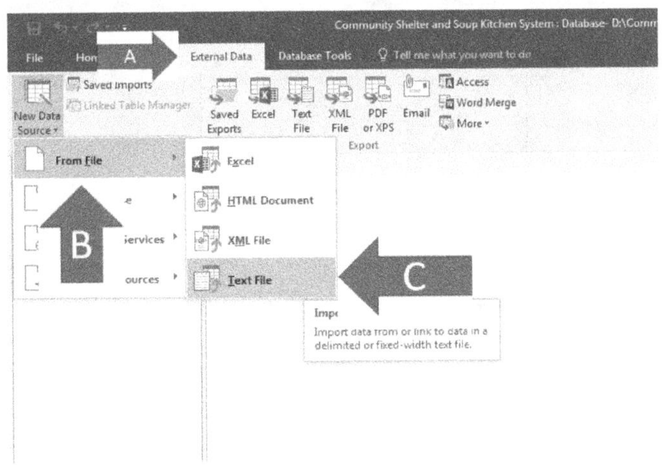

FIGURE 1.28b   Importing a text file (step 2)

Step 3: Click **New Data Source** and then click **From File** (figure 1.28b, arrow B).

Step 4: Click **Text File** in the **drop-down list** that will emerge (figure 1.28b, arrow C). That will give you the menu you see in figure 1.29.

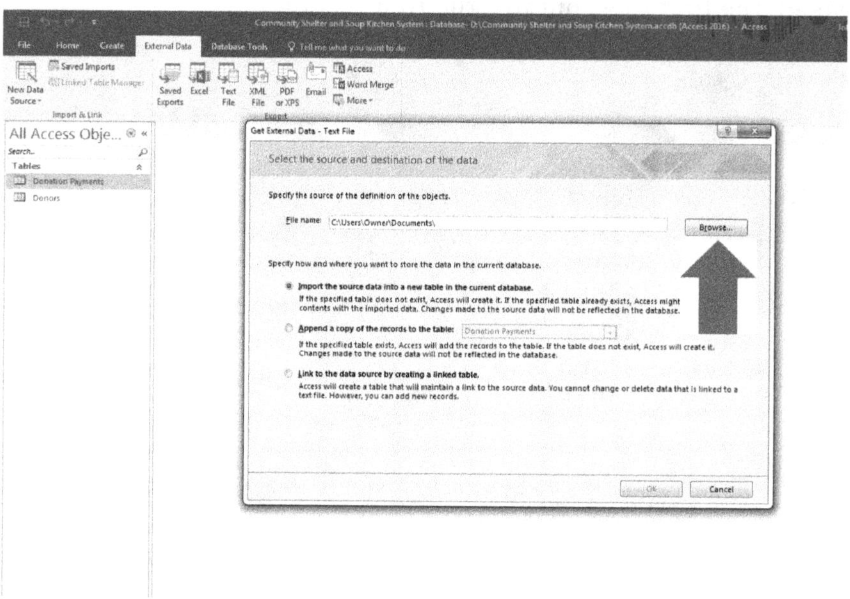

FIGURE 1.29 Importing a text file (step 4)

Step 5: Click **Browse** (figure 1.29, arrow).

Step 6: That will take you to the menu shown in figure 1.30.

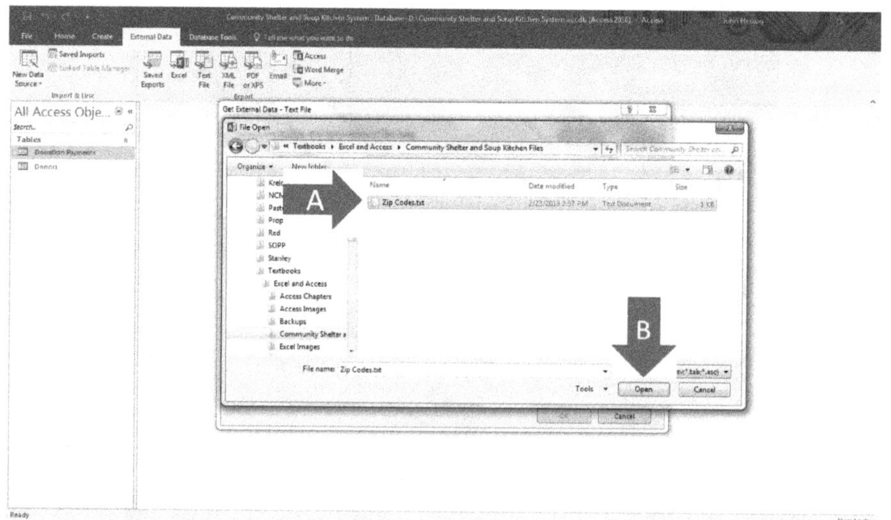

FIGURE 1.30 Importing a text file (step 6)

Notice that Access only displays text files in the folders that we search, because we asked it to look for text files.

Step 7: Click *Zip Codes.txt* (figure 1.30, arrow A) and then click **Open** (figure 1.30, arrow B). When you do it will take you back to the screen you saw earlier (figure 1.31).

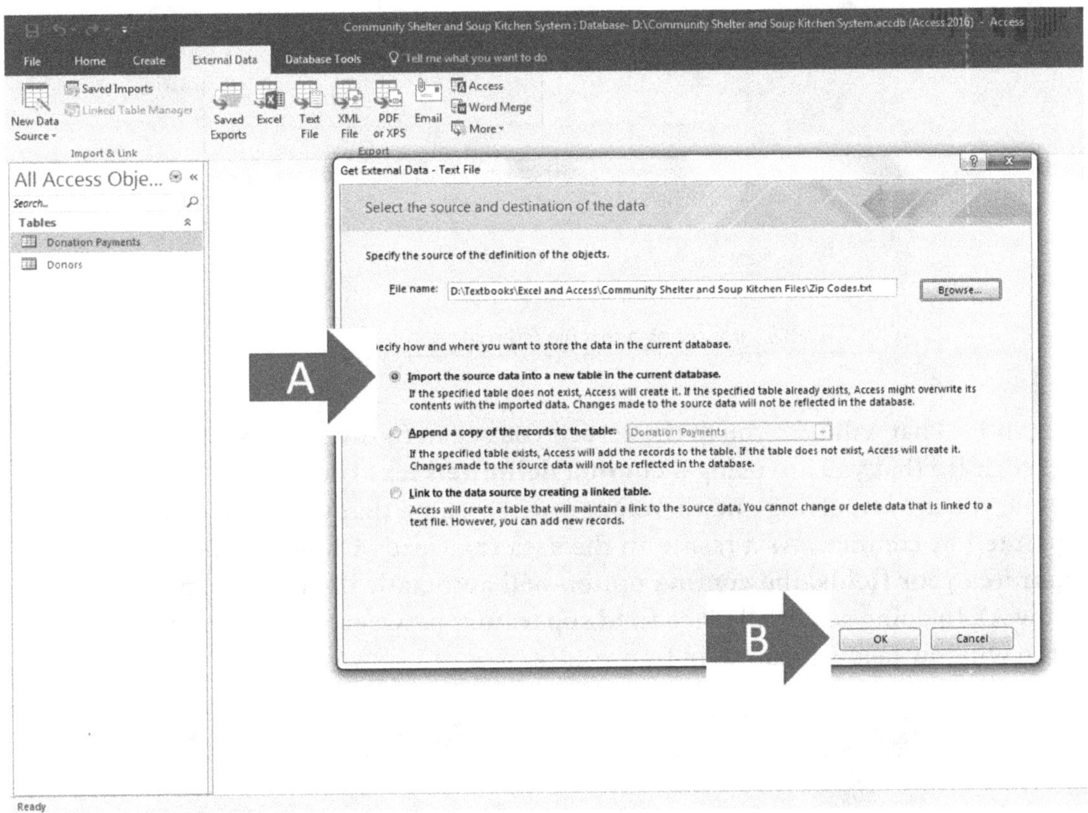

FIGURE 1.31 Importing a text file (step 7 result)

Also, as before, appending a copy of the data (or adding the data) to an existing table is not going to be performed in this example.

Step 8: Make sure that the **Import the source data into a new table in the current database** option button is checked (figure 1.31, arrow A).

Step 9: Click **OK** (figure 1.31, arrow B).

Step 10: That will take you to the menu you see in figure 1.32. As stated before, we are importing a **comma delimited text file**, hence make sure that the option button labeled **Delimited—Characters such as comma or tab separate each field** is selected (figure 1.32, arrow A) and then click **Next** (figure 1.32, arrow B).

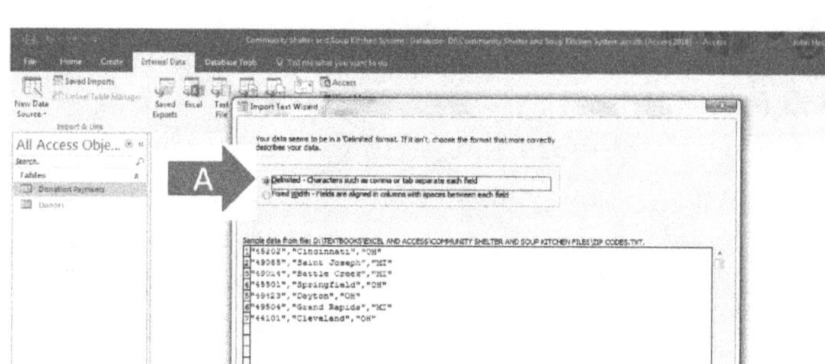

FIGURE 1.32   Importing a text file (step 9)

Step 11: That will take you to the screen you see in figure 1.33. Notice that **Access** will surmise that you are using a **comma delimited text file** due to the structure of the file you are importing and therefore it will assume that you want the fields to be separated by commas. As a result, in the area that reads **Choose the delimiter that separates your fields**, the **comma** option will automatically be chosen (figure 1.34, arrow A). In doing so, it will place **field separators** between the fields, as you can see in the window below arrow A in figure 1.33.

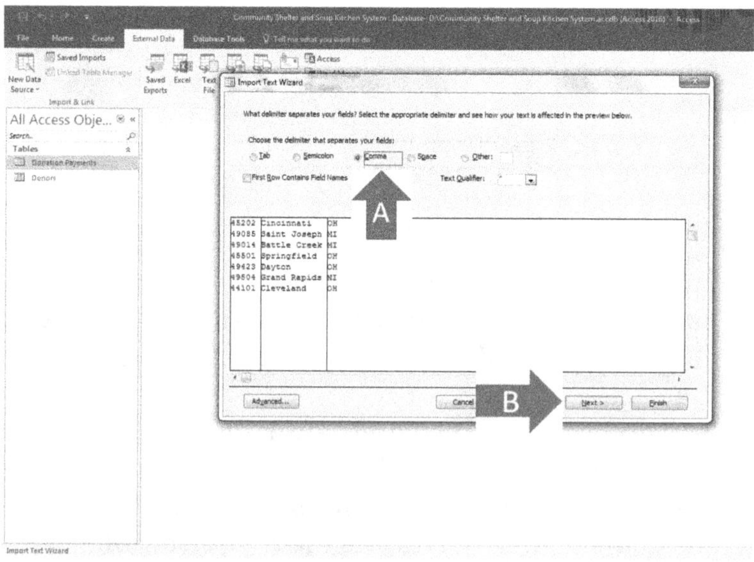

FIGURE 1.33   Importing a text file (step 10)

Step 12. Click **Next** (figure 1.33, arrow B).

Step 13: That will take you to the screen you see in figure 1.34. This screen should look familiar. It works the exact same way that it does for importing an Excel spreadsheet, discussed way back in figure 1.23. However, in this example you will have to change the field names, because the columns had no labels in them like the columns did in the spreadsheet. To name the fields, click the box above their data. For example, click *Field1* (figure 1.34, arrow A) and it will make the column dark.

Step 14: Click inside the box labeled **Field Name** (figure 1.34, arrow B) and then name the first field *Zip Code*. Repeat this step and name *Field2* as *city*, and then name *Field3* as *state*. Do not change their field types in this example.

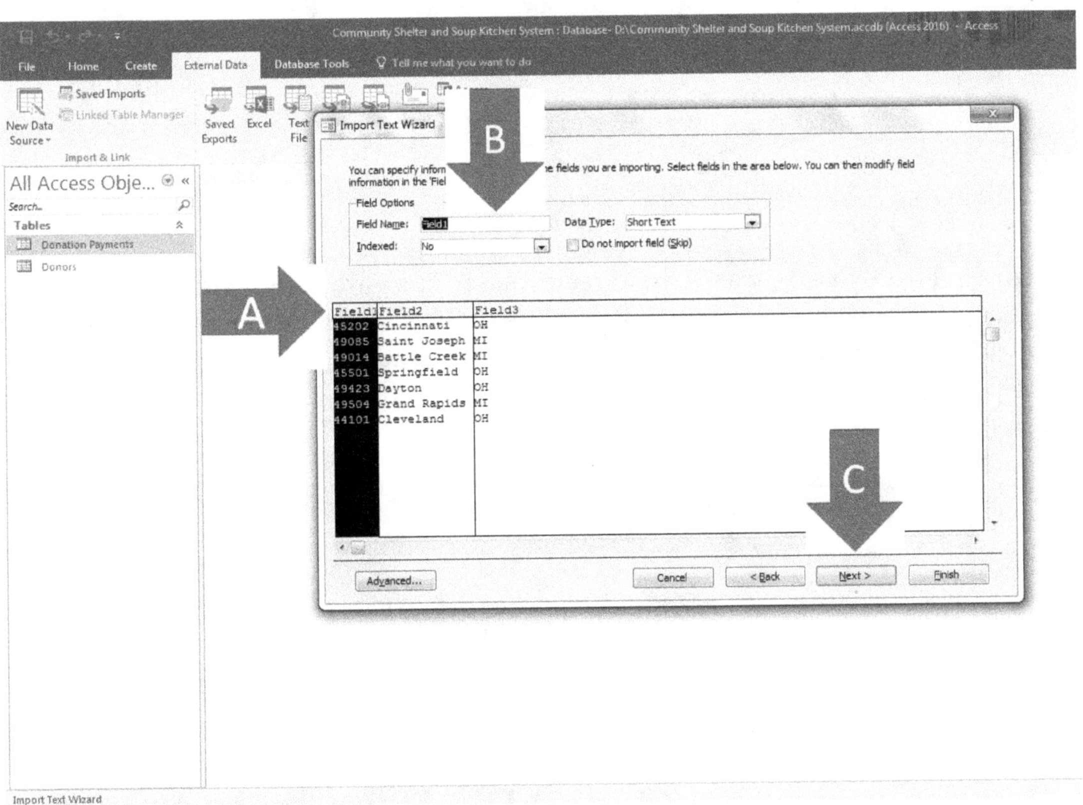

FIGURE 1.34 Importing a text file (step 12)

Step 15: Click **Next** (figure 1.34, arrow C)

Step 16: That will take you to the screen you see in figure 1.35. In this case we will **NOT** allow Access to add a **primary key**. We will choose the *zip code* field in the **drop-down list** box as the **primary key** (figure 1.35, arrow A).

Step 17. Click **Next** (figure 1.35, arrow B).

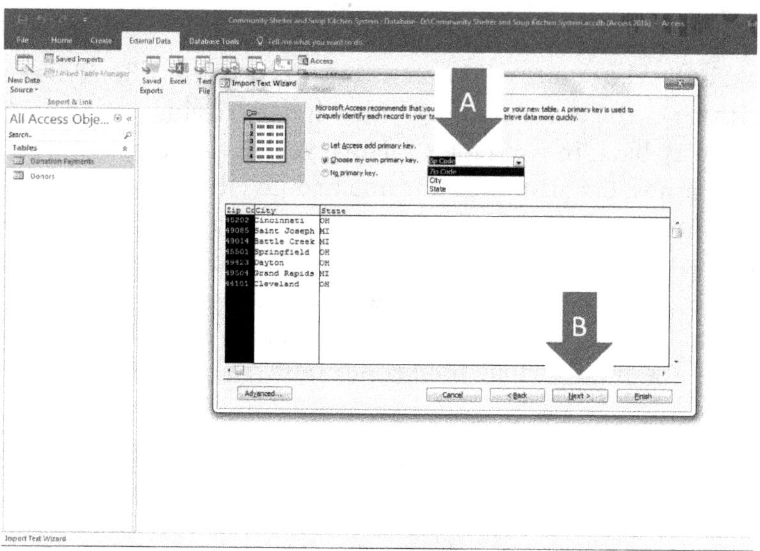

FIGURE 1.35   Importing a text file (step 14)

Step 18: That will take you to the screen you see in figure 1.36. In this case, we are using a text file instead of a spreadsheet, thus there is no referencing of a worksheet tab at this point in order to name the new table. Instead, it uses the name of the text file as your table name. Make sure the table will be named *Zip Codes* and then click **Finish** (figure 1.36, arrow).

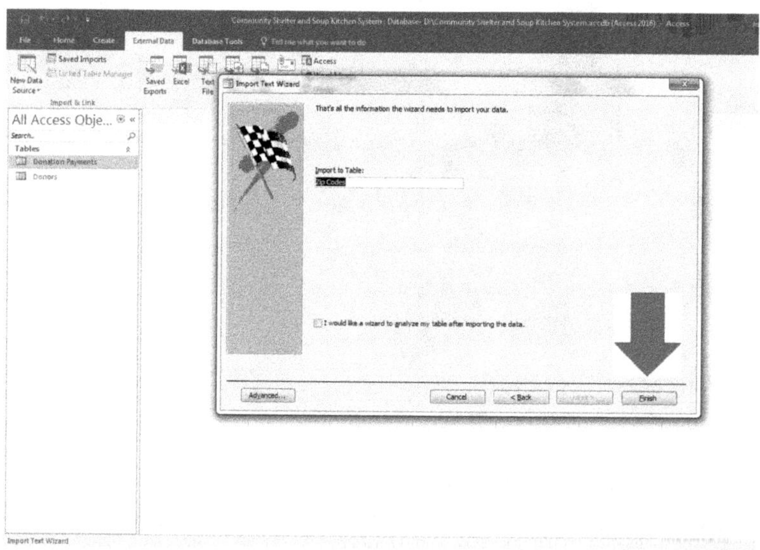

FIGURE 1.36   Importing a text file (step 16)

Step 19: That will take you to the screen you see in figure 1.37. As always, we will not be saving the import steps in this textbook; therefore, click **Close** (figure 1.37, arrow)

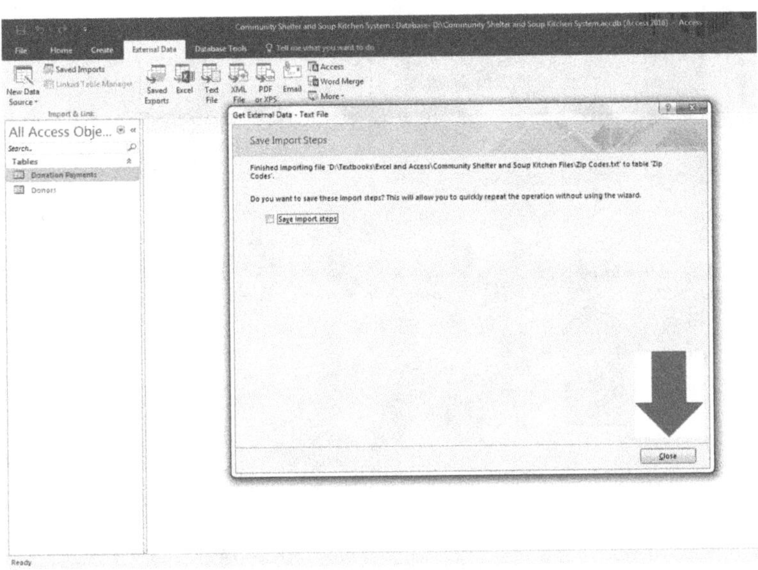

FIGURE 1.37   Importing a text file (step 17)

IMPORTING A TABLE FROM AN EXISTING DATABASE

Why would you need to import a table from another Access database? There are two reasons. First, sometimes people who are amateurs at creating Access databases create tables and find that they have not created the database properly. The problem is that they often enter data into that database and they find that they need that data to be placed into the tables of the new database. Second, it is very common to create a table in Access that is linked to another. For example, if you have databases that use employee information around a large organization, you don't want to have to update all of those employee tables in all of those databases when one employee leaves the organization or when another is hired. If an employee database is created, all of the databases will want to link to it. Thus when one change is made to that employee table, all of the other databases that link to it will instantly see the changes in their systems.

In this example, we will not be linking the table. We will just be importing it.

You can import such a table from an Access database using the following steps:

Step 1: Copy the database named *Donations.accb* that is related to this textbook to your desktop or flash drive.

Step 2: As shown in figure 1.38, click the **External Data** ribbon (figure 1.38, arrow A).
Step 3: Click **New Data Source** (figure 1.38, arrow B).
Step 4: Click **From Database** (figure 1.38, arrow C).
Step 5: Click **Access** (figure 1.38, arrow D).

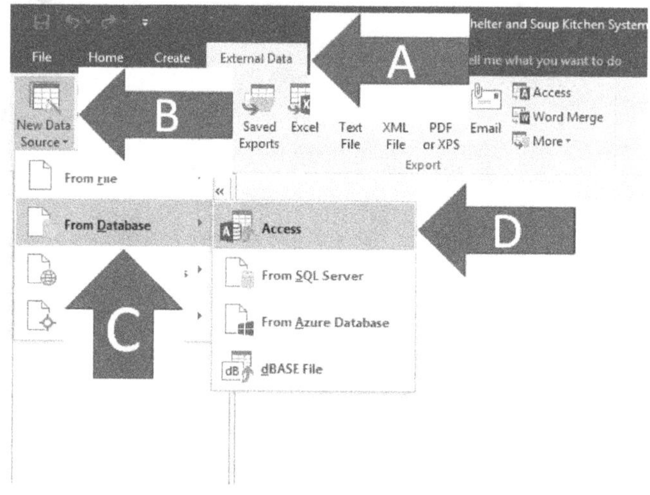

FIGURE 1.38 Importing an Access table (step 2)

Step 6: That will take you to the screen you see in figure 1.39. Click **Browse** (figure 1.39, arrow).

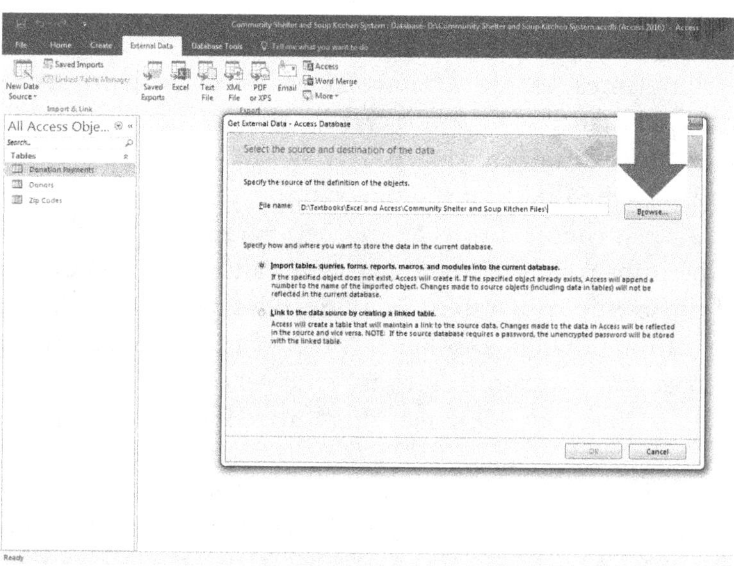

FIGURE 1.39 Importing an Access table (step 3)

Step 7: That will take you to the screen you see in figure 1.40a. As always, you will only see Access database files in the folder, because we have asked Access to import an Access database. In this example, select the *Donations.accdb* database (figure 1.40a, arrow A).

Step 8: Click **Open** (figure 1.40a, arrow B).

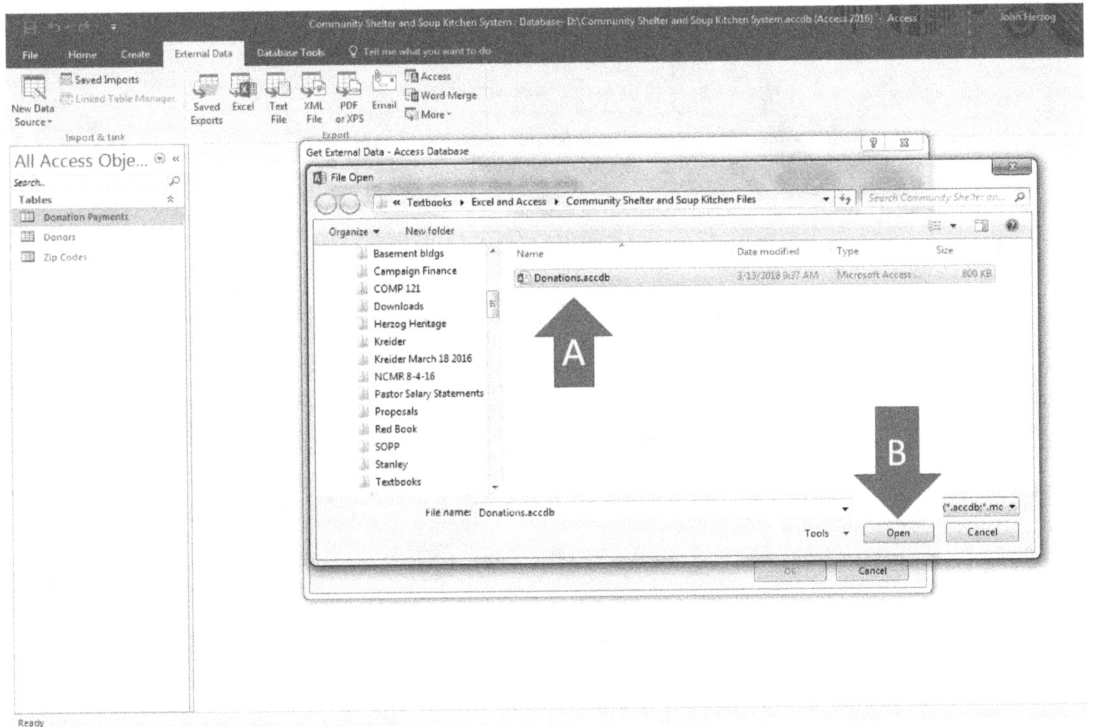

FIGURE 1.40a   Importing an Access table (step 4)

Step 9: That will take you to the screen you see in figure 1.40b.

NOTE: If you wanted to make sure that this table is linked, once again, you can choose the **Link to the data source by creating a linked table**. In this case we don't need to do that, but it is very commonly used. Again, we will import in this case, thus we will choose the option labeled **Import tables, queries, forms, reports, macros, and modules into the current database** (figure 1.40b, arrow A).

Step 10: Click **OK** (figure 1.40b, arrow B).

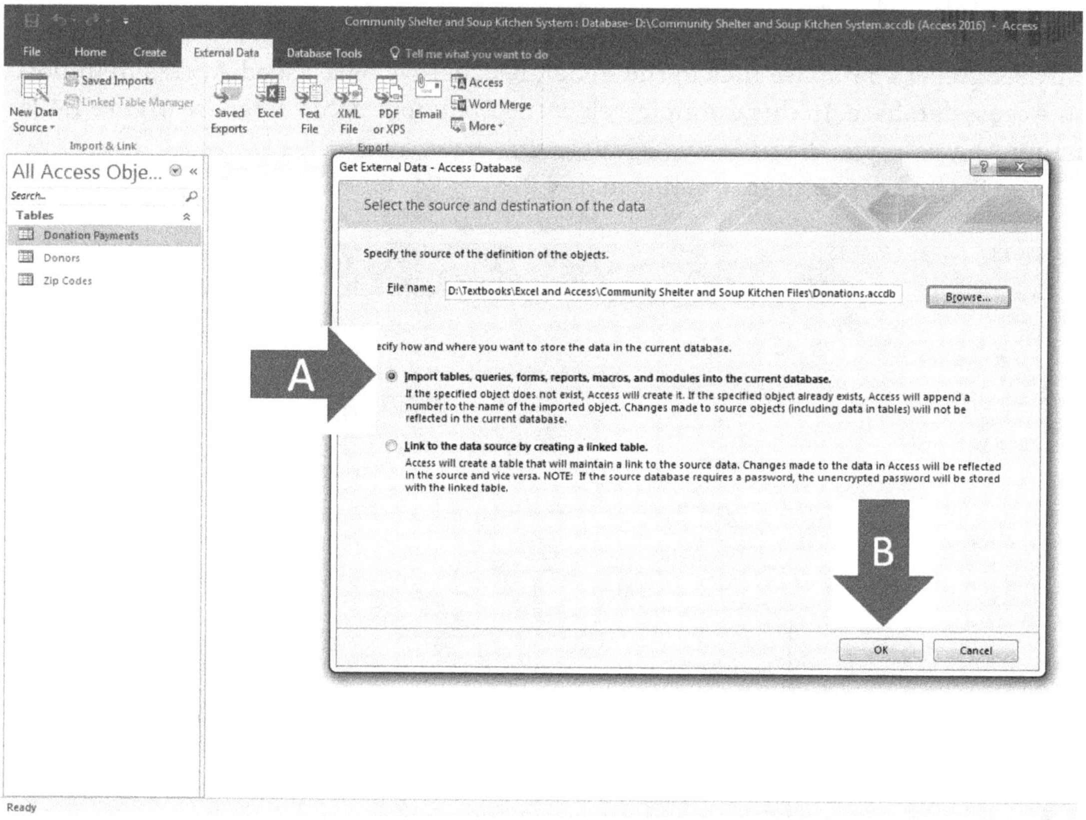

FIGURE 1.40b Importing an Access table (step 5)

Step 11: That will take you to the screen you see in figure 1.41. Click on the table you want to import. In this case it is called *Donations* (figure 1.41, arrow A). After you click it, the line will go dark. If it doesn't, you haven't selected the table, and nothing will be imported.

NOTE: DO NOT JUST CLICK OK WITHOUT SELECTING A TABLE OR OBJECT. IF YOU DO, NOTHING WILL BE IMPORTED.

NOTE: Please notice in figure 1.41, that you can import other **objects** from another Access database, such as the **queries, forms, reports, macros**, and **modules**. We will not be doing that in this example.

Step 12: Click **OK** (figure 1.41, arrow B).

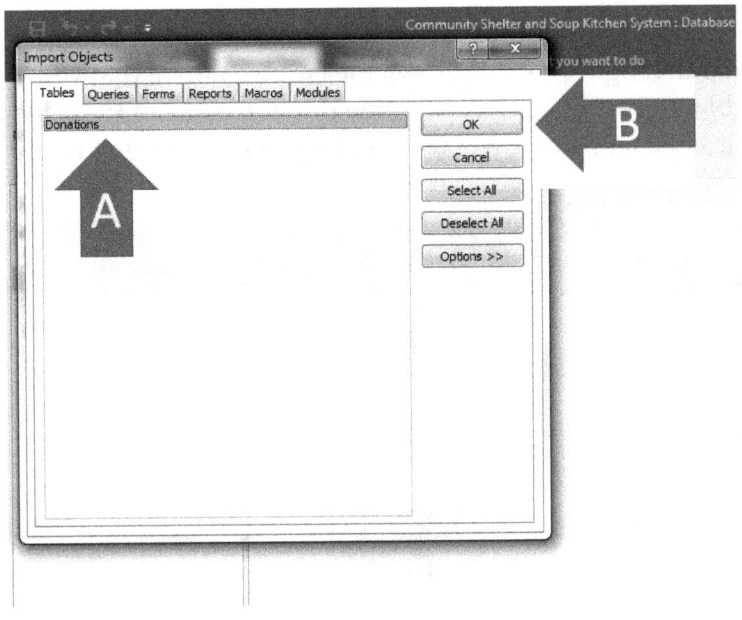

FIGURE 1.41   Importing an Access table (step 6)

Step 13: That will take you to the screen you see in figure 1.42. Do **NOT** save the **import steps**. Click **Close** (figure 1.42, arrow).

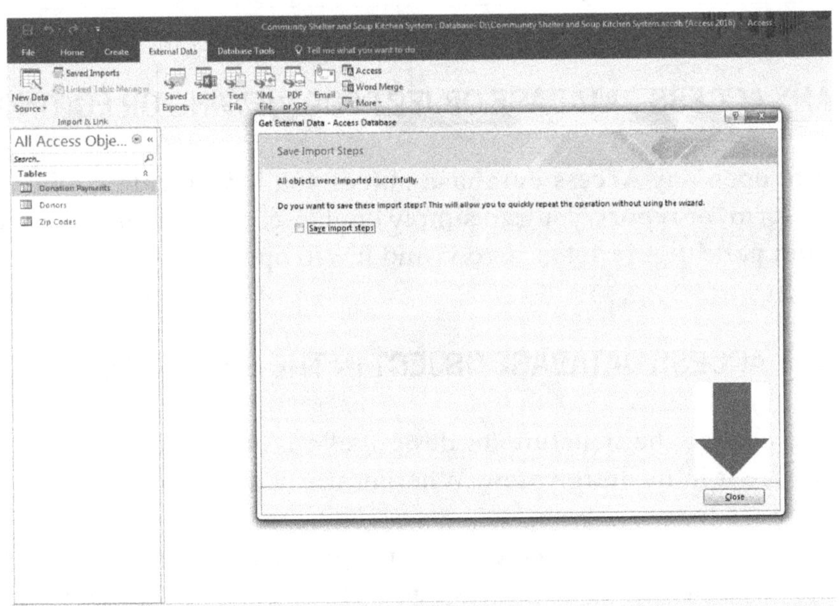

FIGURE 1.42   Importing an Access table (step 7)

Please notice that at this point the new table will be added to your list of tables in the **navigation pane** (figure 1.43a, arrow). Please also notice that you were not asked to name the **table**. Access will always assume that the name of the **table** in the **source database** is the same name that you will use on the new one you have imported.

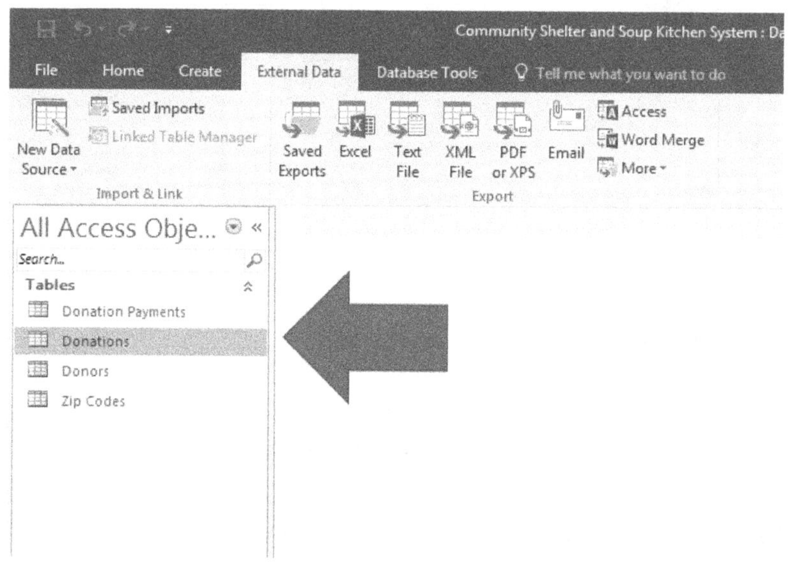

FIGURE 1.43a   Seeing the donations table in the navigation pane

TO OPEN ANY ACCESS DATABASE OBJECT FOR VIEWING OR EDITING

If you want to open any **Access database object** you have created, whether it is a **table**, **query**, **form**, or **report,** you can simply double-click the name of the **object** in the **navigation pane** (figure 1.43a, arrow) and it will open.

TO OPEN ANY ACCESS DATABASE OBJECT IN THE DESIGN VIEW

If you want to change the structure (or design) of any **Access database object**, you will need to open it in its **design view**. Whether it is a **table**, **query**, **form**, or **report** there are two ways to do it.

Method 1: In the **navigation pane** you can right-click on the name of the object and then click **Design View** in the **short-cut menu** (figure 1.43b, arrow).

NOTE: In this same **short-cut menu** you can also open, rename, delete, cut, and copy the object. You can also see its **properties** (which will be discussed later). This menu will also allow you to import or export data as well.

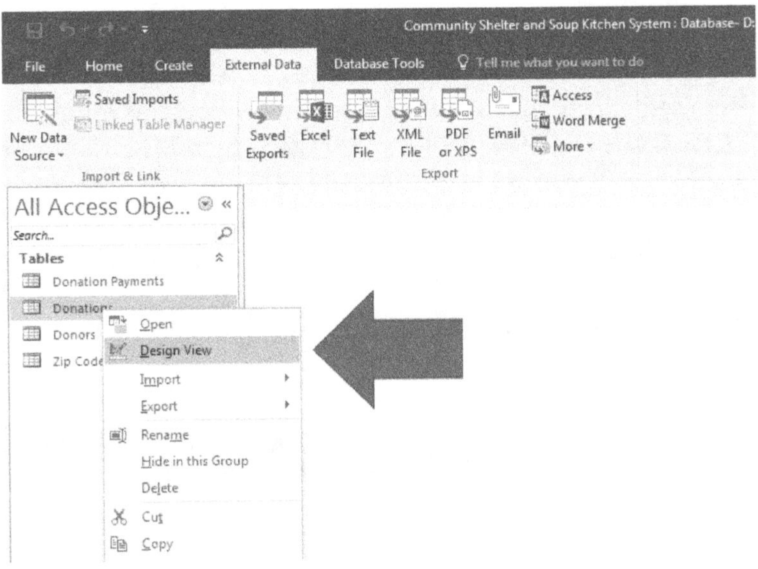

FIGURE 1.43b Going to design view method 1

Method 2: If you have already opened, and are viewing a **table, query, form**, or **report**, you can always click the **View button** in the upper left of your screen and then choose **Design View** (figure 1.43c, arrow).

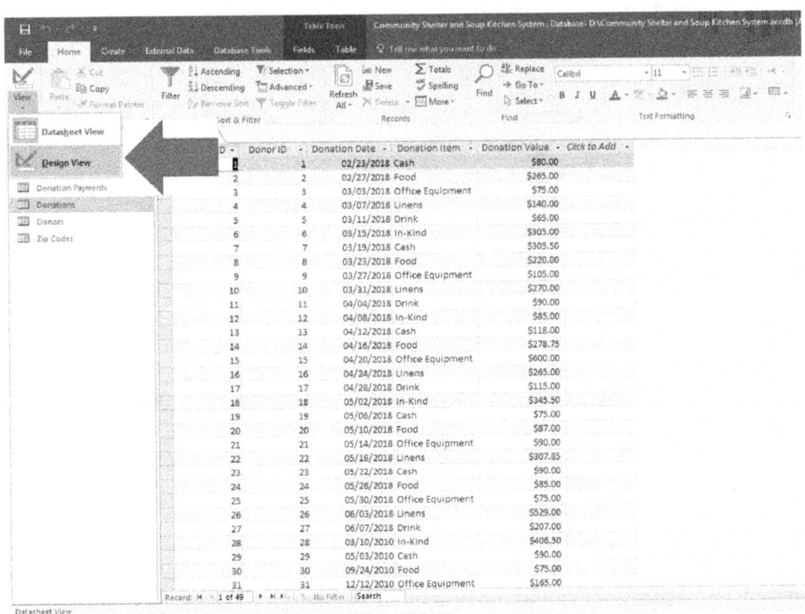

FIGURE 1.43c Going to design view method 2

TO CREATE A SIMPLE FORM OF A TABLE

Forms were discussed earlier and why we have them. You can quickly make a form **WITH ANY TABLE OR QUERY AS ITS SOURCE** by taking the following steps:

Step 1: Open the table you want to use as the source of your form by double-clicking it in the **navigation pane**. In this case it will be the *donors* table (figure 1.44, arrow A). It will open the table to the right of the **navigation pane** as you can see in figure 1.44.

Step 2: Click the **Create** ribbon (figure 1.44, arrow B).

Step 3: Click **Form** (figure 1.44, arrow C).

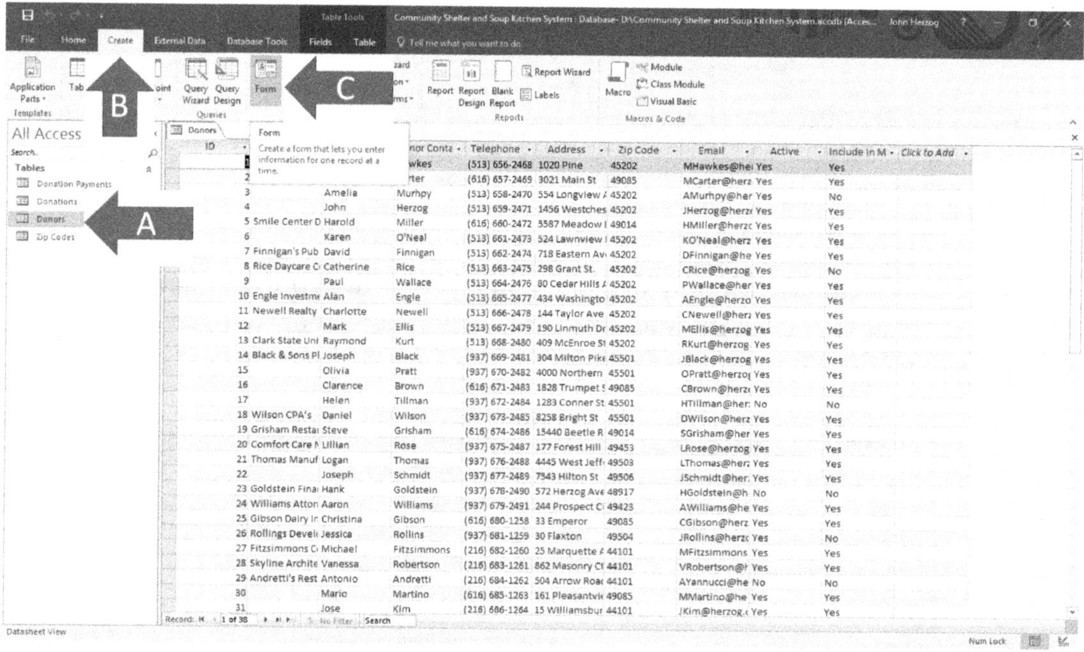

FIGURE 1.44   Creating a simple form (step 1)

Step 4: Your screen will then look like figure 1.45. Click **Save** in the **quick access toolbar** (figure 1.45, arrow).

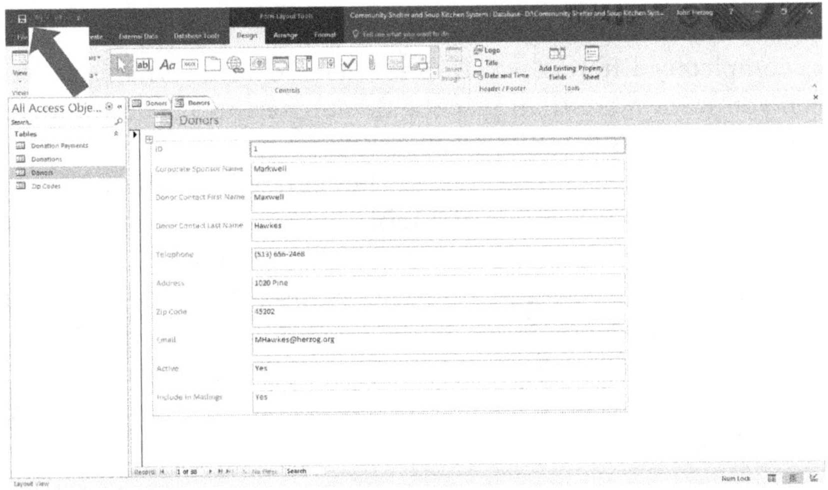

FIGURE 1.45   Creating a simple form (step 2)

Step 5: That will bring you to a screen that is shown in figure 1.46a. Please notice that Access will always assume that you want to name the **form** using the same name as the table. In most cases that is a good idea. In this case we will do just that, therefore all you need to do is click **OK**.

NOTE: The field names in the form will always be the same as the names of the fields in the table, unless you have had to make the captions of the fields different than the field names. In that case, the field captions would then be seen in this form you have just created.

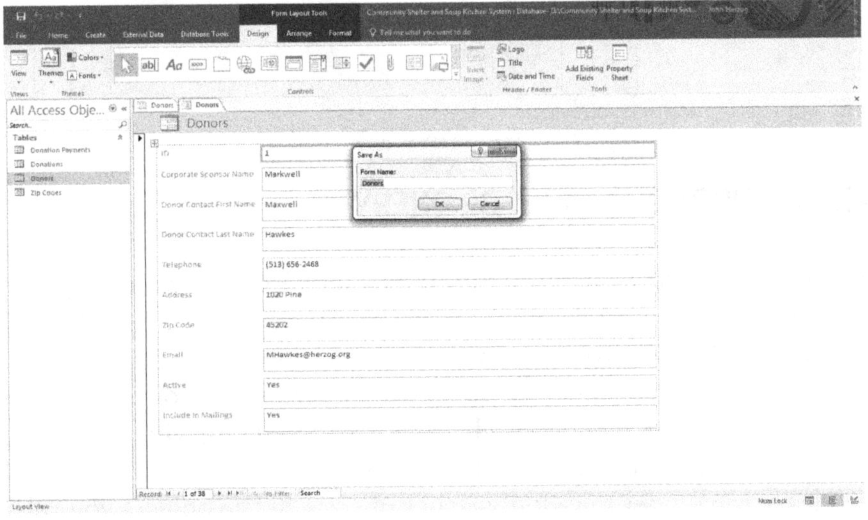

FIGURE 1.46a   Saving a form

You have just created your first **form**!

There is a completely different way to create a form, using what is called the **form wizard**, when you want more than just the default settings. Those procedures will be covered in later chapters.

Data can be entered in these fields just as they can be in a table and whatever you enter into the form will be stored in the table behind the scenes. The **navigation buttons** work the same here as they would in a table or a query.

NOTE: If you are using a report or form you can also open what is called the **layout view**. This allows you to make some design changes while still seeing the data. It gives you an idea as to what your finished form or report will look like as you modify their designs. You can view forms or reports in the **layout view** by choosing it in the **View button** (figure 1.46b, arrow).

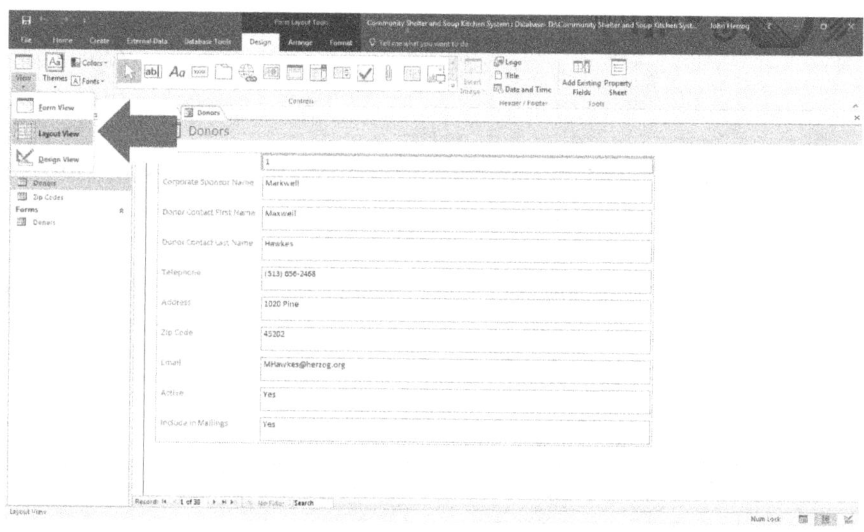

FIGURE 1.46b　Form layout view

TO CREATE A REPORT OF A TABLE OR QUERY

Suppose you want to make a report of a table or query. Reports were discussed earlier as to why we have them. You can quickly make a report **WITH ANY TABLE OR QUERY AS ITS SOURCE**. You can do so by taking the following steps:

Step 1: Open the table or query you want to use as the source of your **report**. In this case open the *Donors* table by double-clicking it in the **navigation pane** (figure 1.47, arrow A). Your screen should look like figure 1.47.

NOTE: Please remember that double-clicking any of the objects you select in the **navigation pane** is one of two different ways to open them. You can also right-click the object and then click **Open** in the **short-cut menu** that would appear.

Step 2: Click the **Create** ribbon (figure 1.47, arrow B).

Step 3: Click **Report** (figure 1.47, arrow C).

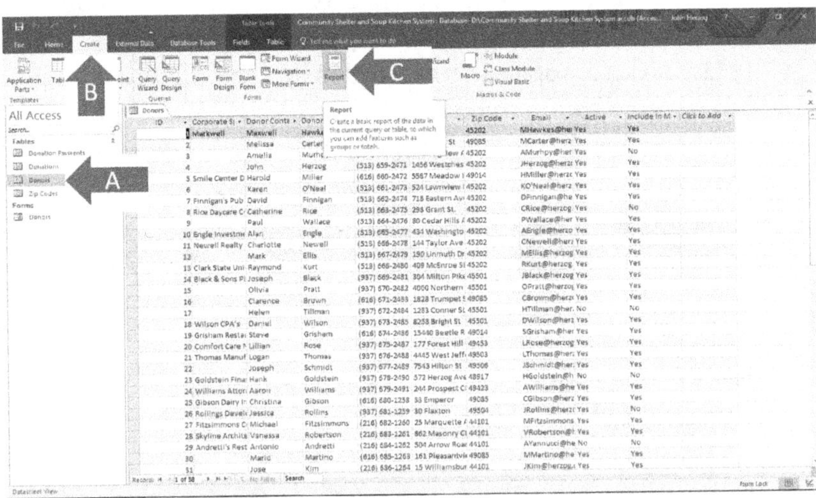

FIGURE 1.47 Opening a table in the navigation pane

Your screen will then look like figure 1.48. You have just created your first **report**!

NOTE: There is a completely different way to create a report, using what is called the **report wizard** when you want more than just the default settings. Those procedures will be covered in later chapters.

Step 4: Click **Save** in the **quick access toolbar** (figure 1.48, arrow).

FIGURE 1.48 Newly created report

Step 5: That will bring you to a screen that is shown in figure 1.49. Please notice, just like with forms, Access will always assume that you want to name the report using the same name as the table or query. In most cases that is a good idea. In this case we will do just that; therefore, all you need to do is click **OK**.

NOTE: The fields in the report will always be the same as the name of the fields in the table unless you have had to make the field captions of the table different than the field names. In that case, the field captions would then be seen in this report you have just created.

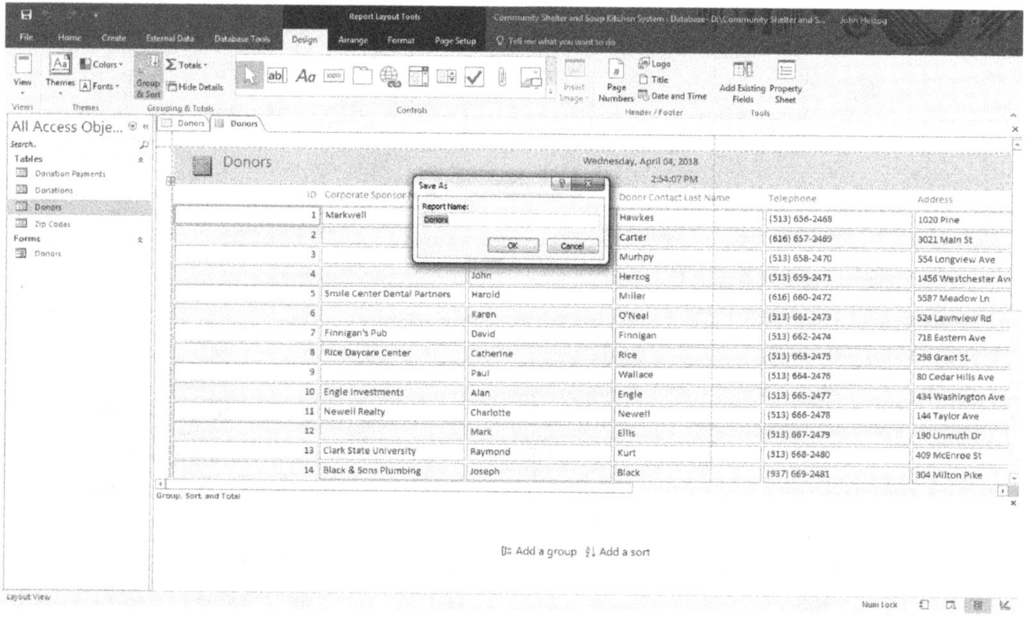

FIGURE 1.49 Saving a report

TO CLOSE THE ENTIRE DATABASE

Click the **upper X** in the upper right of the database window (figure 1.50, arrow).

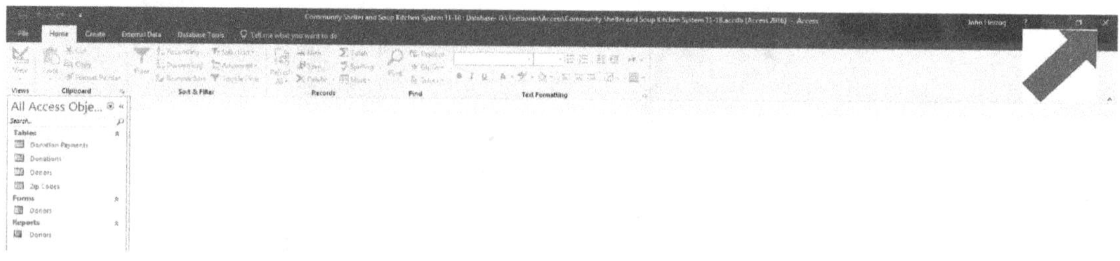

FIGURE 1.50 Close a database

Chapter 1: Assignment (Summary of Tasks)

The local community shelter and soup kitchen wants to keep track of all of their *donors* and the *donations* that they make, in order to give them information for their taxes and to send out mailings when more contributions need to be requested. Please begin to create it by doing the following tasks:

1. Create a new database on your flash drive or desktop. Name it *Community Shelter and Soup Kitchen System*.
2. Create a new table (and when you save it later name it *Donation Payments*). You may use either the **datasheet view** or the **design view** or a combination of both to create this table.
   a. The first field should be called *Payment ID* and it should be an **autonumber** field type. **MAKE THIS FIELD THE PRIMARY KEY.**
   b. The second field should be called *Donation ID* and it should be a **number** field type with a length that is **long integer.**
   c. The third field should be called *Payment Amount* and it should be a **currency** field type.
   d. The fourth field should be called *Check Number* and it should be a **short text** field type with a length of 10 characters.
   e. The fifth field should be called *Payment Date* and it should be a **date/time** field type.
   f. The sixth field should be called *Comments* and it should be a **long text** field type.
3. Enter the following records into the table:

PAYMENT ID	DONATION ID	PAYMENT AMOUNT	CHECK NUMBER	PAYMENT DATE	COMMENTS
Enter no data	1	$50.00	555555	3/05/2018	
in this field,	4	$15.00	8796	3/19/2018	
because this is	3	$25.00	1247	6/25/2018	
an autonumber	4	$25.00	4574	3/10/2018	
field type	7	$15.00	1331	5/06/2018	

4. Import the *Zip Code.txt, Donors.xlsx,*and *Donations.accdb* files associated with this textbook to your desktop. Import these files as new tables into your database (without saving the import steps). Use the following guidelines for each file/table:

    a. *Zip Code*.txt (which is a text file). Name the table *Zip Codes* and name the fields using the following names in the order shown. Do not change any of the field types (thus they will all be text fields):

       i. Field1, zip code. Make this the primary key.

      ii. Field2, city.

     iii. Field3, state.

    b. *Donors.xlsx* (which is an Excel spreadsheet file)

       i. When you import this spreadsheet, use the *Donors* sheet.

      ii. Make sure you indicate that the first row contains column headings by clicking the check box when prompted.

     iii. Do not change any field names or field types.

     iv. Make sure that the option button is selected that is labeled **Let Access add primary key**.

      v. Make sure the table is named *Donors*.

    c. *Donations.accdb* (which is an Access database file)

       i. Import the *Donations* table.

5. Make a simple form of the *donors* table you created earlier. Save the form and name it *Donors*.

6. Make a simple report of the *donors* table you created earlier. Save the report and name it *Donors*.

DO NOT TURN IN THIS DATABASE UNTIL ALL HOMEWORK IS COMPLETED FROM THE REMAINING CHAPTERS.

Chapter 1: Assignment (Alternate) (Summary of Tasks)

The people at the local village bookstore want to keep track of all of their *inventory orders* they have placed with their book publishing *vendors*. They also want to track how many of their book orders they have already received. By doing this, they will know what books they now have on hand. To help them do so, please perform the following tasks:

1. Create a new database on your flash drive or desktop. Name it *Village Bookstore Database System*.

2. Create a new table (and when you save it name it *Shipments Received*). You may use either the **datasheet view** or the **design view** or a combination of both to create this table.

   a. The first field should be called *Shipment ID* and it should be an **autonumber** field type. **MAKE THIS FIELD THE PRIMARY KEY.**

   b. The second field should be called *Order ID* and it should be a **number** field type with a length that is **long integer**.

   c. The third field should be called *Quantity Received* and it should be a **number** field type with a length that is **long integer**.

   d. The fourth field should be called *Freight Bill Number* and it should be a **short text** field type with a length of 10 characters.

   e. The fifth field should be called *Received Date* and it should be a **date/time** field type.

   f. The sixth field should be called *Comments* and it should be a **long text** field type.

3. Enter the following records into the table:

SHIPMENT ID	ORDER ID	QUANTITY RECEIVED	FREIGHT BILL NUMBER	RECEIVED DATE	COMMENTS
Enter no data in this field, because this is an autonumber field type	1	200	555555	10/8/2018	Will send the rest of the order in two weeks.
	4	50	8796	10/19/2018	The remaining will be received soon.
	3	400	1247	10/25/2018	
	2	50	4574	10/21/2018	
	7	500	1331	10/31/2018	

4. Import the following files associated with this textbook to your desktop or to your flash drive: *Zip Code.txt, Vendors.xlsx,* and *Inventory Orders.accdb.* In the

Inventory Orders.accdb file there is a table named *Book Titles* that must also be imported. Import these files as new tables into your database (without saving the import steps). Use the following guidelines for each file/table:

a. *Zip Code*.txt (which is a text file). Name the table *Zip Codes* and name the fields using the following names and do not change any of the field types (thus they will all be text fields):

   i. Field1, zip code. Make this the primary key.

   ii. Field2, city.

   iii. Field3, state.

b. *Vendors.xlsx* (which is an Excel spreadsheet file)

   i. When you import this spreadsheet, use the *Vendors* sheet.

   ii. Make sure you indicate that the first row contains column headings by clicking the check box when prompted.

   iii. Do not change any field names or field types.

   iv. Make sure that the option button is selected that says **Let Access add primary key.**

   v. Make sure the table is named *Vendors*.

c. *Inventory Orders.accdb* (which is an Access database file).

   i. Import the *Inventory Orders* table.

   ii. Import the *Book Titles* table.

5. Make a simple form of the *vendors* table you created earlier. Save the form and name it *Vendors*.

6. Make a simple report of the *vendors* table you created earlier. Save the report and name it *Vendors*.

DO NOT TURN IN THIS DATABASE UNTIL ALL HOMEWORK IS COM-PLETED FROM THE REMAINING CHAPTERS.

Creating Referential Integrity and Queries

IN THIS CHAPTER, you will learn how to do the following:

1. Understand referential integrity (often called one-to-many relationships).
2. Create referential integrity.
3. Create **queries** with different types of criteria using the **query wizard**.
4. Create **queries** with different types of criteria using the **query design view**.
5. Delete unwanted fields or **tables** in a **query**.
6. Apply filters to tables.
7. Rearrange fields in queries and tables.

CHAPTER 2: TASK FINDING CERTAIN DONATIONS AND MAKING SURE NO DONATIONS ARE ADDED WITHOUT A MATCHING DONOR

If you look at the tables we have created so far for the database we are building, you can see that a *donor ID* is entered into each *donation* in order to tell the users who made a given *donation*. The *donor ID*s are used to prevent the need of having to enter **ALL** information about each *donor* into each *Donation*.

Everything will be fine if someone enters a *donor ID* into a *donation* (figure 2.1, arrow B), when in fact that **PARTICULAR** *ID* **DOES** exist in the *donors* table (figure 2.1, arrow A). But what if they **DON'T**'? That *donation* record will have no use!

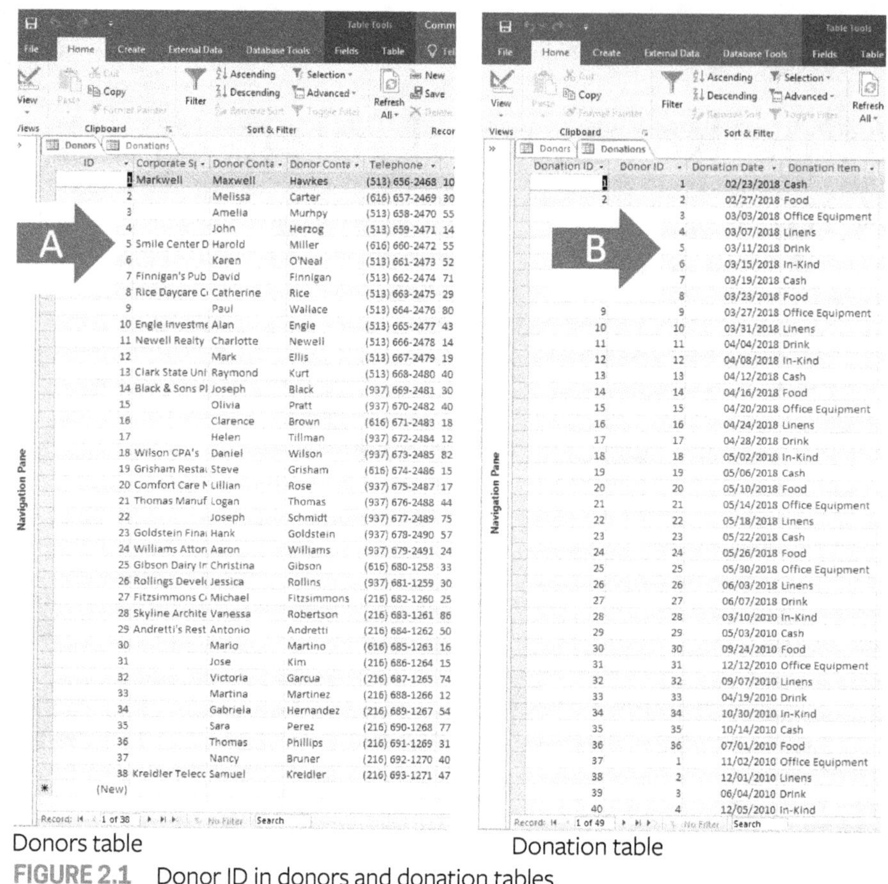

FIGURE 2.1   Donor ID in donors and donation tables.

The *donation* will not be visible in any list that we would want to print in any report or query. They refer to such circumstances with records as being in a **database black hole**. Access will permit you to enter such data unless you use **referential integrity** (often called a **one-to-many relationship**). They need to be established between joined tables to avoid this problem.

CREATING REFERENTIAL INTEGRITY

Setting **referential integrity** is a process that requires a few subprocesses. It can be accomplished by taking the following steps:

Setting referential integrity step 1: Make sure that the tables you will be joining have the following fields:

a. Fields (one in each table) that will link the two tables together
b. Fields that are of the same **field data type**

NOTE: If they are fields with **field data types** that are **numeric** they must have the same **column width**.

In this *community shelter and soup kitchen system*, we have already accomplished this. The *ID* field of the *donors* table will with join with the *donor ID* field in the *donations* table. When you check the **design view** of each table, you can see that they are both **long integer** in length.

Setting referential integrity step 2: Make sure that the two tables you are joining do not already have data that violate **referential integrity rules**. Does the *donations* table (the **child table**) have *donor IDs* that do not exist or has no match in the *donors* table (the **parent table**)? That would take a long time to figure out if you had thousands of records. Therefore, to do this, you can use the find **unmatched query wizard**. This can be done in the following steps:

Finding referential integrity violations in existing tables sub-step 1: Click the **Create** ribbon (figure 2.2, arrow A).

Finding referential integrity violations in existing tables sub-step 2: Click **Query Wizard** (figure 2.2, arrow B).

Finding referential integrity violations in existing tables sub-step 3: When you do, the **New Query Dialogue Box** will appear. Click **Find Unmatched Query Wizard** (figure 2.2, Arrow C) and then click **OK**.

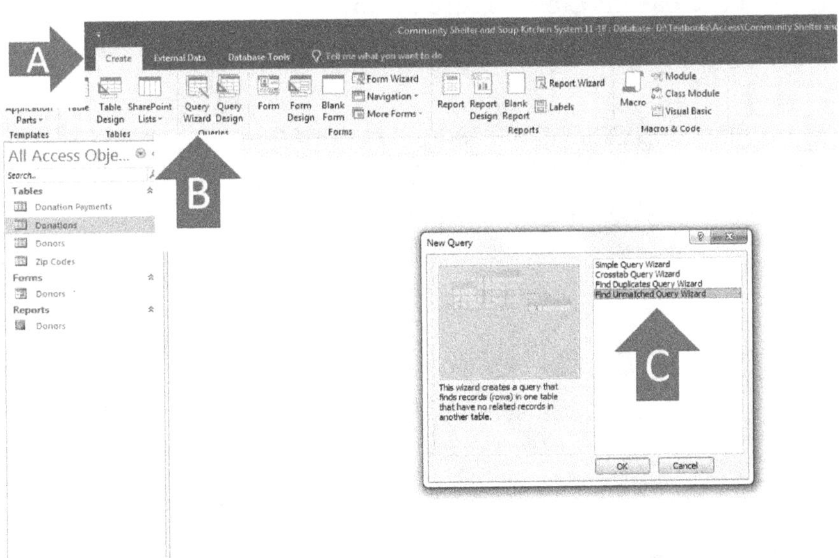

FIGURE 2.2 Creating a find unmatched records query wizard sub-step 1

Finding referential integrity violations in existing tables sub-step 4: That will take you to the screen in figure 2.3. Make sure that the **child table** is selected. In this case the **child table** is the *Donations* table (figure 2.3, arrow A). Then click **Next** (figure 2.3, arrow B).

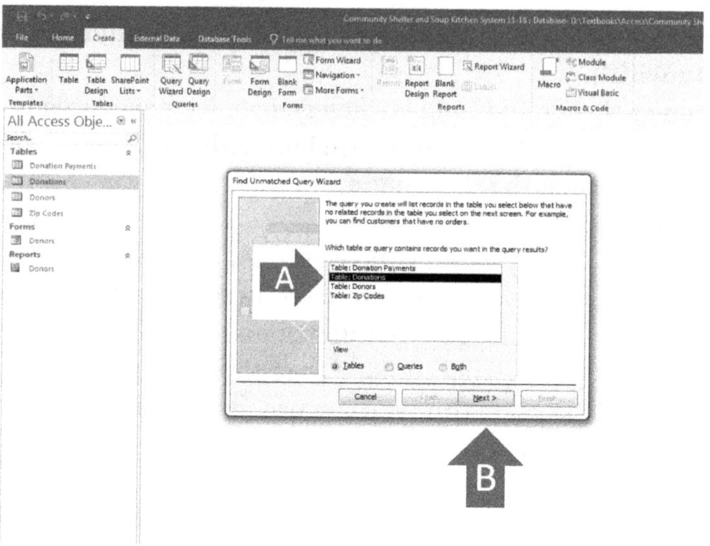

FIGURE 2.3   Creating a find unmatched records query wizard sub-step 2

Finding referential integrity violations in existing tables sub-step 5: That will give you the screen in figure 2.4. Make sure that the **parent table** is chosen. In this example, the *donors* table is the **parent table** (figure 2.4, arrow A).

Finding referential integrity violations in existing tables sub-step 6: Click **Next** (figure 2.4, arrow B).

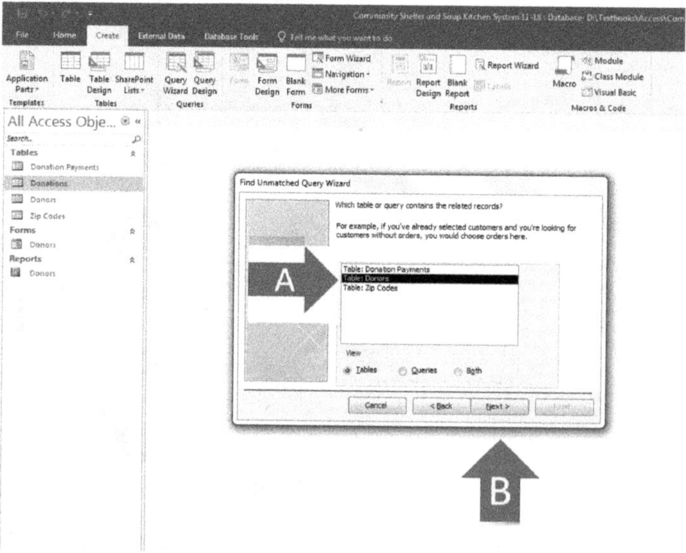

FIGURE 2.4   Creating a find unmatched records query wizard sub-step 3

Finding referential integrity violations in existing tables sub-step 7: That will give you the screen showing in figure 2.5. Make sure that the two fields by which you are joining the tables are selected. In this case, on the *donations* side, it will be the *donor ID* field (figure 2.5, arrow A) and on the *donors* side it will be the *ID* field (figure 2.5, arrow B).

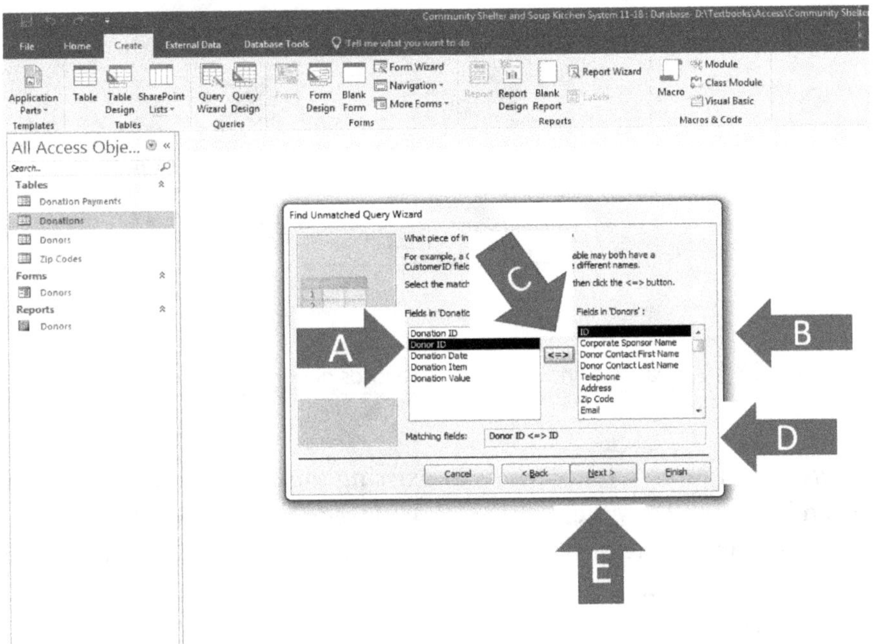

FIGURE 2.5   Creating a find unmatched records query wizard sub-step 4

Finding referential integrity violations in existing tables sub-step 8: Make sure that the **Matching Fields** <=> box is clicked (figure 2.5, arrow C). That will declare the fields you have chosen as the ones you will be checking for unmatched data. It will also enter *Donor ID <=> ID* in the **Matching fields:** box below it (figure 2.5, arrow D).

Finding referential integrity violations in existing tables sub-step 9: Click **Next** (figure 2.5, arrow E).

Finding referential integrity violations in existing tables sub-step 10: That will give you the screen in figure 2.6. Here, Access will ask you to choose the fields to display in the **query** when they find the bogus records that do not have matching data. Click the **Double Arrow** (>>) (figure 2.6, arrow A). That will select all of the fields and it will send them from the **Available fields** window (figure 2.6, arrow B), into the **Selected fields** window (figure 2.6, arrow C).

Finding referential integrity violations in existing tables sub-step 11: Click **Next** (figure 2.6, arrow D).

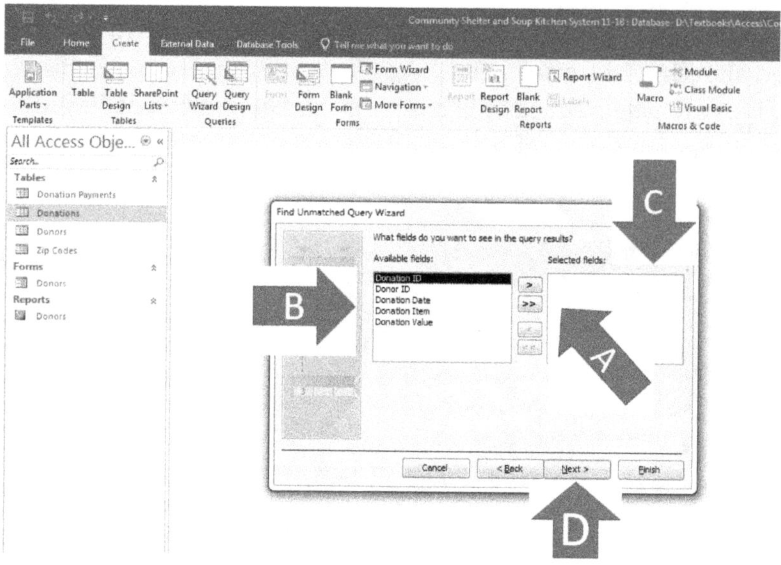

FIGURE 2.6   Creating a find unmatched records query wizard sub-step 5-7

Finding referential integrity violations in existing tables sub-step 12: That will give you the screen in figure 2.7. When you are asked to name the **query** you have just created, the default name is the best name you can give it. In this example it will name it *Donations without Matching Donors* (figure 2.7, arrow A) and that's exactly what we need it to be named.

Finding referential integrity violations in existing tables sub-step 13: Click **Finish** (figure 2.7, arrow B).

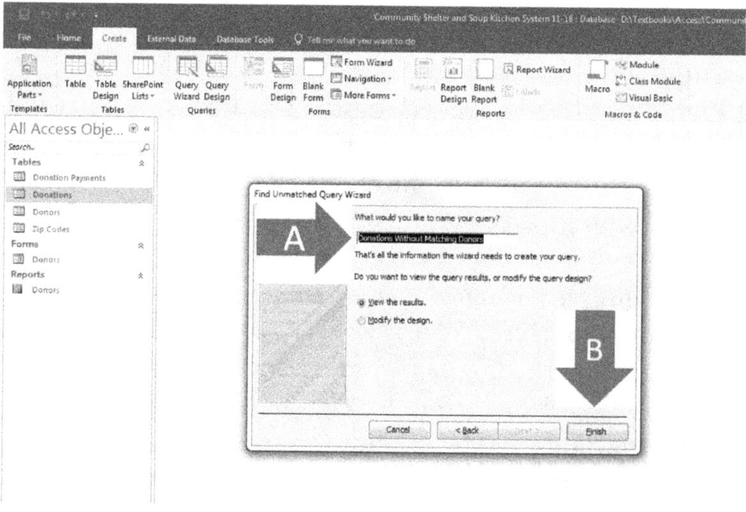

FIGURE 2.7   Creating a find unmatched records query wizard sub-step 8

Finding referential integrity violations in existing tables sub-step 14: The bogus records of *donations* with no matching *donors* are now displayed (figure 2.8, arrow). They are of no use, because the *donor IDs* are bogus; thus, they need to be deleted.

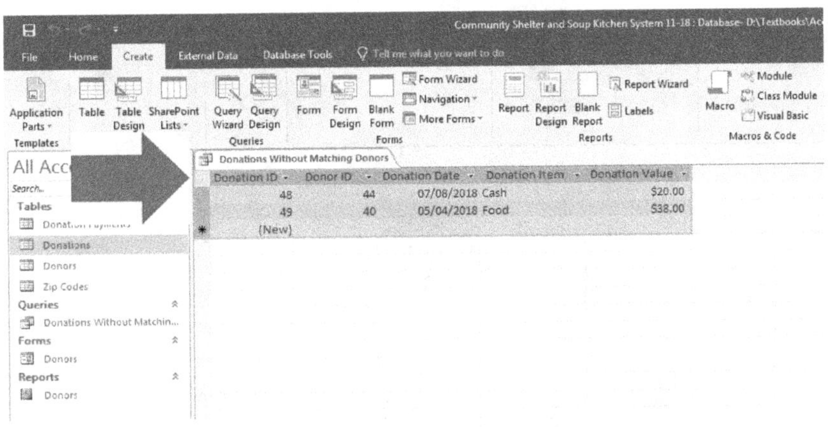

FIGURE 2.8   Creating a find unmatched records query wizard sub-step 10

TO DELETE RECORD(S)

Finding referential integrity violations in existing tables sub-step 15: Referring back to figure 2.8, Click on the **Select All** box at the top left of the records (figure 2.8, arrow).

Finding referential integrity violations in existing tables sub-step 16: When you do, all of the records will be selected. Press the **Delete key** on your keyboard. When asked if you are sure you want to delete the records, click **Yes**.

NOTE: Some ask why the records are to be deleted. The answer is that they do not have any use. As mentioned before, with bogus *donors* entered, there is no way that they can be attributed to a bona fide *donor*. There are those who may try to figure out who the *donors* are to those donations by other data already entered in those records, but in most cases, especially if hundreds of records are displayed, that isn't very practical, and thus the records must be deleted.

JOINING THE TABLES

Setting referential integrity step 3: Resuming now to the step 3, now that the bogus records have been found, CLOSE THE TABLES YOU ARE ABOUT TO JOIN AND CLOSE ALL OF THE OBJECTS (QUERIES, FORMS, AND REPORTS) THAT READ THOSE TABLES IN THE ENTIRE DATABASE!!

Setting referential integrity step 4: Click the **Database Tools** ribbon (figure 2.9, arrow A).

Step 5: Click **Relationships** (figure 2.9, arrow B).

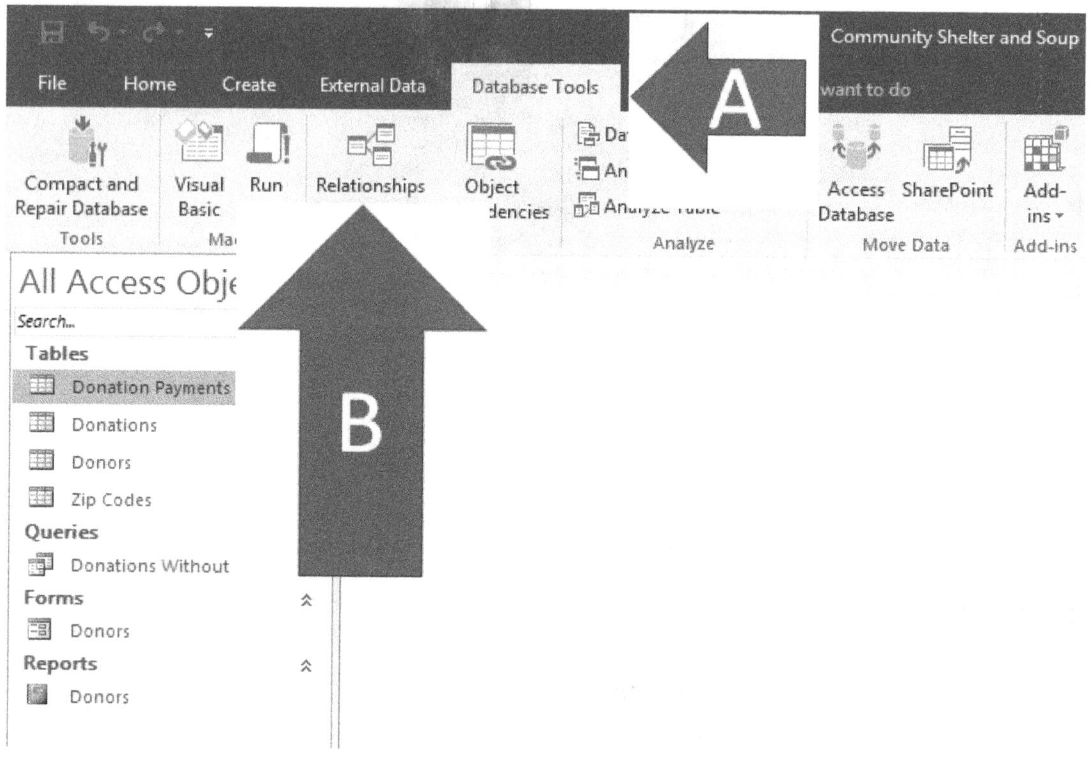

FIGURE 2.9   Creating referential integrity step 4-5

Step 6: That will take you to the screen you see in figure 2.10. The **Show Table Menu** should automatically appear (figure 2.10, arrow A).

NOTE: If the Show Table menu does not appear, click Show Table in the Relationship Tools Design ribbon.

Click the first table for which you would want to set a relationship. In this example, it would be the *donors* table (figure 2.10, arrow B). It is usually easier to understand the table relationships if the **parent** tables are to the left of the **child** tables. That would put the *donors* table first, followed by the *Donations* table.

Step 7: After selecting the *donors* table, click the **Add button** (figure 2.10, arrow C) at the lower part of the **Show Table menu**.

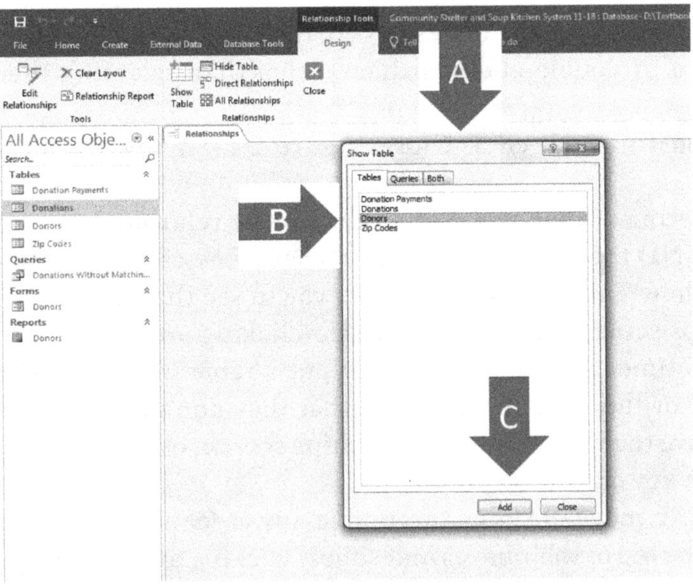

FIGURE 2.10   Show table menu

That will place the *donors* table in the upper left of the **relationships screen** (Figure 2.11). Repeat this process for all of the remaining tables. In figure 2.11, you can see that three tables have been placed in the upper left of the **relationships screen** (figure 2.11, arrow A), and that the *zip codes* table can now be added by clicking on it (figure 2.11, arrow B) and then by clicking **Add** (figure 2.11, arrow C).

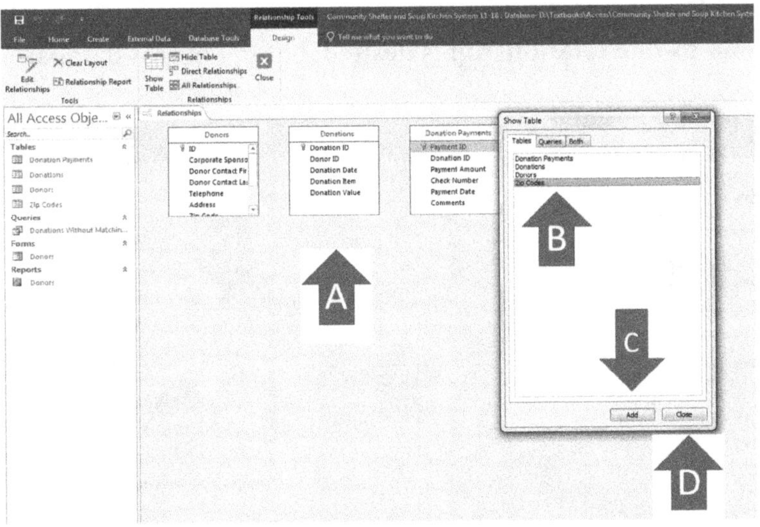

FIGURE 2.11   Setting referential integrity clicking the add button

NOTE: You can also get the tables to take their places in the upper left of the **relationships screen** by double-clicking them rather than selecting them and clicking **Add** each time.

Step 8: When finished, click **Close** (figure 2.11, arrow D) to close the **Show Table menu**.

NOTE: Sometimes when you add a table to the **relationships screen**, they will be placed **BEHIND** the **Show Table menu** where they can't be seen. Click and drag the top of the **Show Table menu** out of the way to see that they have been placed in the **relationships screen**. Do not keep double-clicking until you see the table appear.

NOTE: Sometimes people accidentally put a table in the **relationships screen** more than once or they put objects there that they don't need in the **relationships screen**. To remove them from the **relationships screen**, click on the actual object and press the **Delete** key on your **keyboard**.

NOTE: You can move the tables around in any order you want, by clicking on the table names at the top of their boxes and simply clicking and dragging them wherever you want them.

Step 9: Let's now create the relationship between the *donors* **table** and the *donations* **table**. Click and drag the *ID* field from the *donors* **table** (figure 2.12a, arrow A) over the top of the *donor ID* field in the *donations* table. Make sure that your **mouse arrow** with a **plus sign** (+) attached to its lower right (figure 2.12a, arrow B) will be pointing to the *donor ID* field in the *donations* **table** when you do this. When you let go of the mouse, the **Edit Relationships menu** (figure 2.12a, arrow C) will pop up. It will confirm that you have just begun to join the *ID* field from the *donors* **table** to the *donor ID* field of the *donations* **table** (figure 2.12a, arrow D). It will also show at the bottom of the **Edit Relationships menu** that the **relationship type** is **one-to-many** (figure 2.12b, arrow). This means in this example that one *donor* can have many *donations*. There are occasions when a **one-to-one relationship** is desired, but not in this database.

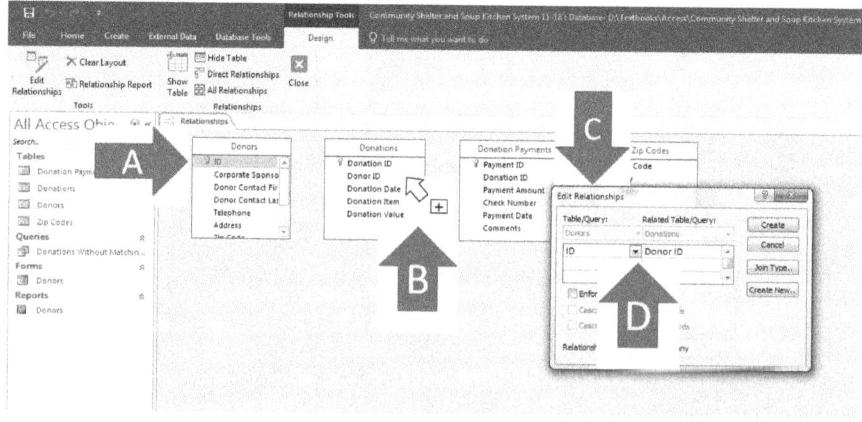

FIGURE 2.12a   Edit relationshps menu

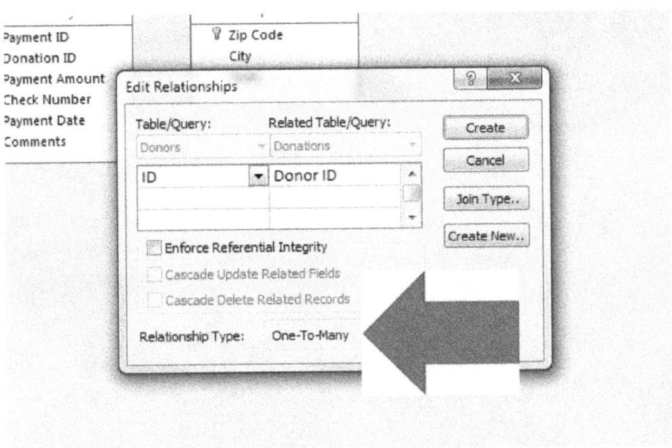

FIGURE 2.12b   One-to-many relationship indicator

Step 10: Click on the Enforce Referential Integrity check box to make sure that it is checked (figure 2.13, arrow A). THIS IS CRITICAL. IF YOU FORGET TO CHECK THAT BOX, REFERENTIAL INTEGRITY WILL NOT BE ESTABLIHSED AND THIS ENTIRE PROCESS WILL BE FOR NAUGHT.

Step 11: Click on the **Cascade Update Related Fields** check box to make sure that it is checked (figure 2.13, arrow B). By doing this, if you change the data in the **primary key** field, it will automatically change the data in the **foreign key** in each of their related records. It's a good habit to form to make this happen. Why? Suppose for a moment that the **primary key** of this database for the *donors* was their *telephone numbers*. When the *donors' telephone numbers* change, you don't want to have to go over to the donations table and change all of their telephone numbers in the **foreign key** field in their donations. In fact, Access will not even allow you to change their *telephone numbers* in that case without clicking on the **Cascade Update Related Fields** check box. That is because changing all of the *telephone numbers* would make all of the telephone numbers in the **child table** bogus data. After clicking the **Cascade Update Related Fields** check box, whenever you change data in a **primary key** field, the data in the **foreign key** field will automatically change as well. In this example, if you change the *telephone number* of a *donor*, the *telephone numbers* in the *donations* table would then be automatically changed instantly to the **NEW** telephone number in all of their *donations telephone number* fields.

NOTE: It is **NOT** recommended that you click the **Cascade Delete Related Records** check box very often (figure 2.13, arrow C). If you do, any time a *donor* would be deleted, all of their *donations* would instantly by deleted as well. If someone would accidentally delete a *donor* that was not intended to be removed, all of their *donations* would be deleted as well. Many records would be lost that could never be recovered.

Step 12: Click the **Create** ribbon (figure 2.13, arrow D).

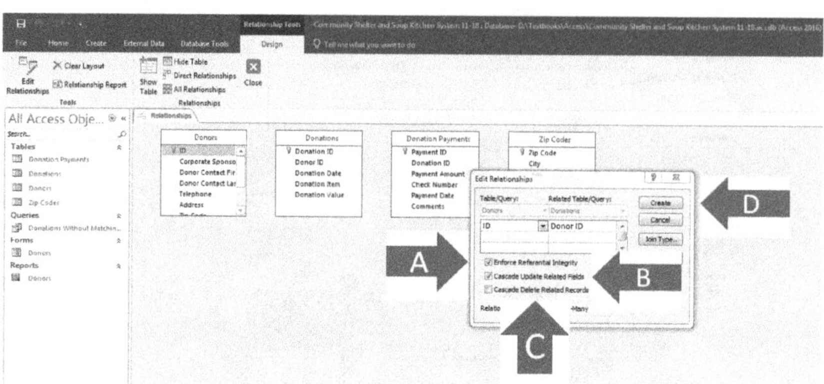

FIGURE 2.13   Edit relationshps menu

When you do, the **Edit Relationships** menu will close and **referential integrity** will now be established.

NOTE: When that happens, a line between the two tables will appear (figure 2.14, arrow A).

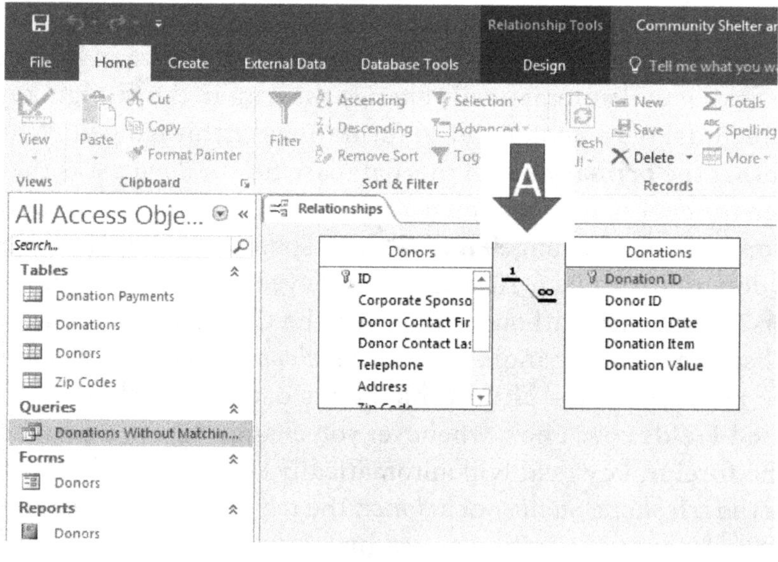

FIGURE 2.14   Relationship established

Where the line connects at the **parent table** (the *donors table*) you will see a 1. Where the line connects at the **child table** (the *donations* table) you will see the infinity ∞ symbol. This signifies that **referential integrity** exists between the **tables** and that one *donor* can have an infinite number of *donations*.

NOTE: Changing of a **field name** or a **field type**, or any other attribute of a field that is used in a **referential integrity relationship** cannot be done after the relationship is set without disconnecting the relationship!

Repeat all of these steps to create **referential integrity** using the following tables:

1. Connect the *donation ID* field from the *donations* **table** to the *donation ID* field of the *donation payments* **table**.
2. Connect the *zip code* field from the *zip codes* **table** to the *zip code* field of the *donors* **table**.

When all tables have been related, your screen should look like figure 2.15.

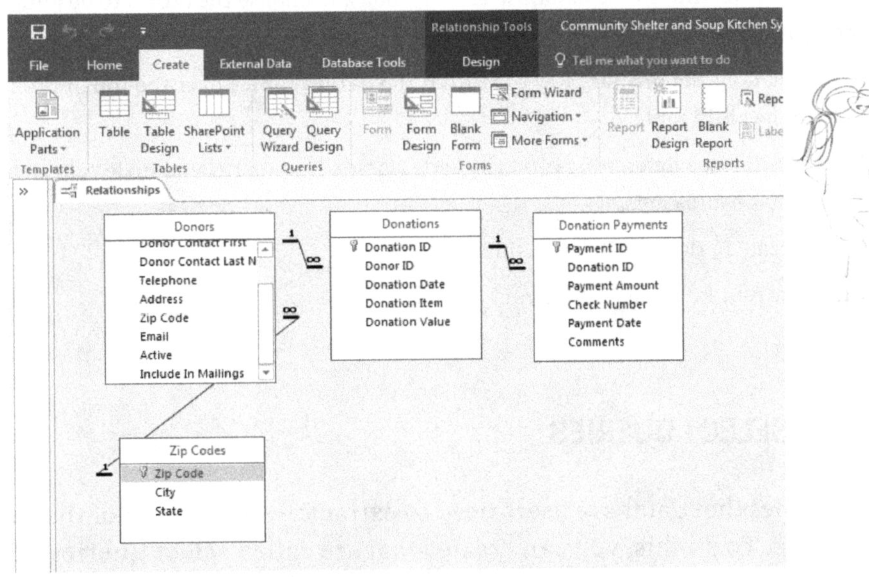

FIGURE 2.15   Relationships established on all tables

This would be a time that you may want to move the *zip codes* table beneath the *donors* table to simply see the relationships better and to understand them better.

NOTE: Sometimes you may have to scroll within the tables in the **Relationships** menu in order to see the fields you want to connect. Scrollbars will appear in the boxes as you click on them. You may also want to resize them for the same purpose. You can do so by placing your mouse on any corner of the **table boxes** and resizing them as you would resize a clipart in Microsoft Word or Microsoft PowerPoint.

When you are finished creating the relationships, close the **Relationships** menu as you would close any object (clicking the **lower X** in the upper right of the window). When asked if you want to save the layout of the **Relationships** menu, click **Yes**.

BELOW IS A SUMMARY OF CREATING REFERENTIAL INTEGRITY, OR A ONE-TO-MANY RELATIONSHIP, BETWEEN TABLES:

1. Make sure that you have a field in each table to be joined (the **primary key** and the **foreign key**).

2. Make sure that the **primary key** is set as such in the linking field of the **parent table** .

3. Make sure that the **primary key** and the **foreign key** are the same field types, and if they are **number** field types, make sure that they are the same width.

4. Make sure that there is not already bogus data in the **tables** you are about to give **referential integrity**.

5. Make sure that both the **parent** and the **child tables** are closed.

6. Go to **relationships** in the **database tools** ribbon and choose the **tables** to be joined and close the **show table** menu.

7. Click and drag the **primary key** field from the **parent tables** over the top of the **foreign key** field in the **child table**.

8. Click the Enforce Referential Integrity and Cascade Update Related Fields check boxes in the Edit Relationships menu.

9. Click Create in the Edit Relationships menu.

10. Close the Edit Relationships menu.

CREATING SELECT QUERIES

There are times that database users need to extract given data out of the **tables** within said database. To do this, you can create what are called **select queries**. As discussed previously, there are two dimensions in which a **select queries** can be created. First, you can show only the fields you need, and second you can display only the records needed. Recall in the example used before, if you want to create a calling list of *donors* of a given *area code*, you don't need to see their other information, and you don't need to see the *donors* who are not of that *area code*. Let's create such a **query**. You can do so by doing the following steps:

Step 1: Click the **Create** ribbon in the **Main Database window** (figure 2.16, arrow A).

Step 2: Click **Query Design** (figure 2.16, arrow B).

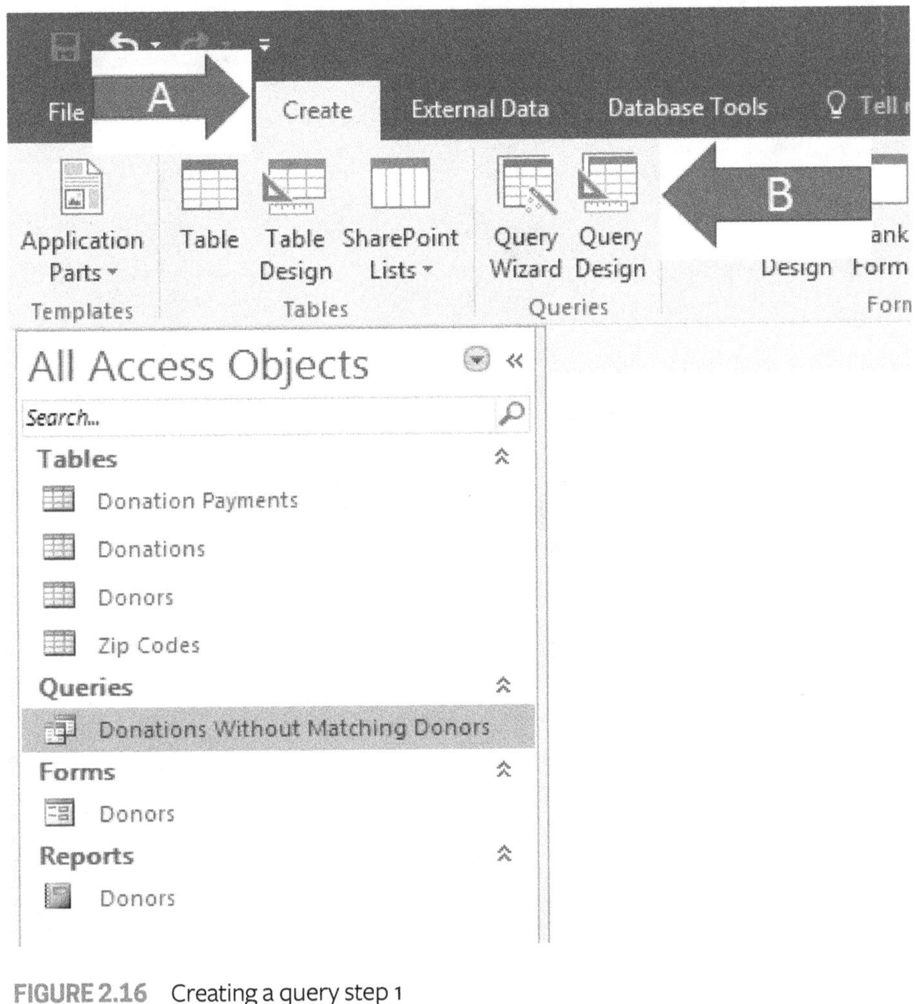

FIGURE 2.16   Creating a query step 1

Step 3: That will take you to the menu seen in figure 2.17. It is **VERY SIMILAR** to the **Relationships** menu that you used to recreate **referential integrity**. There is a **Show Table** menu (figure 2.17, arrow A) displaying the tables from which you will want to extract data. The **Show Table** menu also works the same as the one in the **Relationships** menu, so please do not get them mixed up.

NOTE: YOU CANNOT CREATE REFERENTIAL INTEGRITY IN THE QUERY DESIGN VIEW.

Step 4: Double-click the *donors* **table** (figure 2.17, arrow B) or you can click on the *donors* **table** once and then click **Add** (figure 2.17, arrow C). Doing this will send the **table** to the upper left of the **Query Design View** menu (figure 2.17, Arrow E).

Step 5: Close the **Show Table** menu by clicking **Close** (figure 2.17, arrow D).

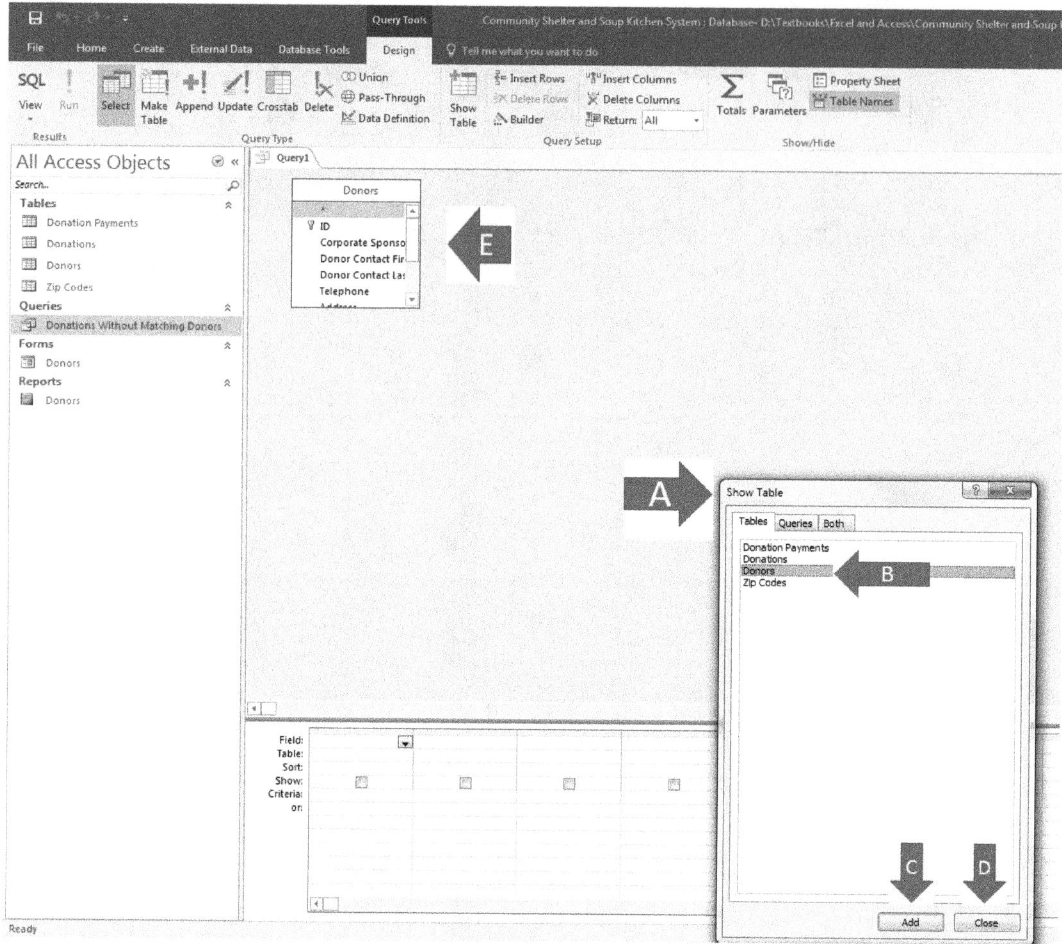

FIGURE 2.17 Creating a query step 3

Step 6: Now you must choose the fields you want to view. To do so, you must send them to the **Query Design grid** (figure 2.18, Arrow C). One way to do so, would be to go to the **table box** of the **table** with the fields you want to view and simply double-click those fields in the order that you want them to be displayed. In this example, double-click the *donor contact last name* (figure 2.18, arrow A) and then double-click the *telephone* field (figure 2.18, arrow B) in the *donors* **table box**. Notice how they automatically with be sent to the **Query Design grid** in the bottom of the **query design view** (figure 2.18, arrow C).

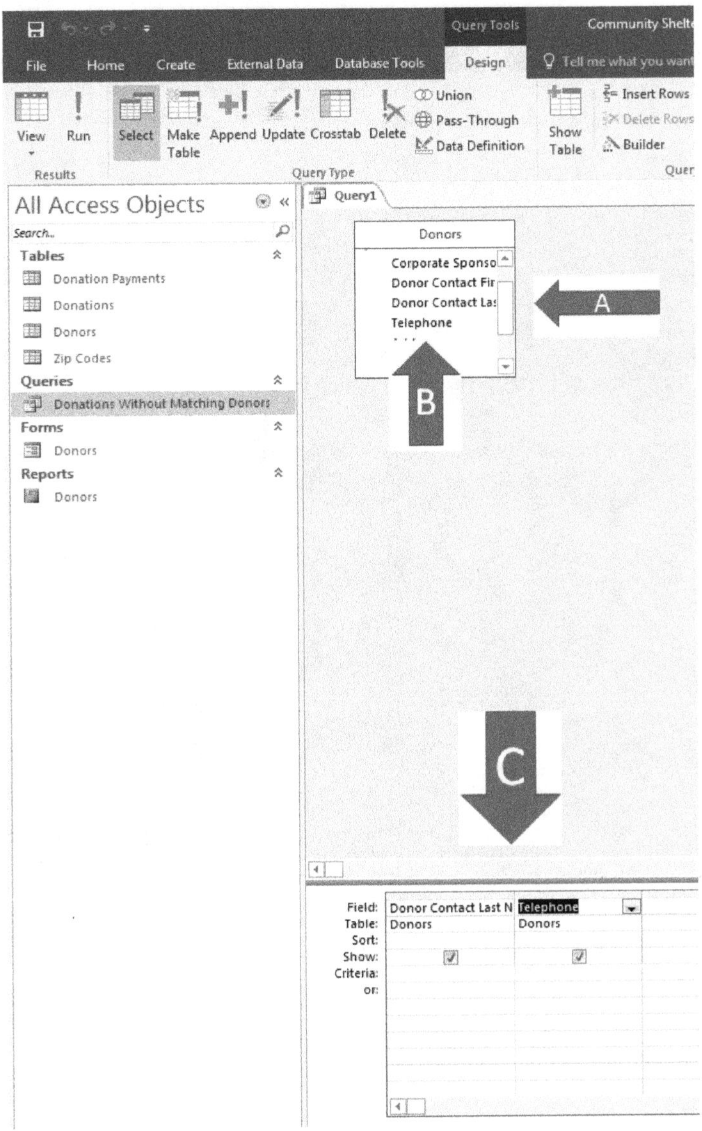

FIGURE 2.18   Creating a query step 6

There is another way to add fields to the **Query Design grid**. Suppose that you wanted to put the *donor contact first name* field **AFTER** the *donor contact last name* field and **BEFORE** the *telephone* field. Rather than having to start over, simply click and drag the *donor contact first name* field down and over the top of the *telephone* field and then let go of the mouse. Doing so will push the *telephone* field to the right and the *donor contact first name* field is now inserted in the middle of the other two fields (figure 2.19, arrow).

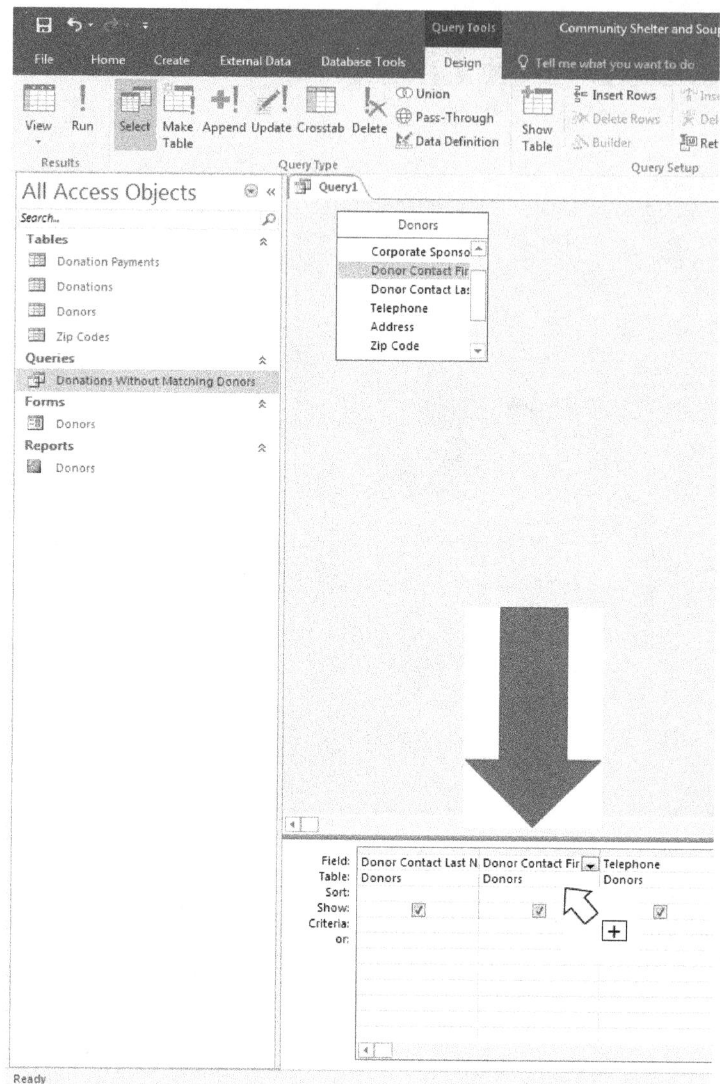

FIGURE 2.19   Inserting a field into a specific part of a query

At this point, the query does no good until we can see the results. We will do so in a moment, but first we want to add more settings to the query.

ADDING CRITERIA TO A SELECT QUERY

Step 1: If you only want to see the people of the 513 area code, type in "(513*" in the **Criteria box** below the *telephone* field (figure 2.20a, arrow). After you do this, Access will automatically convert your criteria to read *Like "(513*"*.

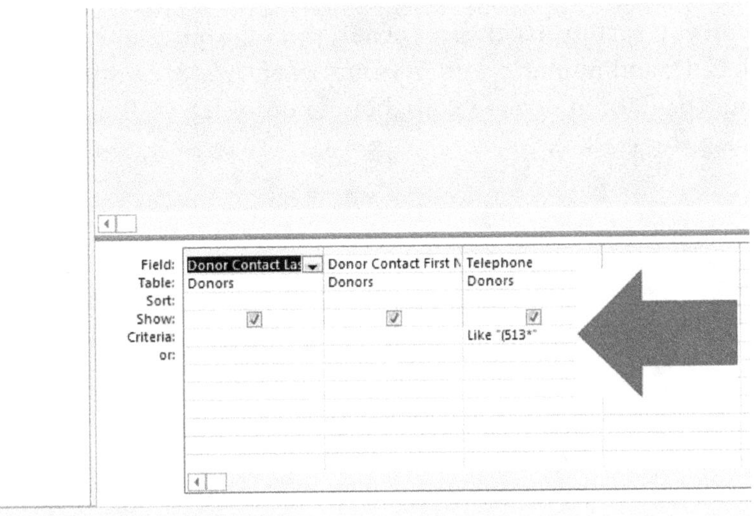

FIGURE 2.20a   Putting criteria into a query

NOTE: The **asterisk** (*) if also called a **wild card**. You **ONLY** need to use the **asterisk** (*) when you want to match ONLY PART of the field's criteria. A summary of rules to follow when using the **asterisk** (*) is next:

USING	DISPLAYS IN THIS EXAMPLE
(513*	All records that begin with (513
513	All records that contain 513 somewhere in that sequence of 513
*513	All records that end with 513

NOTE: If the *telephone* numbers do **NOT** have the parenthesis around the *area code*, you would not need to use the left parenthesis in this criterion. Instead it would then be entered as *"513*"* which would then convert the *"513*"* to *Like "513*"*.

SORTING IN A SELECT QUERY

Step 1: To sort by the *donor contact last name* alphabetically, choose **Ascending** in the **Sort box** under the *donor contact last name* field and do the same for the *donor contact first name* (figure 2.20b, arrow).

NOTE: In the **Query Design grid**, Access will always sort first by the sorted field furthest to the left. It will then sort by the next sorted field on the right and will continue to do so for every field you would sort. If no sort order has been indicated, it will sort by the **primary key** of the **table**.

NOTE: When you sort by **ascending order**, you will put the text data in alphabetical order (A, B, C, D) and numeric data in value order (1, 2, 3, 4). Sorting in **descending order** would put the data in reverse alphabetical order (D, C, B, A) and numeric data in reverse value order (4, 3, 2, 1).

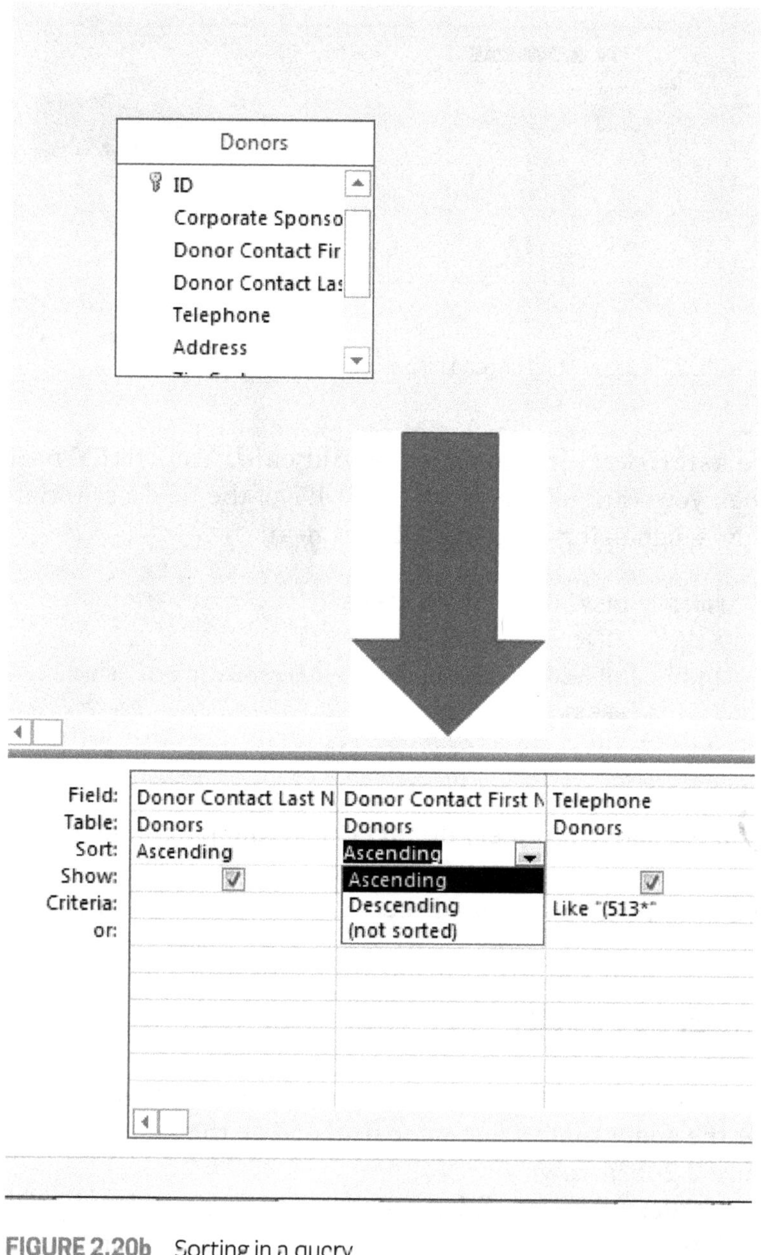

Field:	Donor Contact Last N	Donor Contact First N	Telephone
Table:	Donors	Donors	Donors
Sort:	Ascending	Ascending	
Show:	☑	Ascending	☑
Criteria:		Descending	Like "(513*"
or:		(not sorted)	

FIGURE 2.20b   Sorting in a query

TO RUN (VIEW THE RESULTS OF) A SELECT QUERY TO DISPLAY DATA

To see the results of your query settings:

Step 1: Click the **Run** button in the **Query Tools Design** ribbon (figure 2.20c, arrow).

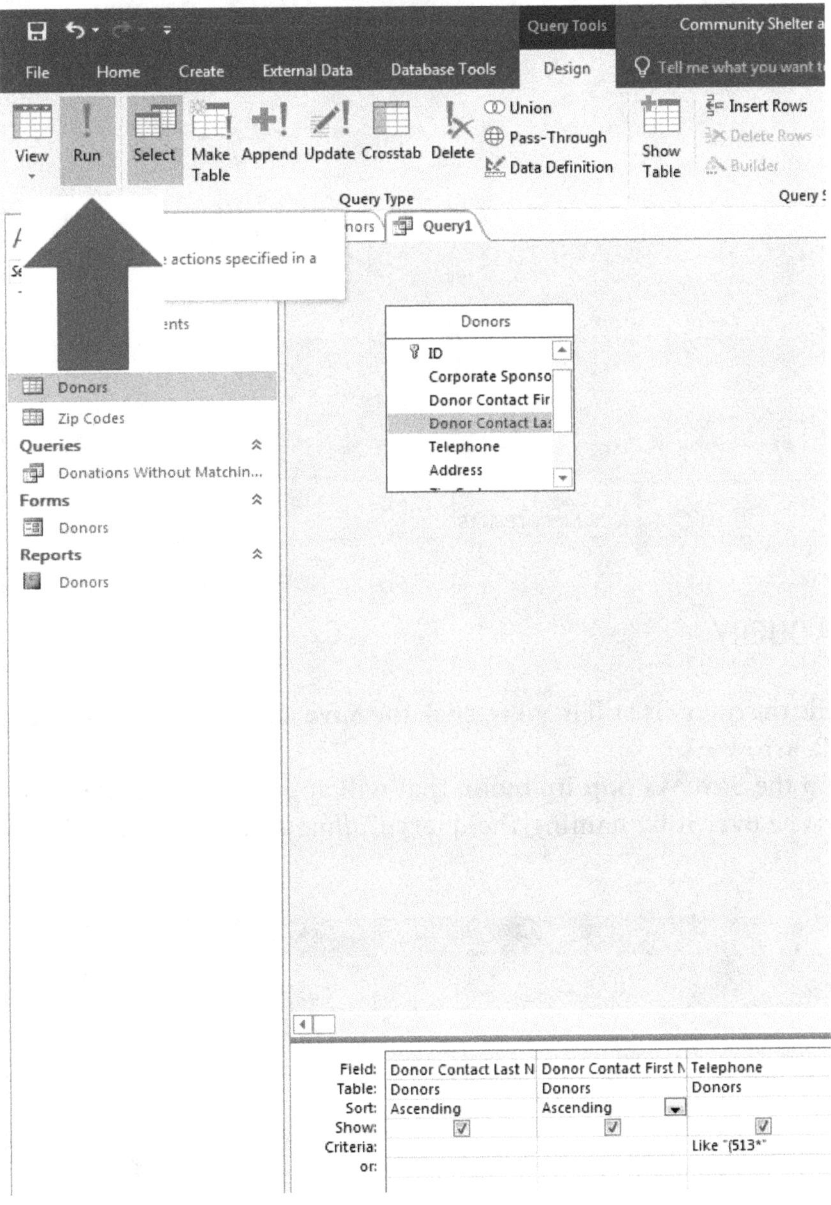

FIGURE 2.20c   Clicking the run button in a query

After you click the **Run button**, the records will display only the fields you have chosen and only the records with the criteria you have entered, and in the **sort order** you have set (figure 2.21).

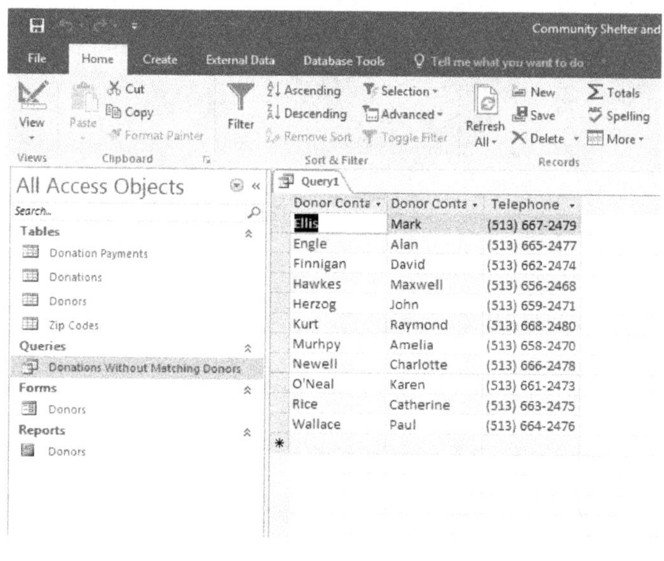

FIGURE 2.21 Query results

TO SAVE A QUERY

Step 1: While the **query** is still in view, click the **Save** icon on the **quick access toolbar** (figure 2.22, arrow A).

Step 2: In the **Save As pop-up menu** that will appear, remove the *Query1* default name and type over it by naming the query *Calling List Area Code 513* (figure 2.22, arrow B).

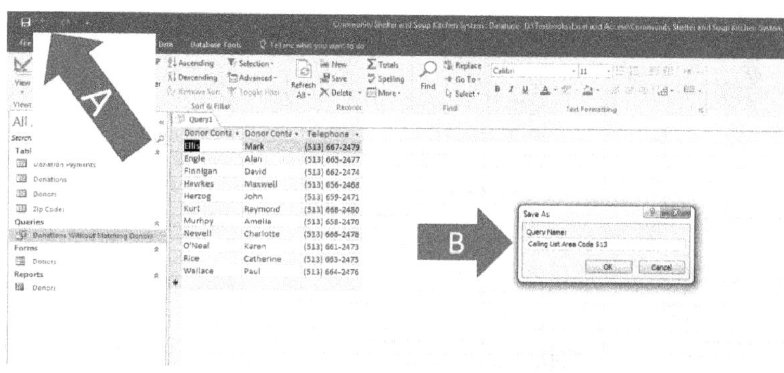

FIGURE 2.22 Saving a query

That will add that new query to the list of objects in the database's **navigation pane** (figure 2.23).

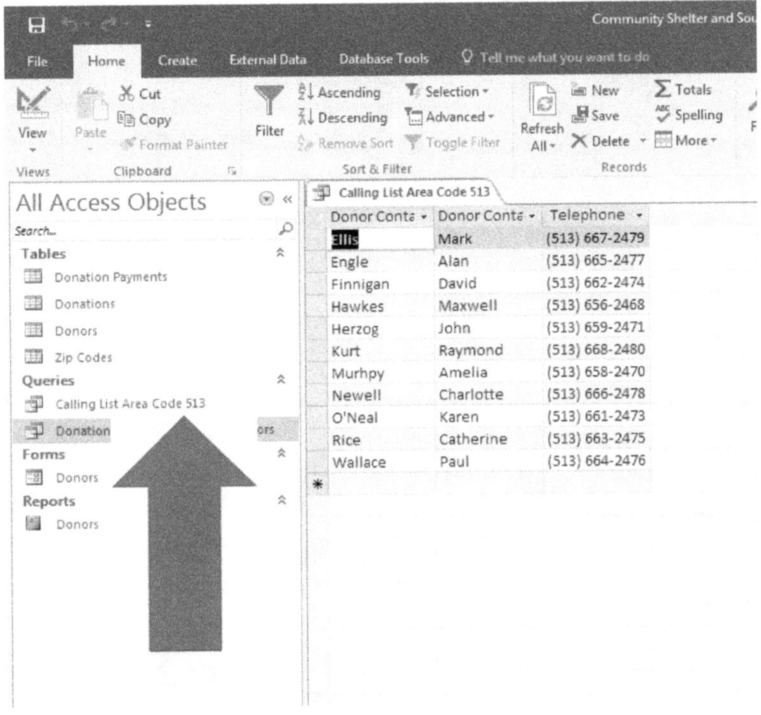

FIGURE 2.23 Seeing the query in the navigation pane

Step 3: Close the **query** by clicking the **lower X** on the right side of the **query**.

CREATING A JOIN-TABLE SELECT QUERY WITH MULTIPLE TABLES AND MULTIPLE CRITERIA

Suppose we now want to know what *Ohio donors* have donated more than *$300*. We will also want to put the **tables** together to see the *donor contact last name* and *donor contact first name* of the *donors* who have contributed to the organization, in what *zip codes, cities,* and *states* they live, the *date* of their *donations,* and the *donation value of their donations.* When you draw data from tables that are linked, you create what is called a **join-table query**. You can make a **query** that will show you this information by taking the following steps:

Step 1: Start a new **query** (using the steps described earlier), but now include the *donors* table, the *zip codes* table, and the *donations* table. Your *query design* should now look like figure 2.24.

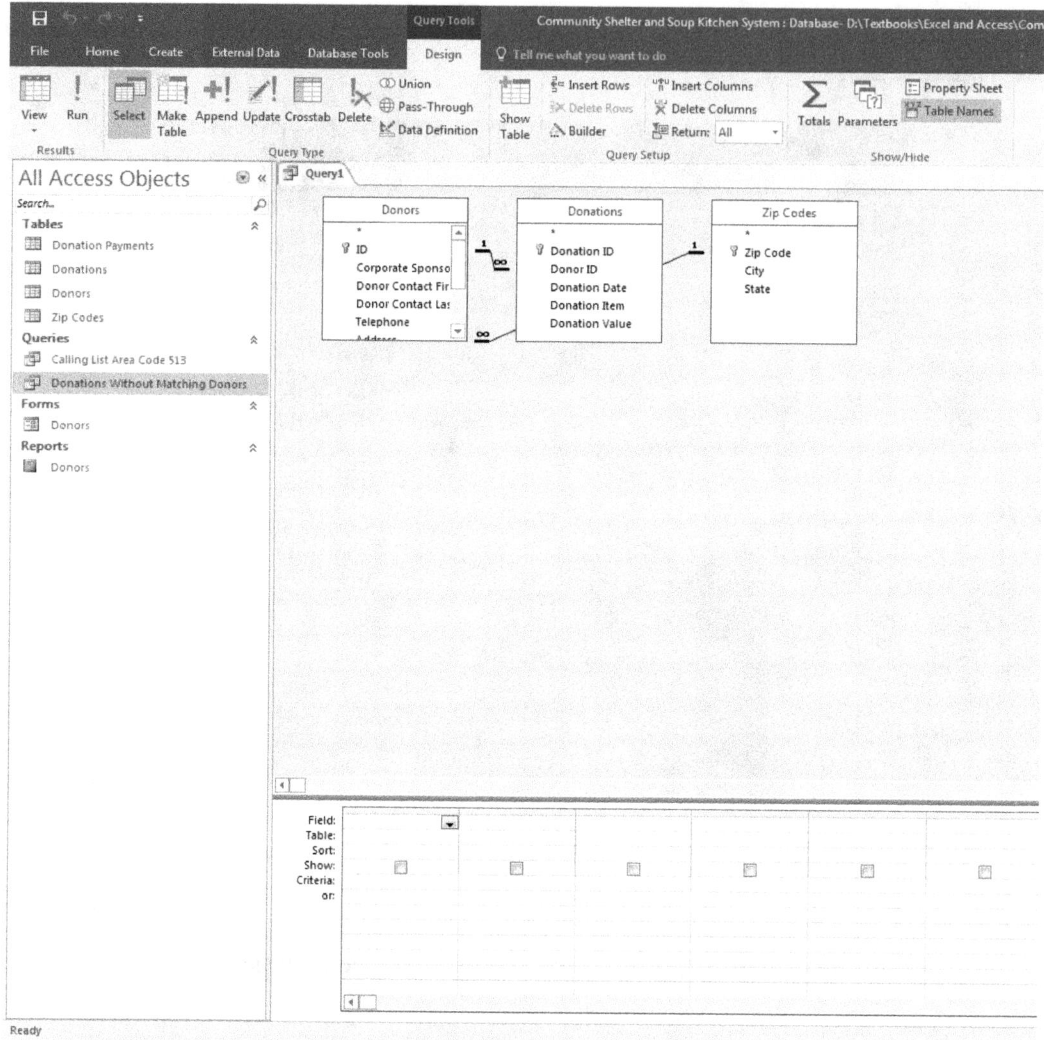

FIGURE 2.24   Starting a new query with different tables

This is one of those times when it might benefit you to rearrange the **table boxes** and resize them as mentioned earlier when we did so in the **Relationship menu**. In this case we will widen and lengthen the *donors* table so that all of the fields can be seen and in their entirety. Then we will move the *zip codes* table to the right of the *donors* table and below the *donations* table (figure 2.25a).

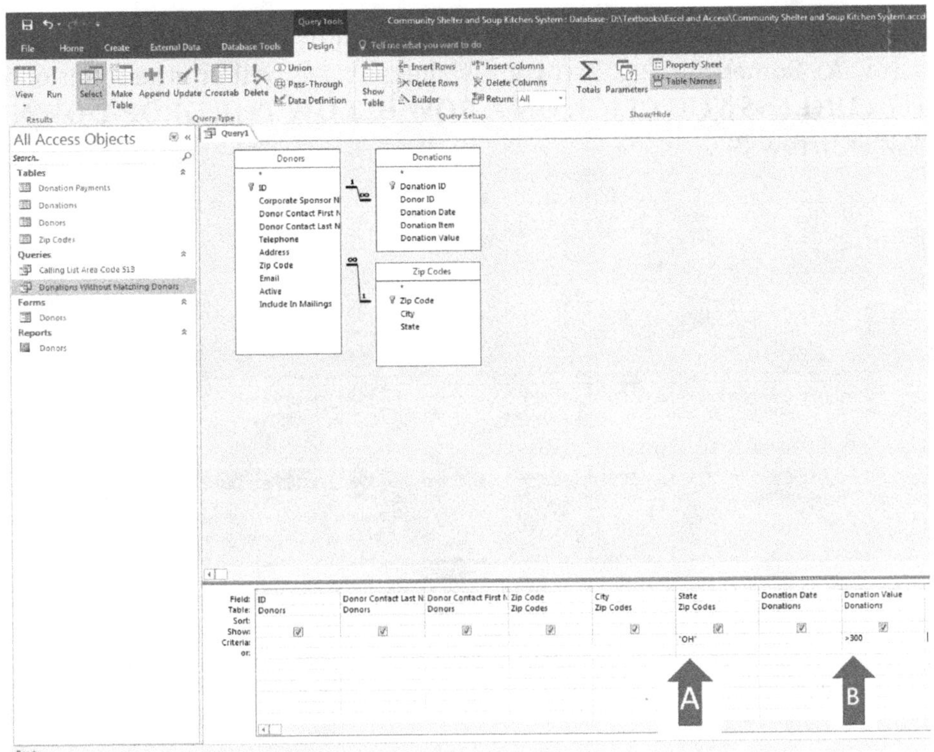

FIGURE 2.25a Adding criteria in a query

Step 2: Add the *ID, donor contact last name, donor contact first name* fields from the *donors* **table**. Add the *zip code, city*, and *state* fields from the *zip codes* **table**. Add the *donation date* and *donation value* fields from the *donations* **table**. Add the criteria of *OH* in the **criteria box** below the *state* field (figure 2.25a, arrow A). After typing it and then pressing **Enter**, Access will convert it to read "*OH.*" It will NOT do that for numeric values.

NOTE: You do **NOT** need to put the **asterisk** (*) in these criteria. In this example, we are looking for data where the **ENTIRE** field will match *OH* exactly. The asterisk (*) is **ONLY** to be used when you are searching for data that matches only **PART** of the data in the field, as you did earlier when you found **PART** of the *telephone numbers of donors* (their *area codes*).

Step 3: Add the criteria of *>300* in the **Criteria box** below the *donation value* field (figure 2.25a, arrow B).

NOTE: By putting these two criteria on the same line, you are basically telling Access that you want **BOTH** criteria to be true for the records that will display.

NOTE: Sometimes when you enter criteria in a **query**, Access will try to anticipate what it is you are trying to enter. For example, when you began to enter the *OH* criteria a moment ago, Access was thinking that you were trying to enter the **oct function**

(which you were not) and it will display a drop-down list for you to choose it (figure 2.25b, arrow A). Sometimes, against your wishes, Access will put the function in the criteria box **UNLESS YOU CLICK ON A ROW BELOW THE DROP-DOWN LIST** (figure 2.25b, arrow B).

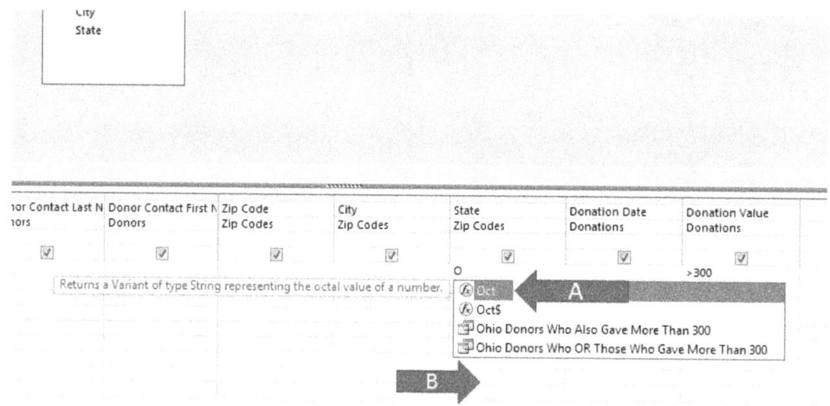

FIGURE 2.25b Watch out for access adding expressions you don't want in criteria

Step 4: Run the **query**. The records in figure 2.27 will be displayed.

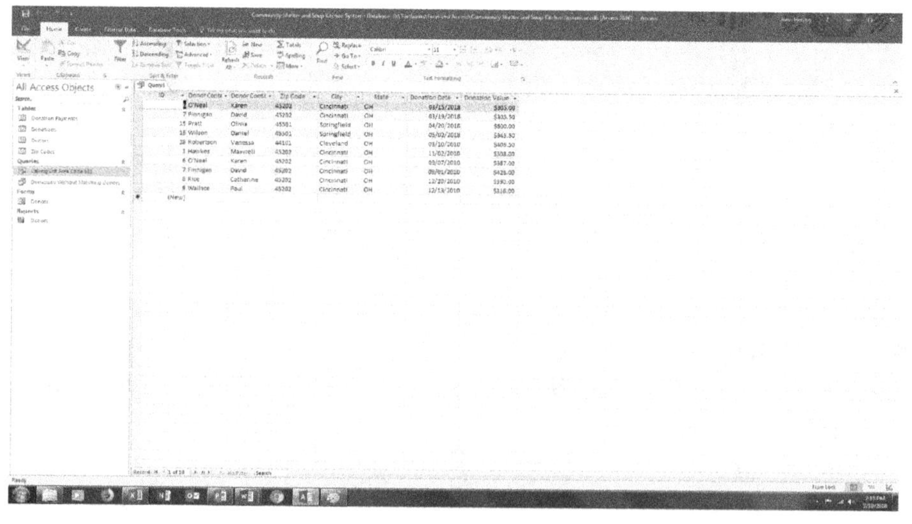

FIGURE 2.26 Query results with criteria

Save and name the **query** *Ohio Donors Who Also Gave More Than 300* and then close the **query**.

COPYING AN ACCESS OBJECT (METHOD 1)

Anytime you want to create a **table**, **query**, **form**, report, or **macro** that is very similar to, but not the same as one of these existing objects, it is best to copy the object and then make the needed minor changes to it. To copy and **access object**, take the following steps.

Step 1. While the query is still open, click **File** in the **Home** ribbon (figure 2.27, arrow).

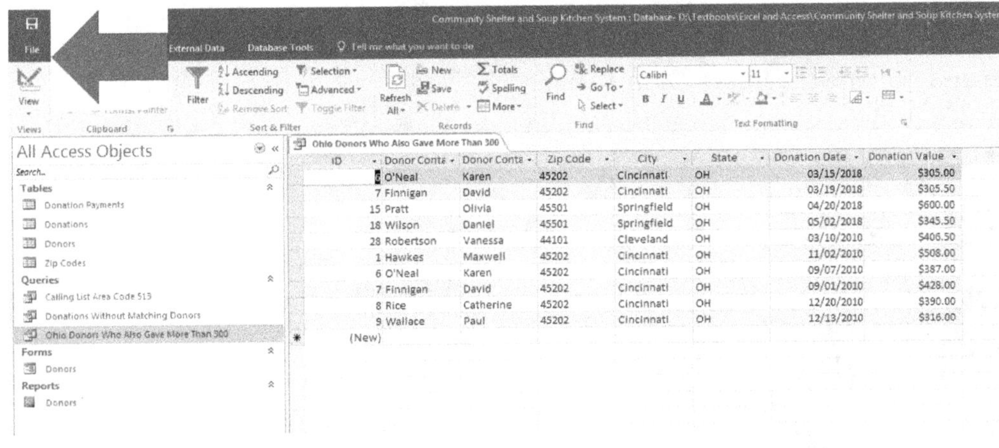

FIGURE 2.27   Copying and access object step 1

Step 2: That will take you to the menu shown in figure 2.29. Click **Save As** (figure 2.28, arrow)

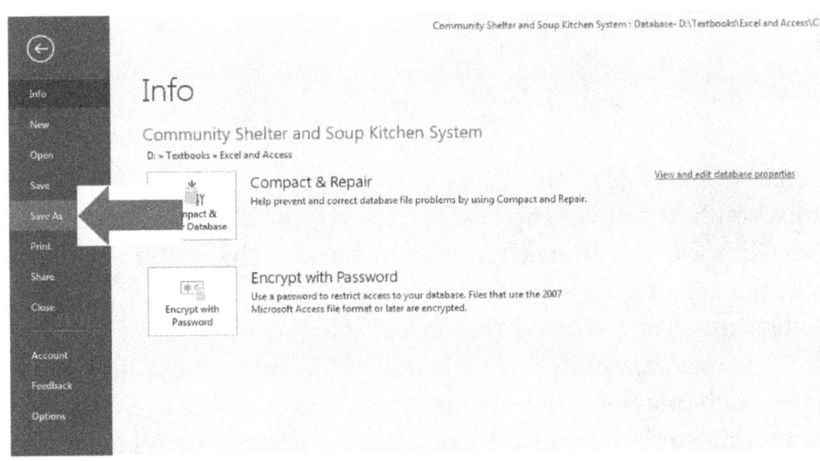

FIGURE 2.28   Copying and access object step 2

Step 3: That will take you to the menu shown in figure 2.29. Click **Save Object As** (figure 2.29, arrow A).

Step 4: Make sure that Save Object As (Save the current database object as a new object) is highlighted (figure 2.29, arrow B).

NOTE: Please notice that you can also make a copy for backup purposes of the entire database at this same screen.

Step 5: Click **Save As** (figure 2.29, arrow C).

Step 6: When you do, a **Save As** dialogue **pop-up menu** will appear. It will have the name of the query you are copying in the **Save to** box, but it will be named *Copy of Ohio Donors Who Gave More Than 300* (figure 2.29, arrow D). Type over that object name and rename the query *Ohio Donors OR Those Who Gave More Than 300*.

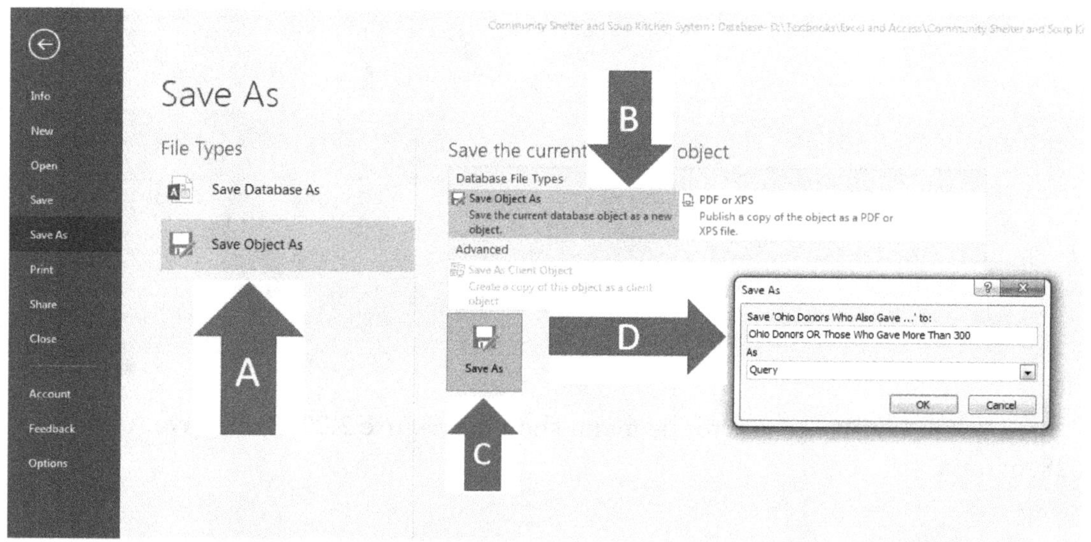

FIGURE 2.29   Copying and access object step 3

Now, let's make a change to the query you just created.

Step 7: Once you have copied the query, Access will take you to the screen you see in figure 2.30. Click **Design View** (figure 2.30, arrow). That will take you back to the **design view** of the **select query**.

NOTE: Select queries are those that select data according the fields you choose and the criteria you use. Later in this textbook, we will be discussing what are called **action queries**. Such **queries** don't display data. They change the data.

NOTE: Any time you are in a **table**, **select query**, **form**, or report if you click the **Design View** button (that looks like a pencil and T-square), it will take you to the **design view** of that object.

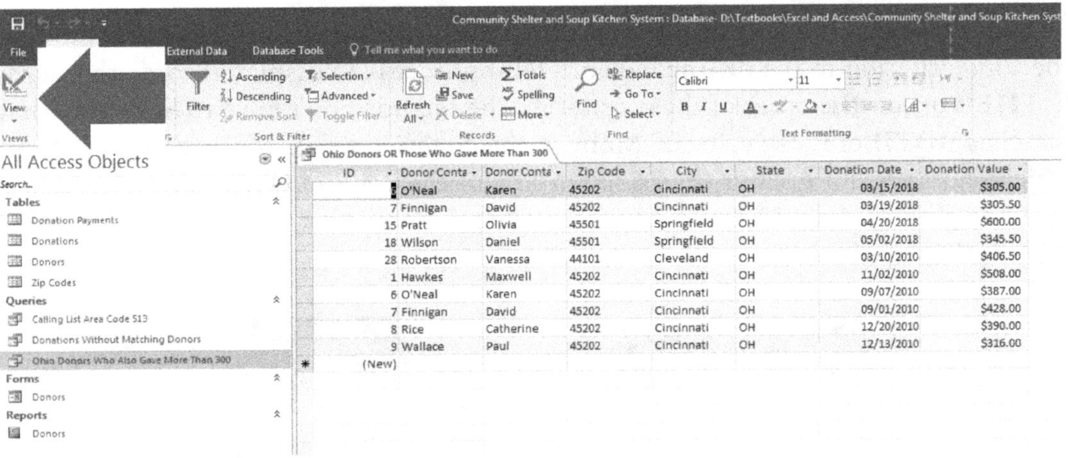

FIGURE 2.30   Copying and access object step 7

Step 8: At this point, you will now see the screen shown in figure 2.31. Move the criteria for the *donation value* that now reads as *>300* down to the row below it in the **Query Design View grid** (figure 2.31, arrow A). That is called the **or line**.

Step 9: Run the query by clicking the **Run** button (figure 2.31, arrow B).

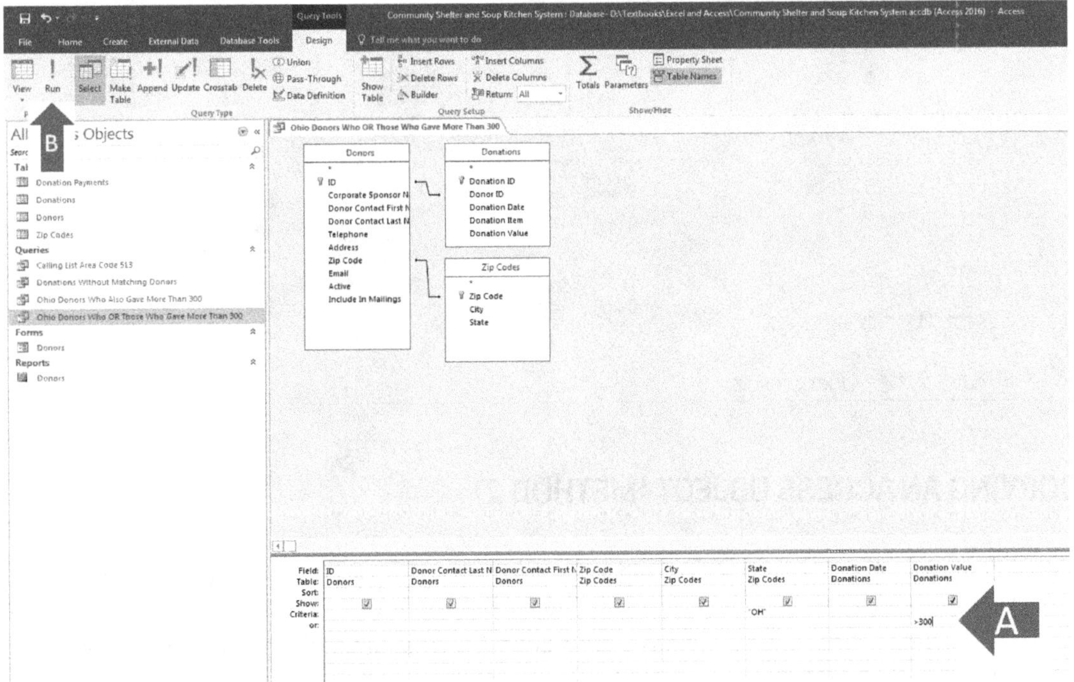

FIGURE 2.31   Copying and access object steps 8 and 9

Step 10: Notice the results of the modified query shown in figure 2.32. By moving the *>300* criterion to the **or line**, you are telling Access to display records that have **EITHER** one or the other of these criteria. The records don't have to have data that matches **BOTH** of those criteria. Many of the *donation values* of the *donations* from *Ohio donors* are not greater than *$300*. Many of the *donations* with *donation values* that are greater than *$300* are not from *Ohio donors*. As a result, there are more records displayed.

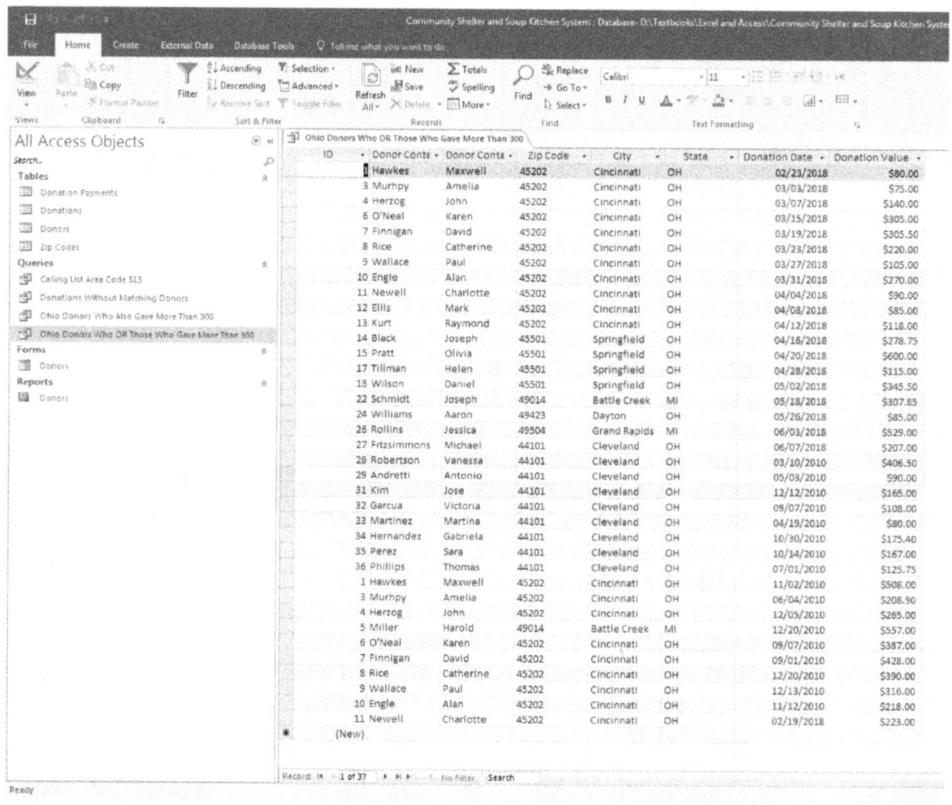

FIGURE 2.32   Query result

COPYING AN ACCESS OBJECT (METHOD 2)

Another way to copy the *Ohio Donors Who Also Gave More Than 300* **query** or copying any Access **object**, would be by taking the following steps:

Step 1: Click on the **object** you want to copy (in this case, choose *Ohio Donors Who Also Gave More Than 300* (figure 2.33, arrow A).

Step 2: Click **Copy** in the **Home** ribbon (figure 2.33, arrow B).

Step 3: Click **Paste** in the **Home** ribbon (figure 2.33, arrow C).

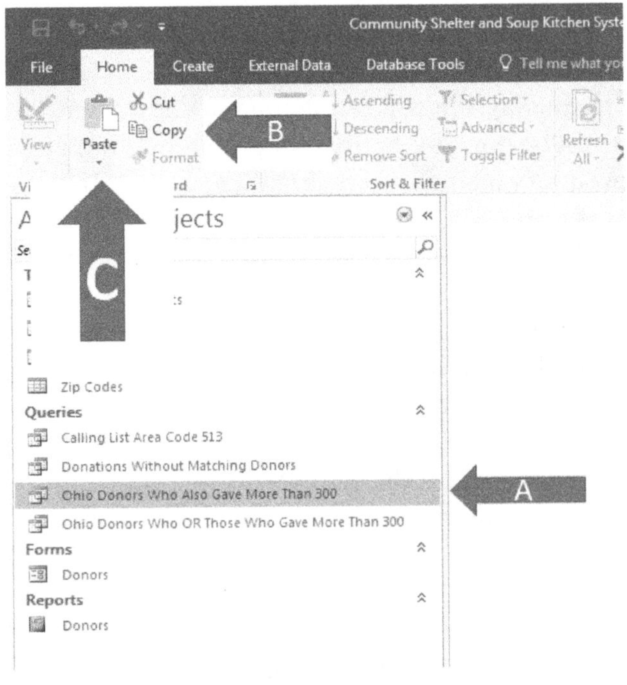

FIGURE 2.33 Copying and object

Step 4: A **dialogue pop-up menu** will appear, with the name *Copy of Ohio Donors Who Also Gave More Than 300* in the **query name** box (figure 2.34). Type over it with the new name you wish to give the **object**.

Step 5: Click **OK**.

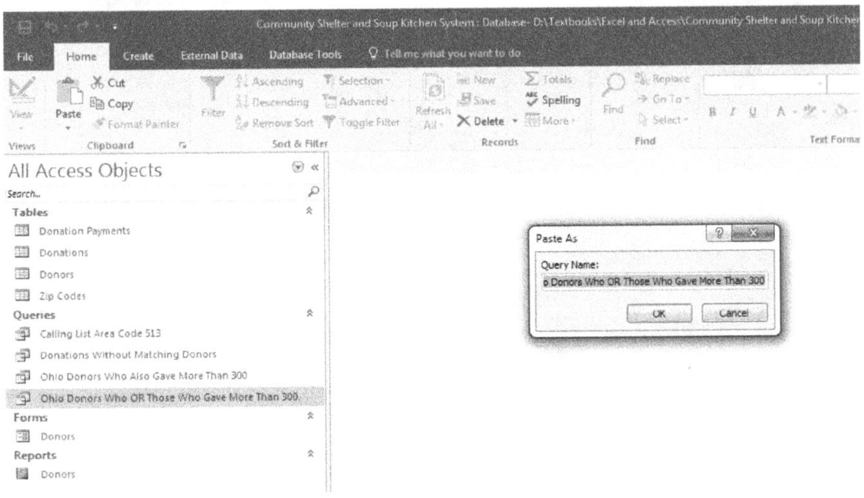

FIGURE 2.34 Saving a copied object

FINDING TOP OR BOTTOM VALUES IN A SELECT QUERY

Suppose you want to display the top ten *donors* of the *community shelter and soup kitchen system*. This can be done by performing the following steps:

Step 1: Start a new **query** (using the steps shown earlier). Include **ONLY** the *donor* and *donations* tables (figure 2.35, arrow A).

NOTE: NEVER add a table to a query **design area** if you don't intend to draw data from it.

Step 2: Make sure that the *donor contact first name* and *donor contact last name* fields from the *donors* table are included in the query and make sure the *donation value* field from the *donations* table is included and sent to the **Query Design View grid** (figure 2.35, arrow B).

Step 3: Sort the *donation value* field in **descending order** (figure 2.35, arrow C).

Step 4: Click the **Return menu** in the **Query Tools Design** ribbon (figure 2.35, arrow D).

NOTE: The default of the number of records that select queries will **return** (or display) or will always be set to display **all** records unless you change it.

Step 5: Type *10* in the **Return menu** box (figure 2.35, arrow E).

NOTE: You can use one of the options in that box (such as 5, 25, 100, 5%, or 15%) but you are not limited to them. If what you want is not in that list, you may type it in as we have just done by entering the *10* (figure 2.35, arrow E). Unfortunately, when you type *10*, it will perform an **autocomplete** and it will enter *100*, because Access assumes that you want to use the *100* that is already in the list. Thus, you must press the **Delete** key on your keyboard to remove the second zero.

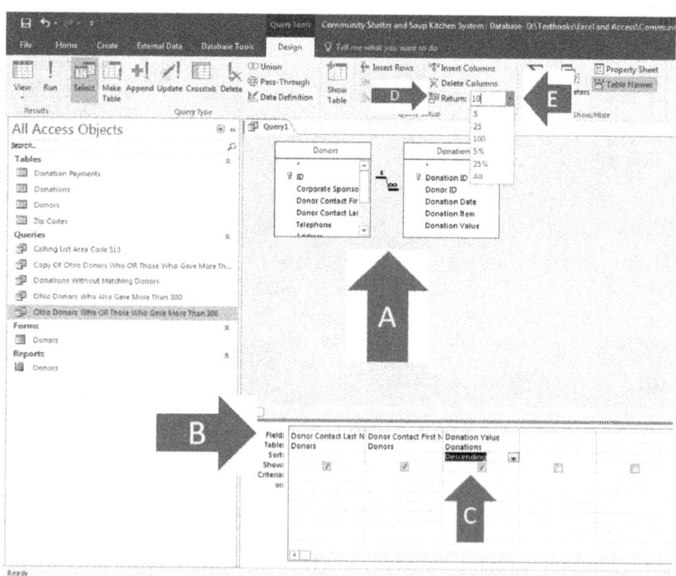

FIGURE 2.35   Finding first, top, or bottom 10 records

Step 6: Run the **query**. The results will be what you see in figure 2.36.

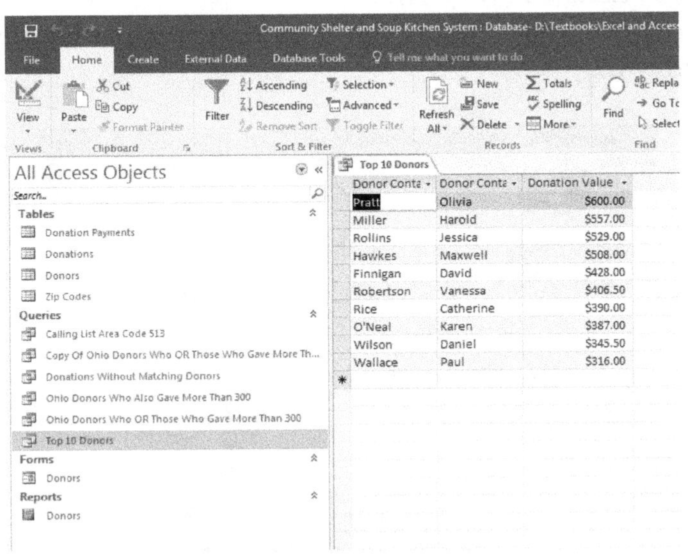

FIGURE 2.36   Finding top 10 donations results

Close and save the **query**. Name it *Top 10 Donors*.

NOTE: If the 10th item has the same value as any other record, all of those records will be displayed, which would give you more than 10 records, even though you asked for only 10.

NOTE: If you wanted to see the bottom (or lowest) *10 donation values*, you would repeat this process, but you would sort the *donation value* field in **ascending order** instead.

CREATING A SELECT QUERY USING DATES AS CRITERIA

Suppose you want to know what *donations* have been made **ON OR AFTER** *January 1 of 2018*. It can be performed with the following steps:

Step 1: Start a new **query** (using the steps shown earlier). Include **ONLY** the *donor* and *donations* tables (figure 2.37a, arrow A).

Step 2: Make sure that the *donor contact first name* and *donor contact last name* fields from the *donors* table are included in the query, and make sure the *donation date* and *donation value* fields from the *donations* table are sent to the **Query Design View grid** (figure 2.37a, arrow B).

Step 3: In the **Criteria box** below the *donation date* field, type *>=1/1/18* (figure 2.37a, arrow C). When you press **Enter**, it will put **pound signs** (#) at the beginning and the end of the date you entered. It will also change the 18 to read 2018, thus the criteria will now have been changed to *>=#1/1/2018#*.

Step 4: Run the query by clicking the **Run** button in the **Design View** ribbon (figure 2.37a, arrow D).

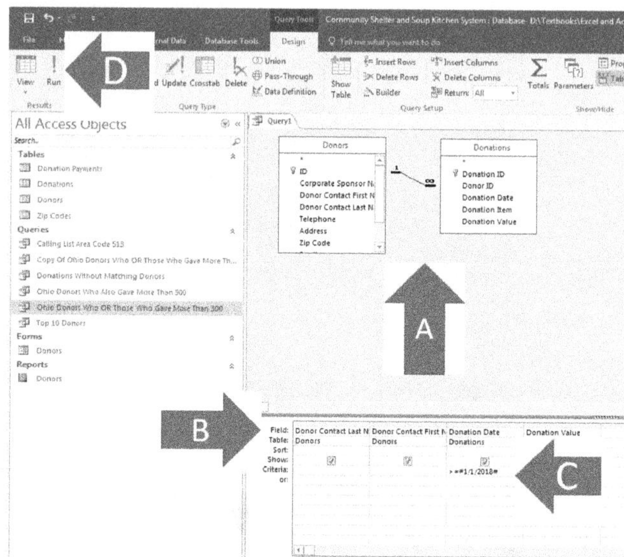

FIGURE 2.37a Querying by date

The results will appear as you see them in figure 2.37b.

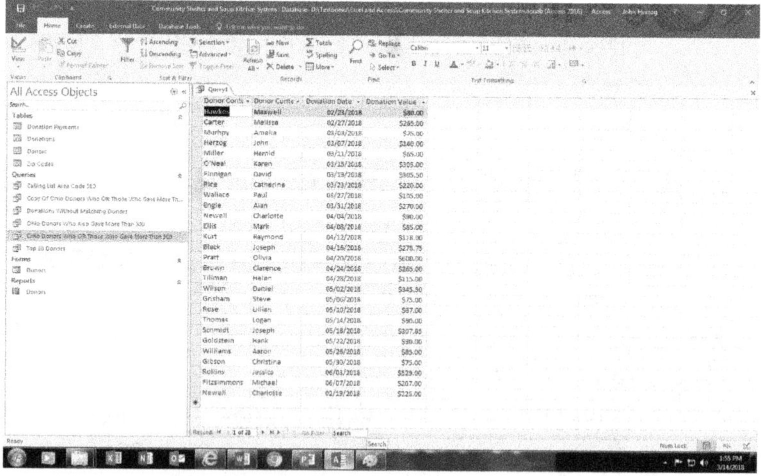

FIGURE 2.37b Querying by date results

Close and save the **query**. Name the it *2018 Donations*.

HIDING A FIELD IN A QUERY

Sometimes you don't need to see a field by which you have set criteria. For example, suppose you just want to see the *donations* after *January 1, 2018*, but you don't need to see the *date* field in the results. You can do so using the following steps:

Step 1: Return to the **design view** of the *2018 donations* **query**.

Step 2: Go to the **Show Row** (figure 2.37c, arrow A) and uncheck the **checkbox** that is below the *donation date* field.

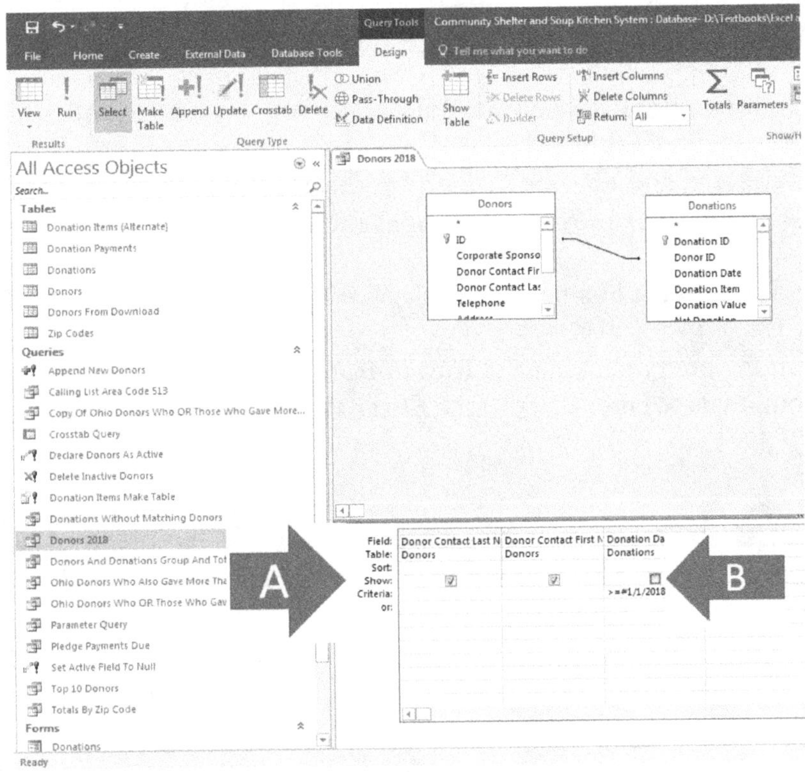

FIGURE 2.37c   Hiding a field in a query

Step 3: Rerun the **query** and you will see that the same records will be displayed, but without the *date* field showing.

Step 4: Resave the **query**.

NOTE: The default for fields in the **Query Design grid** are to be shown. You only need to contend with the **Show box** if you want to hide a field as you just did.

NOTE: If you wanted to find *donations* given **BEFORE** 2018, you would need to enter *<1/1/18* into the **Criteria box**. If you wanted to find *donations* given **BETWEEN** and **INCLUDING** *1/1/2018* **AND** *3/1/18* you would need to enter *>=1/1/18 and <=3/1/18* into the *donation date* **criteria box**.

TO CREATE A PARAMETER QUERY

Suppose you want to be able to display the *donors* of a given *city* without having to make a separate query for each *city*. You can have the computer prompt you for the city you want to use as criteria. This is called a **parameter query**. Let's create such a **select query** using the following steps:

Step 1: Start a new **query** (using the steps shown earlier). Include **ONLY** the *donors, donations,* and *zip codes* tables (figure 2.39a, arrow A).

Step 2: Make sure that the following fields from the following **tables** are included in the **query**:

1. The donor contact first name, donor contact last name, and address fields from the Donors **table**.
2. The *city, state,* and *zip code* fields from the *zip codes* **table**.
3. The *donation value* from the *donations* **table** (figure 2.38a, arrow B).

Step 3: In the **Criteria box** below the *city* field, type *[Please Enter the City]* (figure 2.38a, arrow C).

Step 4: Run the **query** clicking the **Run** button (figure 2.38a, arrow D). When you do, it will prompt you to enter a *city* in the **Enter Parameter Value** menu (figure 2.38b, arrow).

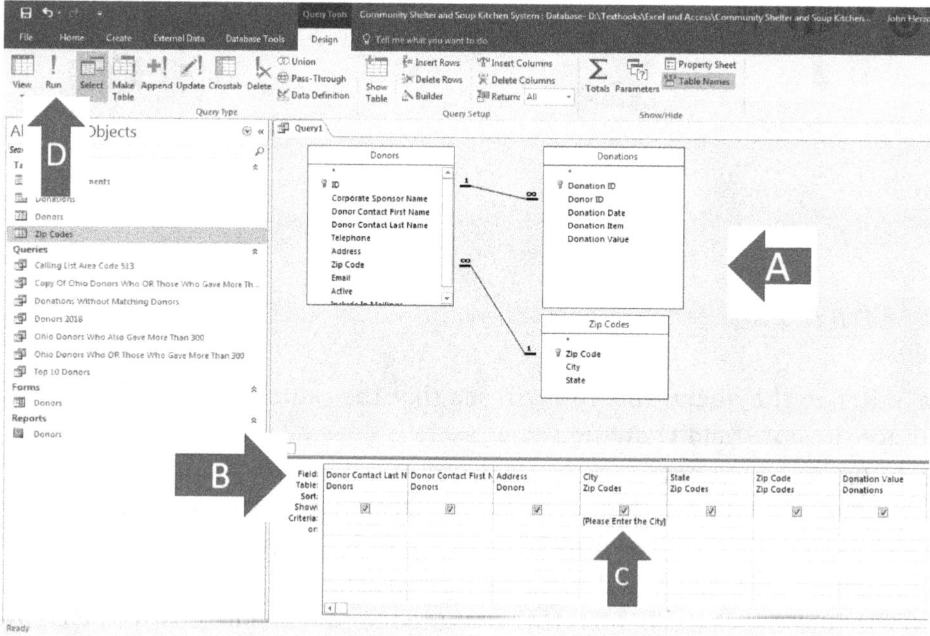

FIGURE 2.38a Querying with a parameter query

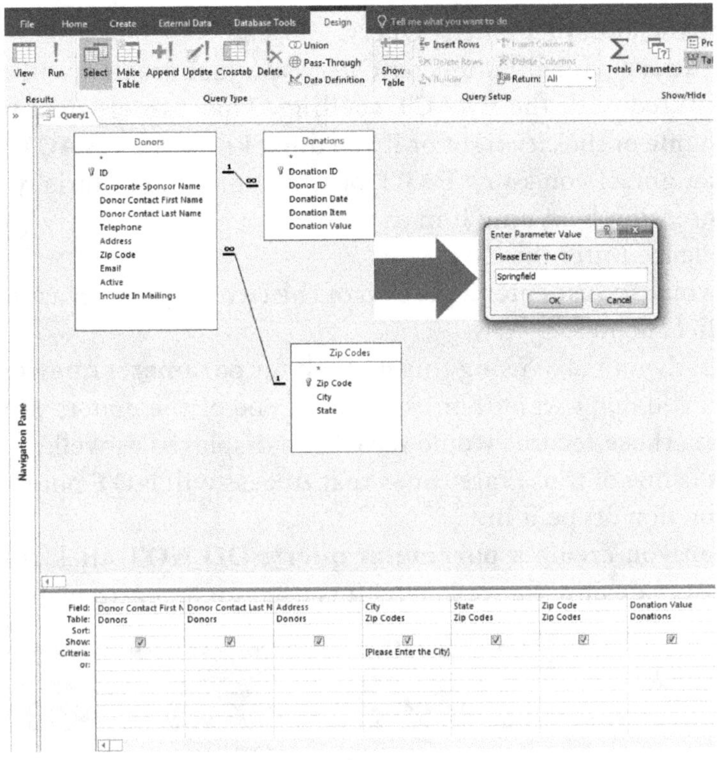

FIGURE 2.38b Querying with a parameter query

Step 5: For this example, enter *Springfield* (figure 2.38b, arrow) into the **Enter Parameter Value prompt** and you should see the results shown in figure 2.39a.

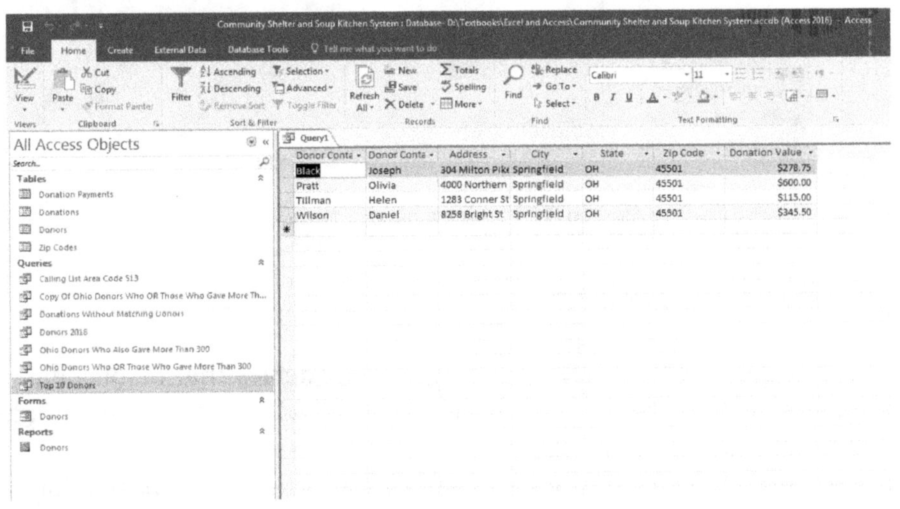

FIGURE 2.39a Querying with a parameter query

Step 6: Name the **query** *Parameter Query.*

NOTE: The **query** named *Parameter Query* you just created will require you to have the *city* entered with the **EXACT** spelling. That can create a problem if you don't spell the name of the *city* right or if you don't know the **EXACT** spelling of the criteria being sought. If you know **PART** of spelling of the criteria, you can modify the **criteria expression** by typing it in as follows:

Like "*" & [Please Enter the City] & "*"

In doing so, you can just enter a portion of the *city* name, such as *spring* or *field* to find the *Springfield donors.*

NOTE: A disadvantage of using this method for **parameter queries** is that it may give you results you don't want. For example, if one of the *donors* was from a town called *Springboro*, those records would have been displayed as well.

NOTE: This is one of those rare times that Access will **NOT** put the word **like** in the criteria if you don't type it in.

NOTE: When you create a **parameter query, DO NOT** click the **Parameters button** in the **query design** (figure 2.39b)! It is used for an entirely different purpose.

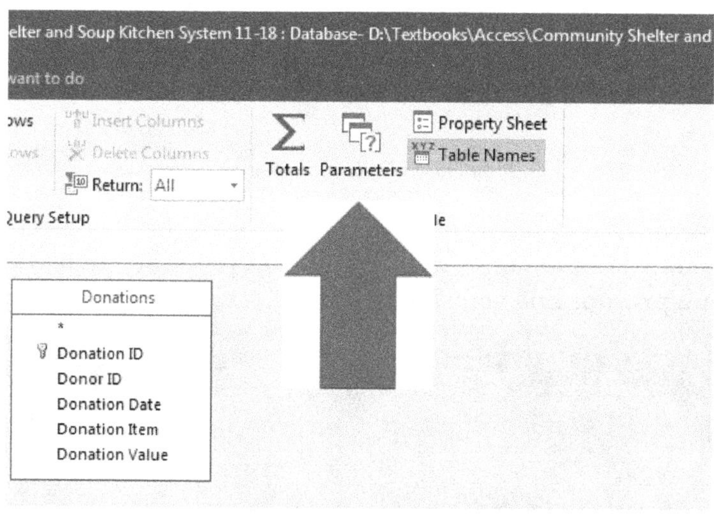

FIGURE 2.39b   How to NOT create a parameter query

CREATING CROSSTAB QUERIES

As mentioned earlier, we have been creating **select queries** that display data. When you choose the type of **queries** you want to create, by default, Access will make them **select queries,** because they are the most commonly used **queries.**

Sometimes you will want to see data on a three-dimensional basis. For example, suppose you want to know the total *donation values* of given *donation items* from *donors* living in certain *zip codes.* If you use Microsoft Excel, you can create what is referred to as a **pivot table**. In Access the equivalent of a **pivot table** is called a **cross tab query**. To create one, use the following steps:

Step 1: Create a new **query** using the *donors* and *donations* **tables**.

Step 2: Send the *zip code* field from the *donors* **table** into the **Query Design View grid**.

Step 3: Send the *donation item* field and the *donation value* field from the *donations* **table** into the **Query Design View grid** (figure 2.40, arrow A).

Step 4: Click the **Crosstab** icon in the **Query Design** ribbon (figure 2.40, arrow B) to select the **query type**. Notice that it will add a **total line** (figure 2.40, arrow C) and **crosstab row** (figure 2.40, arrow D) in the **Query grid**.

Step 5: In the **total line** (figure 2.40, arrow C) use the **Group By** setting beneath the *zip code* and *donation item* fields (they are the default settings). Use the **SUM** option beneath the *donation value* field (figure 2.40, arrow C).

Step 6: In the **crosstab row** select **Row Heading** beneath the *zip code* field (figure 2.40, arrow D).

Step 7: In the same **crosstab row** select **Column Heading** beneath the *donation item* field (figure 2.40, arrow D).

Step 8: In the same **crosstab row** select **Value** beneath the *donation value* field (figure 2.40, arrow D).

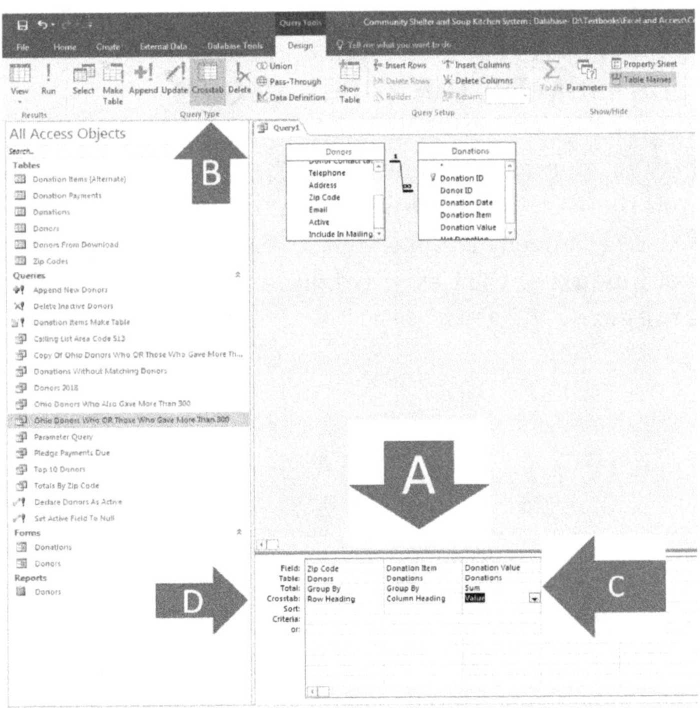

FIGURE 2.40   Parameter query results

Step 9: Click the Run button in the Query Design ribbon.
Step 10: The results should look like what you see in figure 2.41.

Zip Code	Cash	Drink	Food	In-Kind	Linens	Office Equipment
44101	$257.00	$287.00	$125.75	$581.90	$108.00	$165.00
45202	$819.50	$298.90	$1,048.00	$655.00	$800.00	$1,116.00
45501		$115.00	$278.75	$345.50		$600.00
49014	$632.00	$65.00				
49085			$340.00		$405.00	$75.00
49423			$85.00			
49504					$529.00	

FIGURE 2.41 Crosstab query design grid

Step 11: Save the query as **Crosstab Query**.

REMOVING FIELDS FROM THE QUERY DESIGN GRID

Sometimes you may accidentally put a field into the **Query Design View grid** that you didn't intend to put there In the example shown in figure 2.42, the second *donation value* field needs to be removed. To remove it, use the following steps:

Step 1: Point your mouse to the tiny gray box above the field name and a tiny black arrow will appear (figure 2.42, arrow).

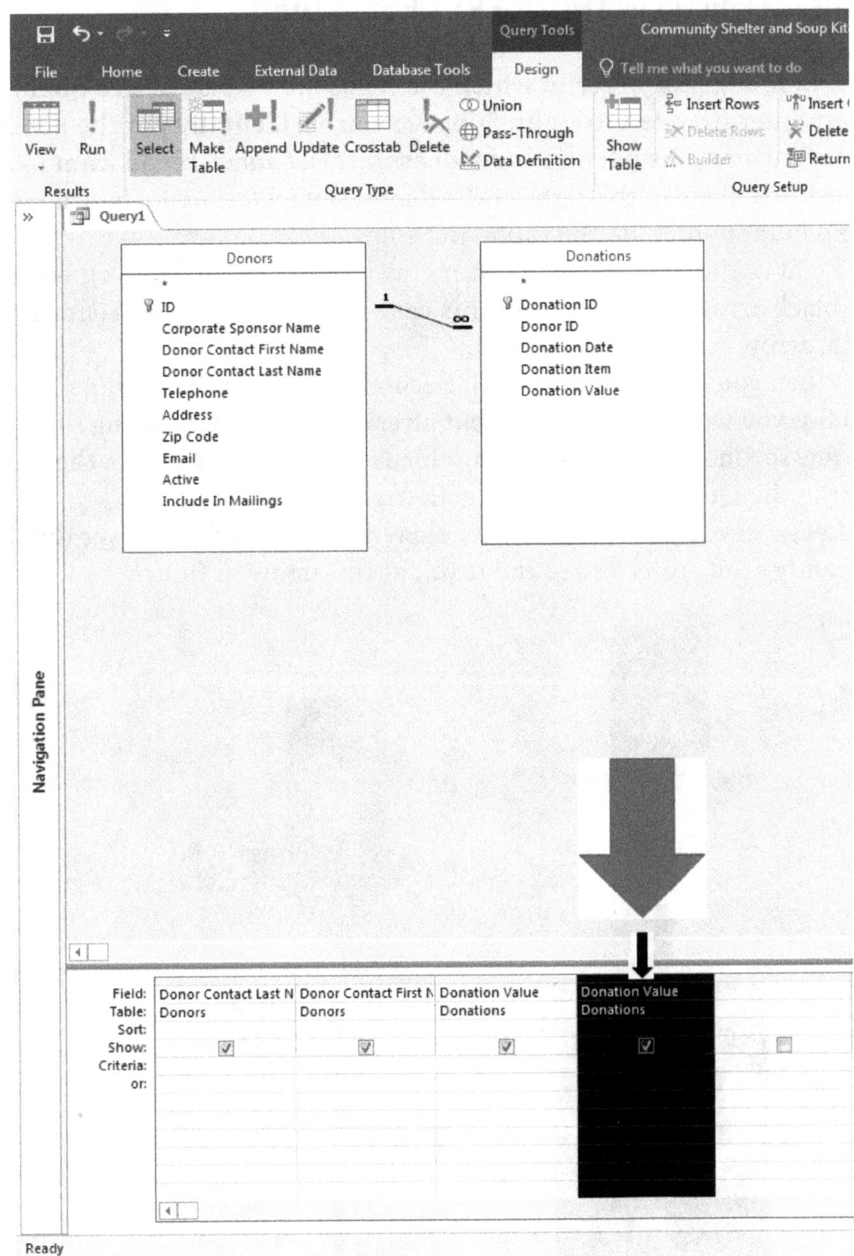

FIGURE 2.42   Crosstab query results

Step 2: When you do, the entire column will become dark. Press the **Delete** key on your keyboard and the field will be removed.

REARRANGING FIELDS IN THE QUERY DESIGN GRID

If you want to change the order in which the fields are displayed in a **query**, you can change their order from left to right. Suppose you accidentally put the *donor contact first name* before, or to the left of, the *donor contact last name*. If you want to move the *donor contact first name* to the right of the *donor contact last name* in order to be able to sort them in that order, do the following steps:

Step 1: Point to the tiny, wide, and thin gray box above the field you want to move, and a tiny black arrow will appear. In this case it will be the *donor contact first name* (figure 2.43, arrow A).

Step 2: When you do, the column will become dark. Click and drag the box to the right. Initially, you won't see a change, but after clicking and dragging the field to the right for a few seconds, you will see a thin black line appear between the field names to the right in the **Query Design View grid** (figure 2.43, arrow B).

Step 3: Let go of your mouse, and the field will be dropped wherever the thin black line is currently seen. You can see the result at the arrow in figure 2.44.

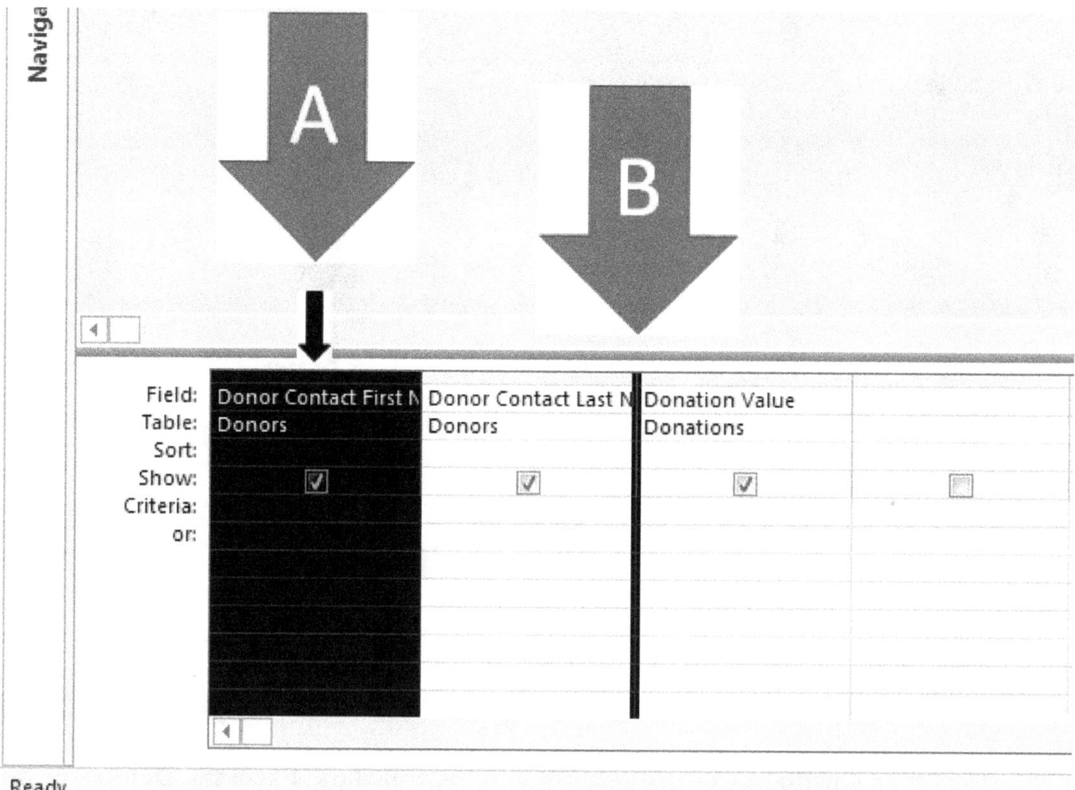

FIGURE 2.43 Selecting a column to be moved

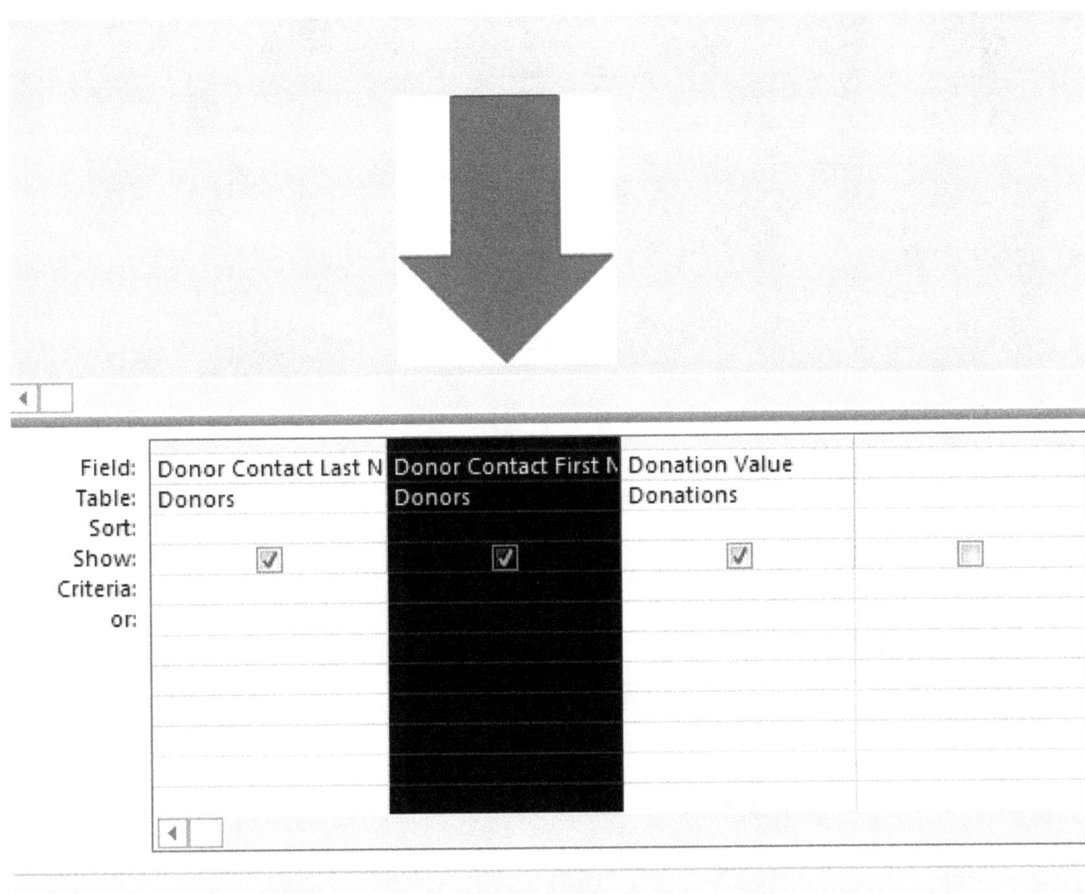

FIGURE 2.44 Selecting a column to be moved (Result)

REMOVING UNWANTED TABLES IN THE QUERY DESIGN VIEW

There will be times when you may add a table above the **Query Design View grid** while you are in **query design view**. For example, if you accidentally send the same table into the **query design view** twice, it will give the table a new name. In the example in figure 2.45, you will see that the *donations* **table** has been chosen twice. Access will automatically name the second one *Donations_1*. To remove it, use the following steps:

Step 1: Click on the table you want to remove. In this case it will be the one named *Donations_1* (figure 2.45).

Step 2: Press the **Delete** key on your keyboard and the table will disappear.

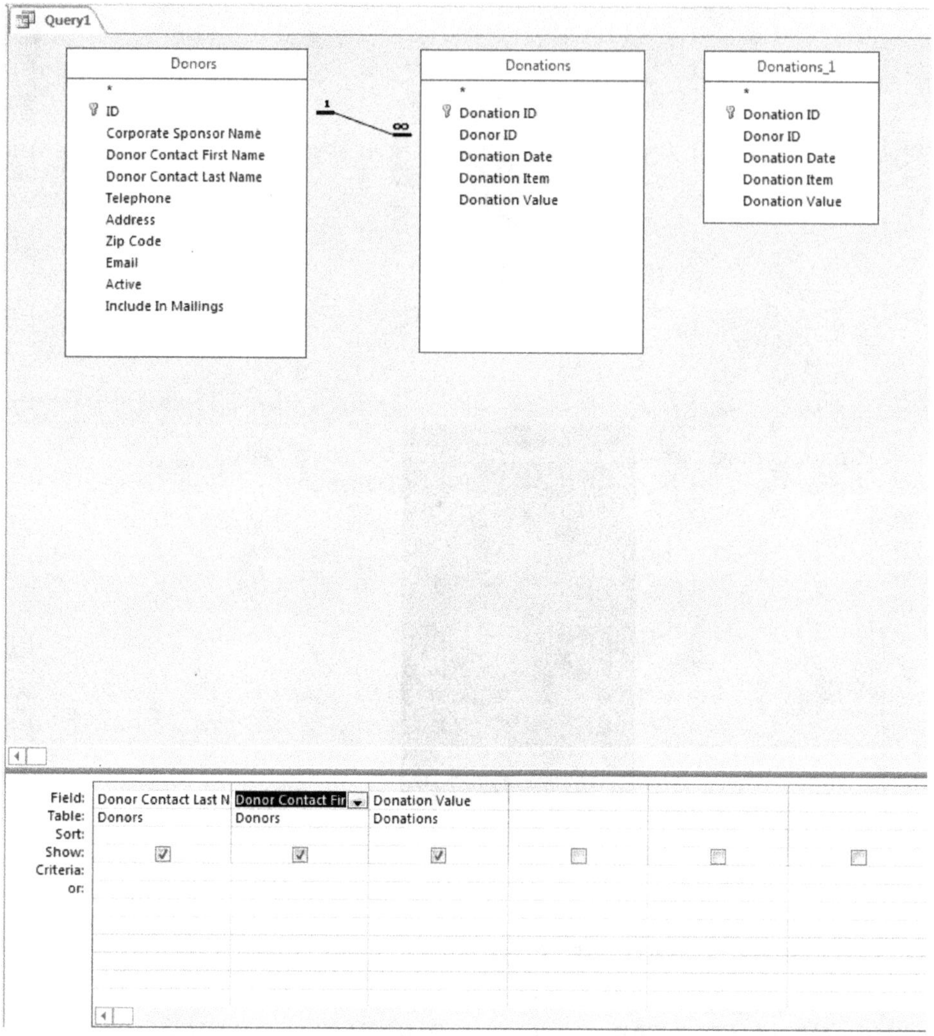

FIGURE 2.45   Unwanted tables in the query design view

CREATING FILTERS IN A QUERY OR A TABLE IN THE DATASHEET VIEW (METHOD 1)

When you are using a table or viewing a query, sometimes you may want to use a **filter** to **TEMPORARILY** remove records that are not needed and to view only those that you **DO** need to see.

NOTE: Once you create a **filter** in a **table** or **query**, it will be removed when you close it. You can **filter** records by using the following steps:

Step 1: Click on the tiny arrow at the top right of the field you are wanting to **filter**. In this case it will be the *zip code* field (figure 2.46, arrow A).

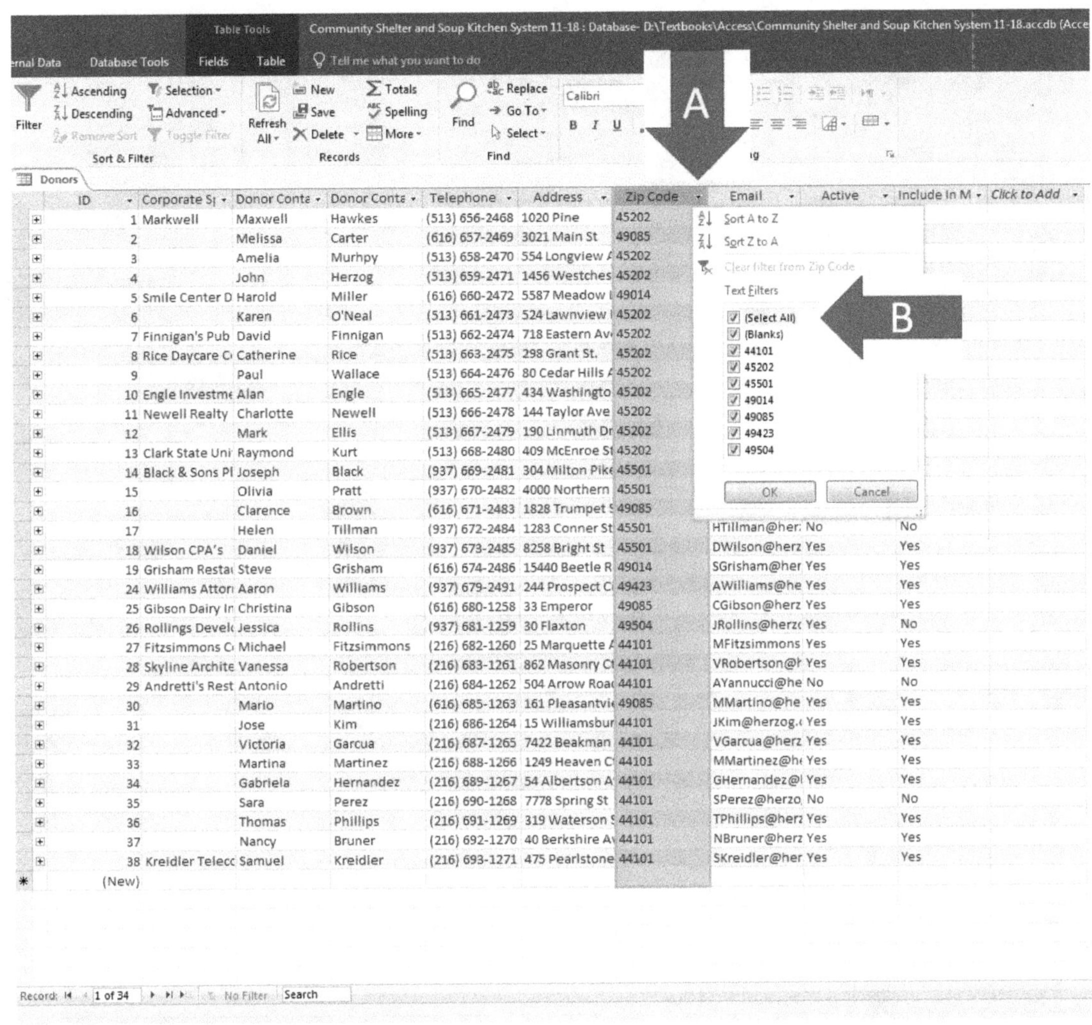

FIGURE 2.46 Filtering method 1 (step 1)

Step 2: There will be a **filter drop-down menu** that will appear (figure 2.46, arrow B). Please notice that you can use this menu to sort records as well as filter them. To **filter**, however you should first remove the checkmark from the **Select All option** in the menu (figure 2.46, arrow B). That will remove the checkmarks to all of the *zip codes* that are displayed.

NOTE: This procedure is virtually the same as the one you could use to **filter** data in an **Excel spreadsheet**. In Excel, however the **filter** is **NOT** removed upon closing the spreadsheet.

Step 3: Put a checkmark in each box of the records you want to see in the **filter** by clicking them. In this case, select *zip codes 45501* and *49423* (figure 2.47, arrow A).

Step 4: Click **OK** (figure 2.47, arrow B).

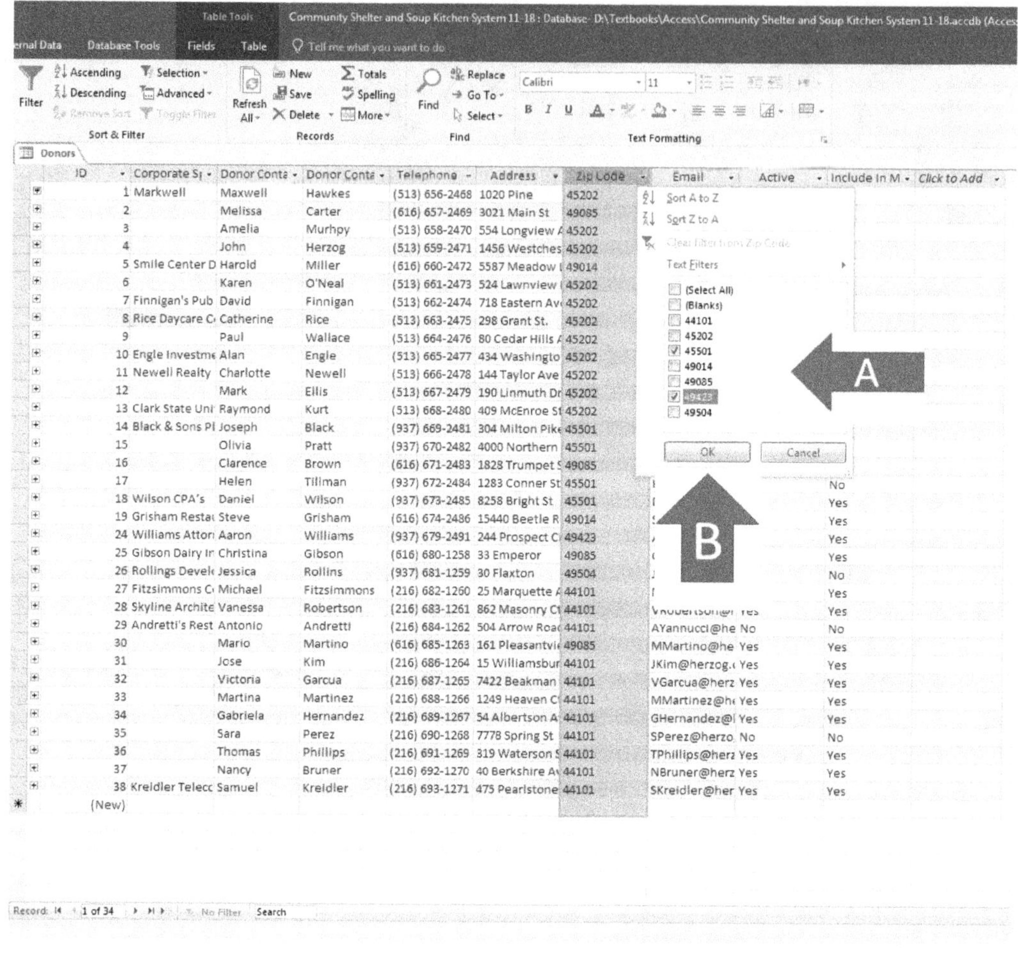

FIGURE 2.47   Filtering method 1 (step 2)

You will see the results shown in figure 2.48.

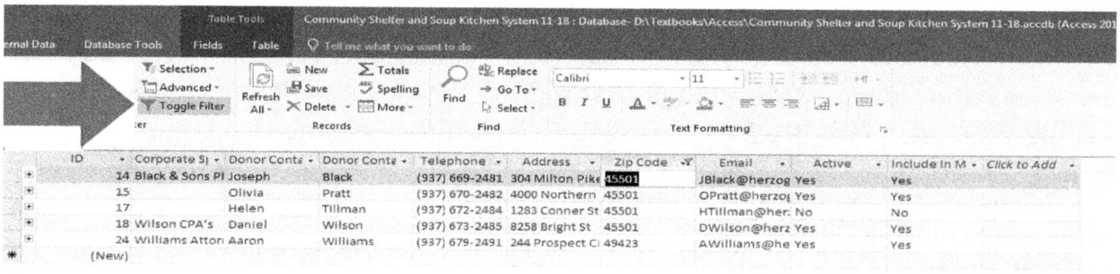

FIGURE 2.48   Filtering method 1 (Results)

REMOVING A FILTER

To remove the **filter**, click the **Toggle Filter** command in the **Home** ribbon (figure 2.48, arrow). **ALL** of the records in the table with be displayed again.

CREATING FILTERS IN A QUERY OR A TABLE IN THE DATASHEET VIEW (METHOD 2)

Suppose you see the *zip code* of *45202* in one of the records of the *donors* table. Also suppose that you decide that you want to see only *donors* of that particular *zip code*. If you right click on a cell of one of the records that holds that *zip code*, you will get a **context-sensitive** (or **shortcut**) **menu** (figure 2.49a). Please notice all of the options from which you can choose:

a. Copy that *zip code* (figure 2.49a, arrow A).
b. Sort data in the *zip code* field (figure 2.49a, arrow B).
c. Display only records that are equal to/are **NOT** equal to or contains/does **NOT** contain the same *zip code* as the *45202* that you selected (figure 2.49a, arrow C).
d. Search for *zip codes* that are/are not equal to, begin with/do not begin with, contain/do not contain, end with/do not end with certain data (figure 2.49a, arrow D).

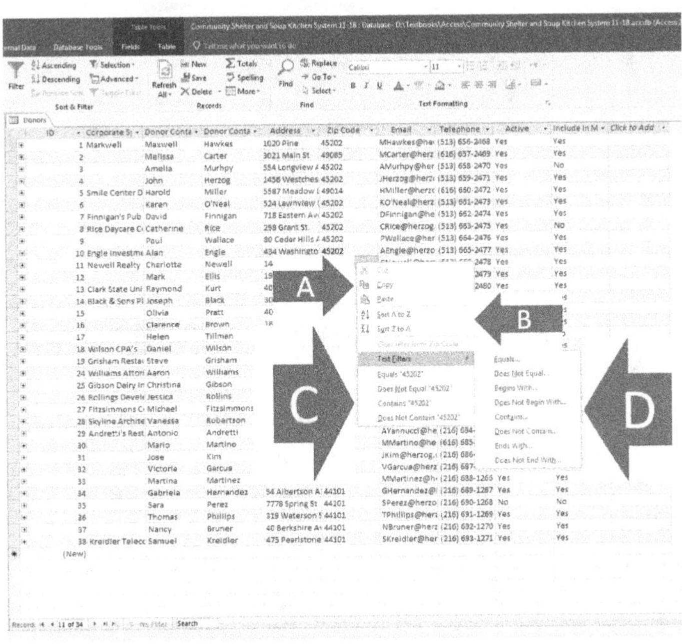

FIGURE 2.49a   Filtering method 2 (step 1)

If you choose to search, for example, for a *zip code* that equals 45889, you can click **Equals** and then place that *zip code* in the **Dialogue box** that appears (figure 2.49b, arrow). Then click **OK**, and only records with that *zip code* will be displayed.

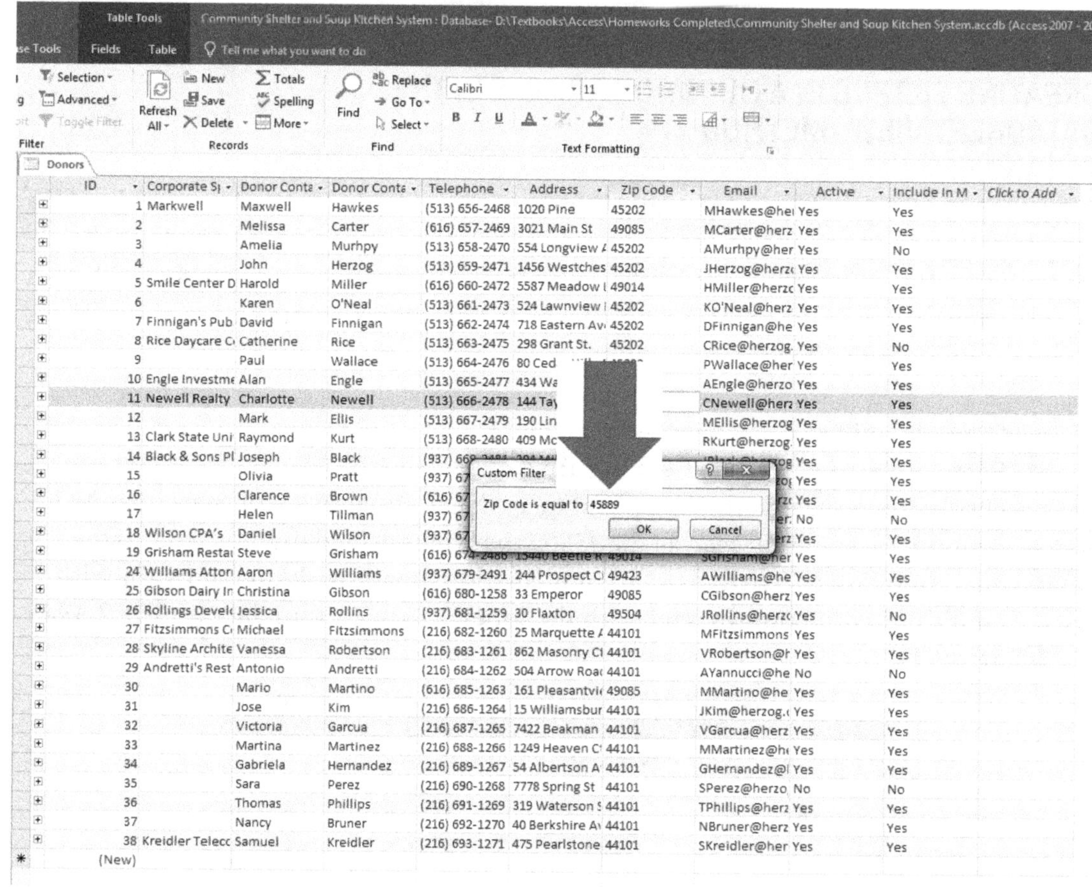

FIGURE 2.49b   Filtering method 2 (step 2)

REARRANGING COLUMNS IN THE DATASHEET VIEW OF A TABLE OR QUERY

Sometimes you will want to change the order that columns appear in a **query**, because the sort order must be from left to right in the **query design**, which may be different than the order you desire. To do this in either a **query** or **table** while in the **datasheet view**, do the following steps:

Step 1: Point the mouse on the field name of the field you wish to move. In this case it will be the *telephone* field of the *donors* table. When you click the field name, a tiny black arrow will appear (figure 2.50, arrow A) and the entire column will become dark.

Step 2: Click and drag the field to the desired location. In this case, we will put it between the *email* and *active* fields. When you do this, a black line will show where the field will be when you release the mouse button (figure 2.50, arrow B).

NOTE: After you do this, Access will ask you if you want to save this field position change. Click **Yes** if you want the new field order to remain permanent.

NOTE: When you rearrange fields in a **query datasheet view**, it does not change their order in the **query design view**. That can cause problems if you are trying to find fields in the **query design view** when the order of them differs from that of the **query datasheet view**.

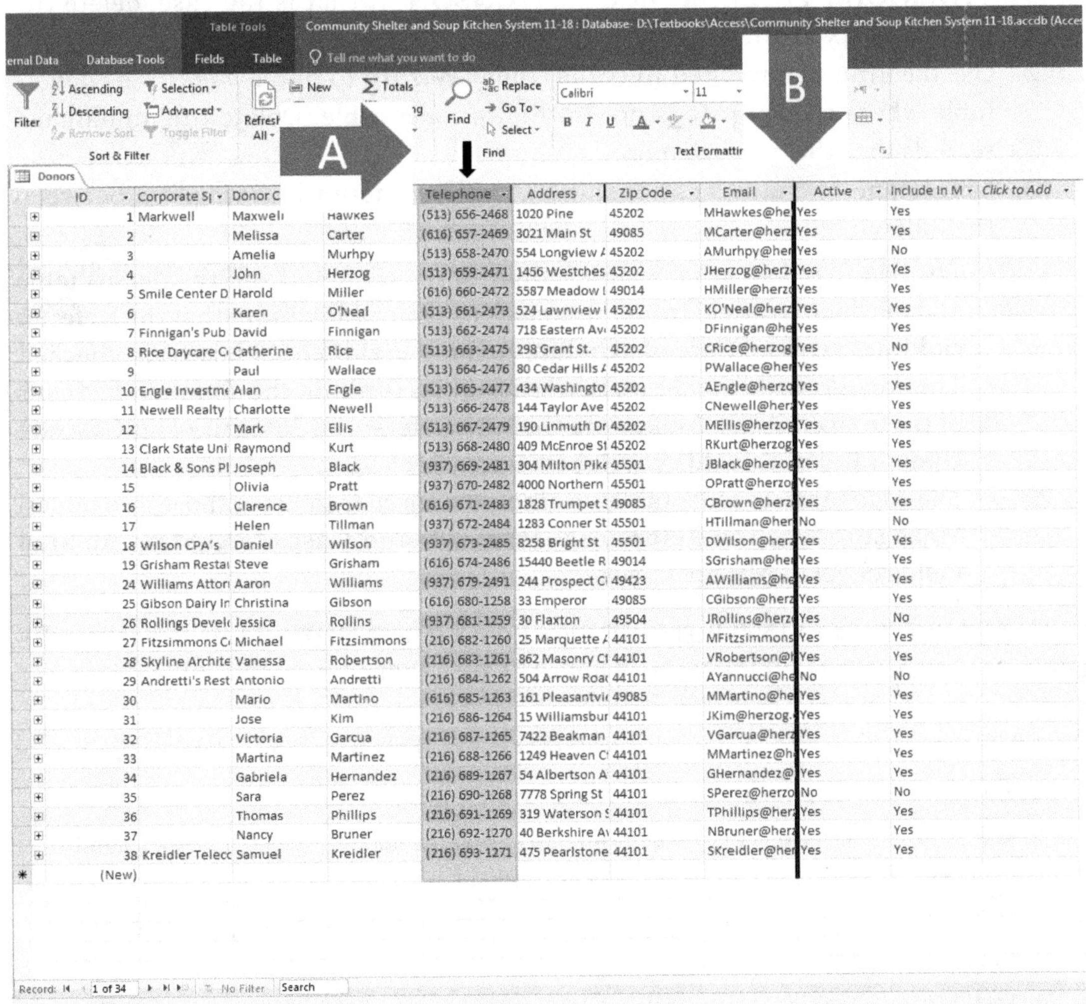

FIGURE 2.50   Moving a column (Datasheet view)

Chapter 2: Assignment (Summary of Tasks)

1. Open the *Community Shelter and Soup Kitchen System* database. Use the **Find Unmatched Records Query Wizard** to find *donations* in the *donations* table where the *donor ID* is not in the *donors table*. Delete the unmatched records that the query displays. **YOU SHOULD ONLY HAVE A FEW RECORDS THAT ARE BOGUS. IF YOU HAVE MORE THAN A FEW, DON'T DELETE THEM. THAT WOULD BE AN INDICATION THAT YOU DID NOT USE THE RIGHT FIELDS FOR COMPARISON.** If that is the case, delete the query and try again.

2. Use the **Find Unmatched Records Query Wizard** to find *donors* in the *donors* table where the *zip code* is not in the *zip codes* table. Delete the unmatched records that the query displays.

3. Define one-to-many relationships (creating referential integrity) between the following:

   - The *donor* and *donations* tables (linking the *ID* field from the *donors* table and the *donor ID* field from the *donations* table). Make sure that the **Referential Integrity** and **Cascade Update Related Fields** check boxes are checked.
   - The *donations* and *donation payments* tables (linking the *donation ID* fields of those two tables). Make sure that the Referential Integrity and **Cascade Update Related Fields** check boxes are checked.
   - The *donor* and *zip codes* tables (linking the *zip code* fields of those two tables). Make sure that the **Referential Integrity** and **Cascade Update Related Fields** check boxes are checked.

USE THE FOLLOWING REMINDER REGARDING THE STEPS TO CREATE ONE-TO-MANY RELATIONSHIPS (OR REFERENTIAL INTEGRITY):

1. Make sure that you have a field in each table to be joined (the **primary key** and the **foreign key**).

2. Make sure that the **primary key** is set as such in the linking field of the **parent** table.

3. Make sure that the **primary key** and the **foreign key** are the same types, and if they are **number** field types, make sure that they are the same width.

4. Make sure that there is not already bogus data in the tables you about to give **referential integrity**.

5. Make sure that both the **parent** and the **child** tables are closed.

6. Go to **Relationships** in the **Database Tools** ribbon and choose the tables to be joined and close the **Show Table** menu.

7. Click and drag the **primary key** field from the **parent** table over the top of the **foreign key** field in the **child** table.

8. Click the Enforce Referential Integrity and Cascade Update Related Fields check boxes in the Edit Relationships menu.

9. Click Create in the Edit Relationships menu.

10. Close the Edit Relationships menu.

4. Create a **select query** using the following specifications:
   a. Use the donor contact last name, donor contact first name, and telephone fields from the donors table.
   b. Set criteria in the **Query Design View grid** to display only the records with a *513* area code.
   c. Sort the query by donor contact last name and then by donor contact first name.
   d. Name the query Calling List Area Code 513.

5. Create a **NEW select query** using only the *donors* table, *zip codes* table, and the *donations* table, and use the following specifications:
   a. Use only the donor contact last name and donor contact first name fields from the donors table.
   b. Use only the *city, state,* and *zip code* fields from the *zip codes* table .
   c. Use only the *donation date* and *donation value* fields from the *donations table.*
   d. Set criteria in the **Query Design View grid** so that only the *Ohio (OH) donors* who have donated more than *300* dollars will be displayed.
   e. Hide the *donation value* field.
   f. Name the query Ohio Donors Who Also Gave More than 300.

6. Copy the *Ohio Donors Who Also Gave More Than 300* query that you just created and give the copy the name of *Ohio Donors OR Those Who Gave More Than 300* and then make the following changes:
   a. Move the *>300* criterion to the or line in the Query Design View grid.
   b. Save the query.

7. Create a **NEW select query** using only the *donors table* and the *donations table* and use the following specifications:
   i. Use the donor contact last name and donor contact first name fields from the donors table.
   ii. Use the *donation value* field from the *donations table.*
   iii. Show the records with the Top 10 donation values.
   iv. Save the query as Top 10 donors.

8. Create a **NEW select query** using only the *donors table* and the *donations table* and use the following specifications:
   a. Use the donor contact last name and donor contact first name fields from the donors table.
   b. Use the *donation date* and the *donation value* fields from the *donations* table.
   c. Using criteria in the **Query Design View grid**, show only the records with donations given **ON OR AFTER** January 1 of 2018.
   d. Name the query *2018 Donations*.
9. Create a **NEW select query** using only the *donors* table, *donations* table, and the *zip codes* table, and use the following specifications:
   a. Use the donor contact last name and donor contact first name fields from the donors table.
   b. Use the *donation date* and the *donation value* fields from the *donations* table .
   c. Use the *city, state,* and *zip code* fields from the *zip codes* table.
   d. Use criteria in the **Query Design View grid** beneath the *city* field that will prompt the user to enter a *city* by using the formula criterion *Like* "*" & *[Please Enter the City]* & "*".
   e. Name the query Parameter Query.
10. Create a **crosstab query** using the *zip code* field from the *donors* table as **row headings**, using the *donation item* field from the *donations* table as the **column heading**, and the *donation value* from the *donations* table as the *value*. Name the query *Crosstab Query*.

Chapter 2: Assignment (Alternate) (Summary of Tasks)

1. Open the *village bookstore database system.* Use the **Find Unmatched Records Query Wizard** to find *inventory orders* in the *inventory orders* table where the *vendor ID* is not in the *vendor* table. Delete the unmatched records that the query displays. **YOU SHOULD ONLY HAVE NONE OR A FEW RECORDS THAT ARE BOGUS. IF YOU HAVE MORE THAN A FEW, DON'T DELETE THEM. THAT WOULD BE AN INDICATION THAT YOU DID NOT USE THE RIGHT FIELDS FOR COMPARISON.** If that is the case, delete the query and try again.

2. Use the **Find Unmatched Records Query Wizard** to find *vendors* in the *vendors* table where the *zip code* is not in the *zip codes* table. Delete the unmatched records that the query displays.

3. Define one-to-many relationships (creating referential integrity) between the following:

   - The *vendors* and *inventory orders* tables (linking the *ID* field from the *vendors* table and the *vendor ID* field from the *inventory orders* table). Make sure that the **Referential Integrity** and **Cascade Update Related Fields** check boxes are checked.
   - The *inventory orders* and *shipments received* tables (linking the *order ID* fields of those two tables). Make sure that the **Referential Integrity** and Cascade **Update Related Fields** check boxes are checked.
   - The *inventory orders* and *book titles* tables (linking the *book title* fields). Make sure that the **Referential Integrity** and **Cascade Update Related Fields** check boxes are checked.
   - The *vendors* and *zip codes* tables (linking the *zip code* fields of those two tables). Make sure that the **Referential Integrity** and **Cascade Update Related Fields** check boxes are checked.

USE THE REMINDER BELOW REGARDING THE STEPS TO CREATE ONE-TO-MANY RELATIONSHIPS (OR REFERENTIAL INTEGRITY):

1. Make sure that you have a field in each table to be joined (the **primary key** and the **foreign key**).

2. Make sure that the **primary key** is set as such in the linking field of the **parent** table.

3. Make sure that the **primary key** and the **foreign key** are the same types, and if they are **number** field types, make sure that they are the same width.

4. Make sure that there is not already bogus data in the tables you about to give **referential integrity**.

5. Make sure that both the **parent** and the **child** tables are closed.

6. Go to **Relationships** in the **Database Tools** ribbon and choose the tables to be joined and close the **Show Table** menu.

7. Click and drag the **primary key** field from the **parent** table over the top of the **foreign key** field in the **child** table.

8. Click the Enforce Referential Integrity and Cascade Update Related Fields check boxes in the Edit Relationships menu.

9. Click Create in the Edit Relationships menu.

10. Close the Edit Relationships menu.

4. Create a **select query** using the following specifications:
   a. Use the vendor contact last name, vendor contact first name, and telephone fields from the vendors table.
   b. Set criteria in the **Query Design View grid** to display only the records with a *937* area code.
   c. Sort the query by vendor contact last name and then by vendor contact first name.
   d. Name the query Calling List Area Code 937.

5. Create a **NEW select query** using only the *vendor* table, *zip codes* table, and the *inventory orders* table, and use the following specifications:
   a. Use the vendor contact last name and vendor contact first name fields from the vendor table.
   b. Use the *city, state,* and *zip code* fields from the *zip codes table*.
   c. Use the date of order and quantity ordered fields from the inventory order table.
   d. Set criteria in the **Query Design View grid** so that only the *Ohio (OH) vendors* where the *quantity ordered* is greater than *800* will be displayed.
   e. Hide the *quantity ordered* field.
   f. Name the query Ohio Vendors with Orders Greater than 800.

6. Copy the *Ohio vendors with Orders greater than 800* query that you just created and give the copy the name of *Ohio Vendors or Orders Greater Than 800* and then make the following changes:
   a. Move the *>800* criterion to the or line in the query design view.
   b. Save the query.

7. Create a **NEW select query** using only the *vendors* table and the *inventory orders* table and use the following specifications:
   i. Use the vendor contact last name and vendor contact first name fields from the vendors table.
   ii. Use the quantity ordered fields from the inventory orders table.

      iii. Show the records with the top 10 quantity ordered.

      iv. Save the query as Top 10 orders.

8. Create a **NEW select query** using only the *vendors table* and the *inventory orders* table and use the following specifications:

   a. Use the vendor contact last name and vendor contact first name fields from the vendors table.

   b. Use the date of order and the quantity ordered fields from the inventory orders table.

   c. Using criteria in the **Query Design View grid**, show only the records with *date of order* dates **ON OR AFTER** January 1 of 2018.

   d. Name the query *2018 Orders.*

9. Create a **NEW select query** using only the *vendors* table, *inventory orders* table, and the *zip codes* table, and use the following specifications:

   a. Use the vendors contact last name and vendors contact first name fields from the *vendors* table.

   b. Use the *date of order* and the *quantity ordered* fields from the *inventory orders* table.

   c. Use the *city, state,* and *zip code* fields from the *zip codes* table.

   d. Use criteria in the **Query Design View grid** beneath the *city* field that will prompt the user to enter a *city* by using the formula criterion *Like* "*" & *[Please Enter the City]* & "*".

   e. Name the query Parameter Query.

10. Create a **crosstab query** using the *zip codes* from the *vendors* table as **row headings**, using the *book title* field from the *inventory orders* table as the **column heading**, and the *quantity ordered* field from the *inventory orders* table as the *value*. Name the query *Crosstab Query.*

Calculations, Action Queries, and Group/Total Reports

IN THIS CHAPTER you will learn how to do the following:

1. Create calculations in **tables**
2. Create calculations in **queries**
3. Create **queries** that perform actions such as updating, adding records, deleting records, and creating new **tables**
4. Create **graphs**
5. Create group and total reports
6. Create **labels**

CHAPTER 3 TASK: CALCULATING, UPDATING, DELETING, APPENDING DATA IN MASS, AND CREATING A GROUP AND TOTAL REPORT OF DONORS

Another nice thing about databases is that they will also make calculations for you. They can be created in the design of a **table**, **query**, **form**, or **report**. We will now discuss the different types of calculations that are commonly created and then you will create them to fit the needs of this scenario.

CREATING A CALCULATING FORMULA (OR EXPRESSION) IN A TABLE WITHIN EACH RECORD (LATERALLY)

Suppose the *community shelter and soup kitchen* has administrative costs of *$3.75* for every donation they receive. Also assume that they would like to show what the *net donation* would be after those costs are subtracted to a given *donation*. You can set up a **calculated field** in the *donations* **Table** that will do the math for you on each *donation*. You can do so using the following steps:

Step 1: Open the *donations* table in the datasheet view by double-clicking it in the navigation pane.

Step 2: Click the **Design View** icon in the **Home** ribbon in order to switch to the **design view** (figure 3.1, arrow).

NOTE: Remember, when a **table**, **query**, **form**, or **report** are open, you can always get to their **design views** by clicking the **Design View** icon in the upper left of your screen.

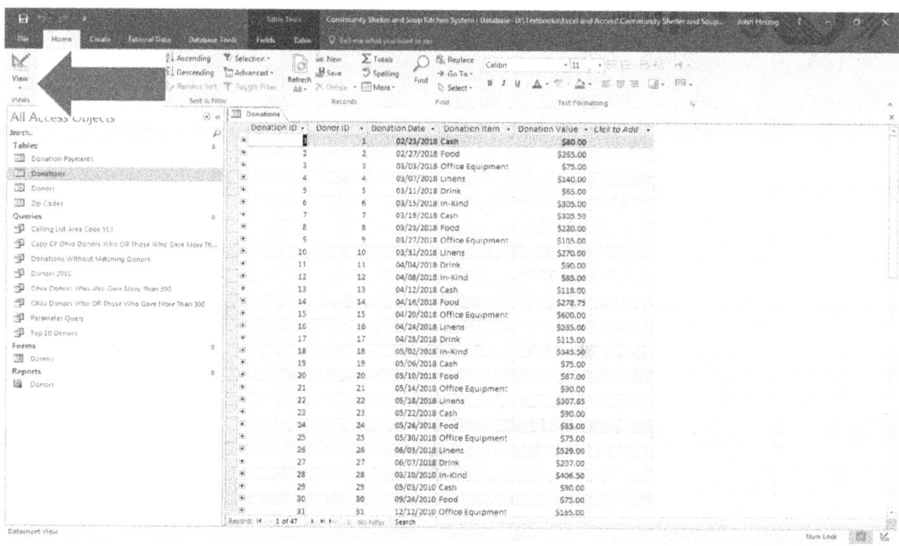

FIGURE 3.1   Creating a calculated field in a table

Step 3: That will take you to the menu you see in figure 3.2. Add a new field in the first blank row below the existing fields and name it *Net Donation*, with a field type of **calculated** (figure 3.2, arrow A).

Step 4: That will display a **pop-up menu** also seen in figure 3.2. You can enter the **calculation formula** (or **expression**) in the box labeled **Enter an expression to calculate the value of the calculated column** (figure 3.2, arrow B). In this example *3.75* will be subtracted from whatever value is in the *donation value* field for each given record. Formulas/expressions are entered into what is called the **Expression Builder window**. In order to designate that a field will be referenced in a calculation, just as cells are referenced in a spreadsheet, you must put square brackets [] around the name of the field; **NOT PARENTHESIS** () and **NOT SET SIGNS** {}. Therefore, at this point, type the formula that reads *[Donation Value]-3.75* in the **expression builder** window (figure 3.2, arrow B).

NOTE: The field name typed between the brackets must be exactly the same as the field name it is referencing. If there is a space in the field name, a space must also be placed in the expression.

NOTE: Do **NOT** put commas or dollar signs in such a formula.

NOTE: Instead of typing out the *donation value* field name between the square brackets to reference it, you can also double click the *Donation Value* field in the **Expression Categories** box (figure 3.2, arrow C) and it will place the field with the brackets in your expression in the **Expression Builder** window for you.

NOTE: You can also add, multiply, and divide fields by other fields or fixed numbers. For example, if you added a field to create a *due date* (assuming that payments are due thirty days after the *donation date*) you could create an **expression** in that **calculated field** that reads

[Donation Date]+30

Please also remember that the you must **NEVER** put square brackets around the value of a fixed number (or constant). In this example, *30* is the constant. Remember, you can only put square brackets around an actual field name (or control which will be discussed later).

Step 5: Click **OK**.

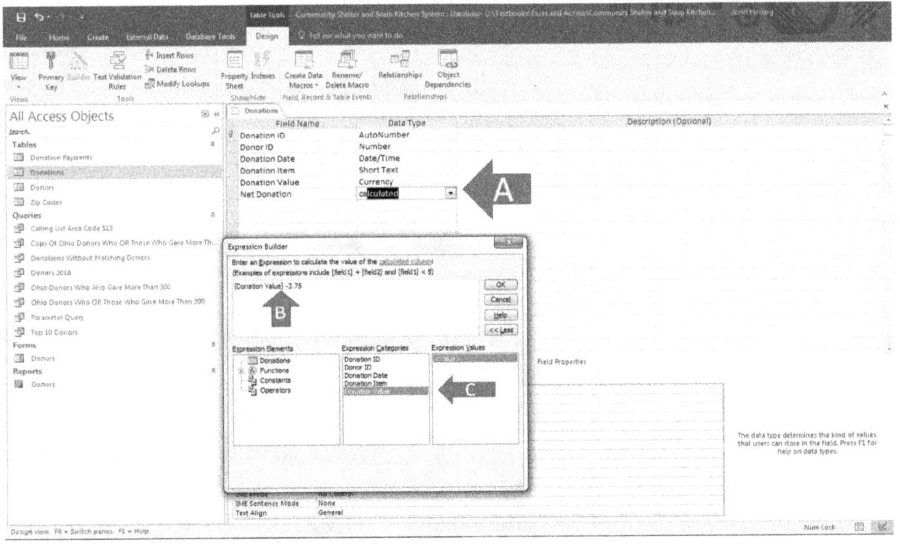

FIGURE 3.2   Creating a calculated field in a table

Step 6: That will display the menu in figure 3.3. Notice that the formula/expression has automatically been placed into the **Expression box** (figure 3.3, arrow A) in the **field properties** section of the **table design view**. **Properties** are characteristics or attributes of things in a database as well as many other types of documents. **Formats** are examples of **properties**. In this case, enter the **property** (or the **format**) of the field to be **currency** in the **Format box** (figure 3.3, arrow B) to make sure that the *net donation* field will be displayed with dollar signs.

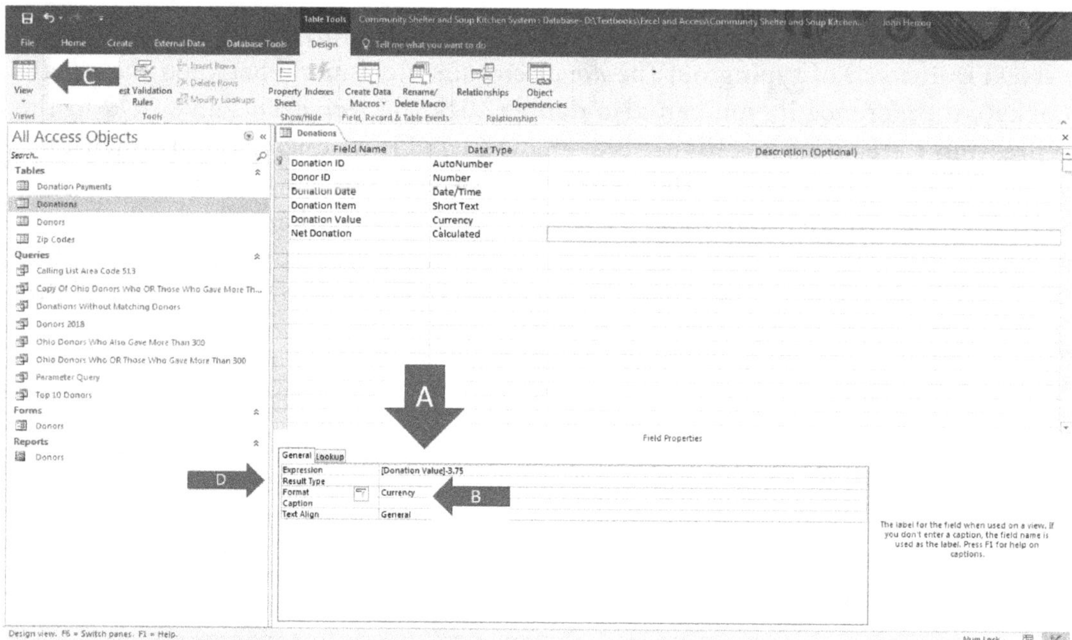

FIGURE 3.3   Creating a calculated field in a table

NOTE: If you want to choose a **result type** (figure 3.3, arrow D) in **ANY** calculated field, it will let you choose among the numeric types discussed earlier (such as **integer**, **long integer**, and the like). If you don't choose one of them Access will automatically set the **result type** as a **decimal** type.

Step 7: Click the **View icon** (figure 3.3, arrow C) to return to the **datasheet view** of the table. You will be asked if you want to save your design changes, and when prompted as such, click **Yes**. Your table will look like what you see in figure 3.4. Please notice that the *net donation* field is showing the correct calculations for each record.

If you change any of the *donation value* numbers, the table will automatically change the *net donation* field just as a formula would do in a spreadsheet!

NOTE: If you attempt to open this database in **Microsoft Office 2007**, or an earlier version with this calculated field, the table will not open.

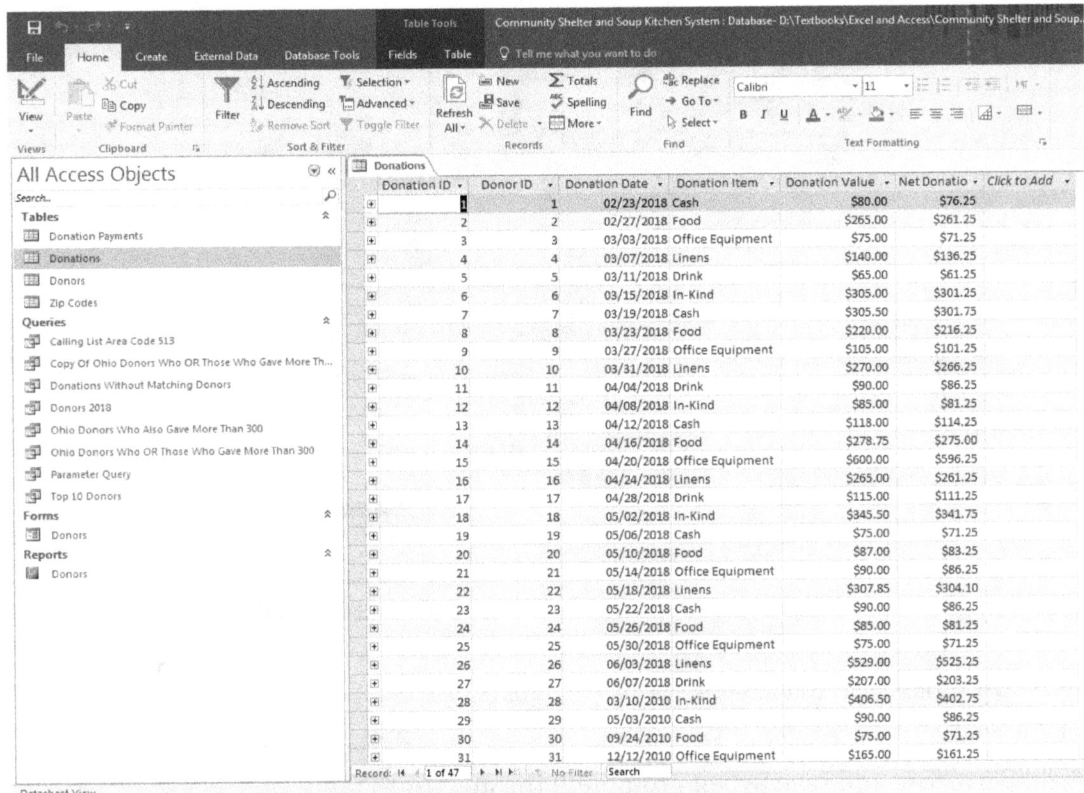

FIGURE 3.4   Creating a calculated field in a table results

TO GET CALCULATIONS OF TOTALS/AVERAGES, ETC. OF MANY RECORDS (AGGREGATELY) IN A TABLE

You can also get totals, averages and more from a field in **tables** or **queries** just as you can in spreadsheets. You can do so by doing the following steps:

Step 1: While remaining in the **datasheet view**, click the **Totals** icon (Σ) in the **Home** ribbon (figure 3.5, arrow A).

Step 2: Notice that a **total row** is then automatically displayed at the bottom of the **tables** (figure 3.5, arrow B). Click on the cell in the **total row** at the bottom of the *donation value* field.

Step 3: Click the tiny arrow on that box (or cell). It is a **drop-down list box** button in (figure 3.5, arrow C) and a list will be displayed (figure 3.5, arrow D).

Step 4: Choose the math you want. In this case choose **SUM** (figure 3.5, arrow D).

NOTE: The list also shows that you can get the **average**, **count**, **maximum**, **minimum, standard deviation**, or **variance** of a field using the **total row**.

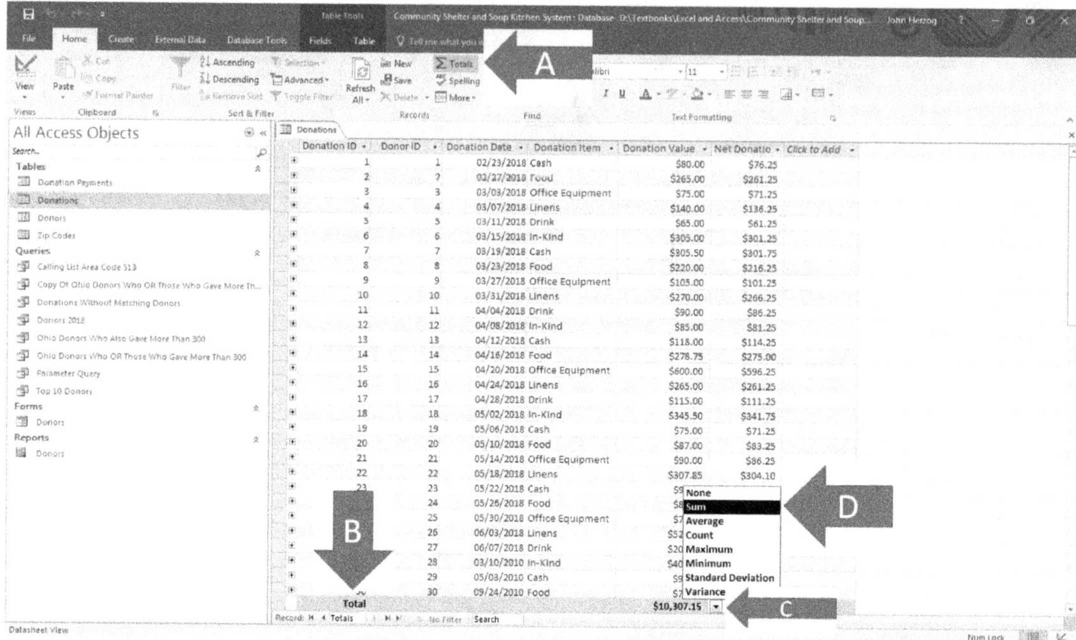

FIGURE 3.5   Totals in the datasheet view of many records (aggregate)

Step 5: Close and save (when prompted) the *donations* table.

CREATING CALCULATIONS IN A FORM

Calculating things as you enter data is also possible in **forms**. Suppose you want to display the tax savings of each donor by each donation that they give, and that you want to see it on the form as you enter data. Also assume that the tax savings will be equal to *25%* of each donation. This can be done by using the following steps:

Step 1: Create a simple form of the *donations* table (as was discussed in chapter 1), and it should look something like what you see in figure 3.6.

Step 2: Go to the **design view** of the form by clicking the **View** icon (figure 3.6, arrow) in the upper left of your screen.

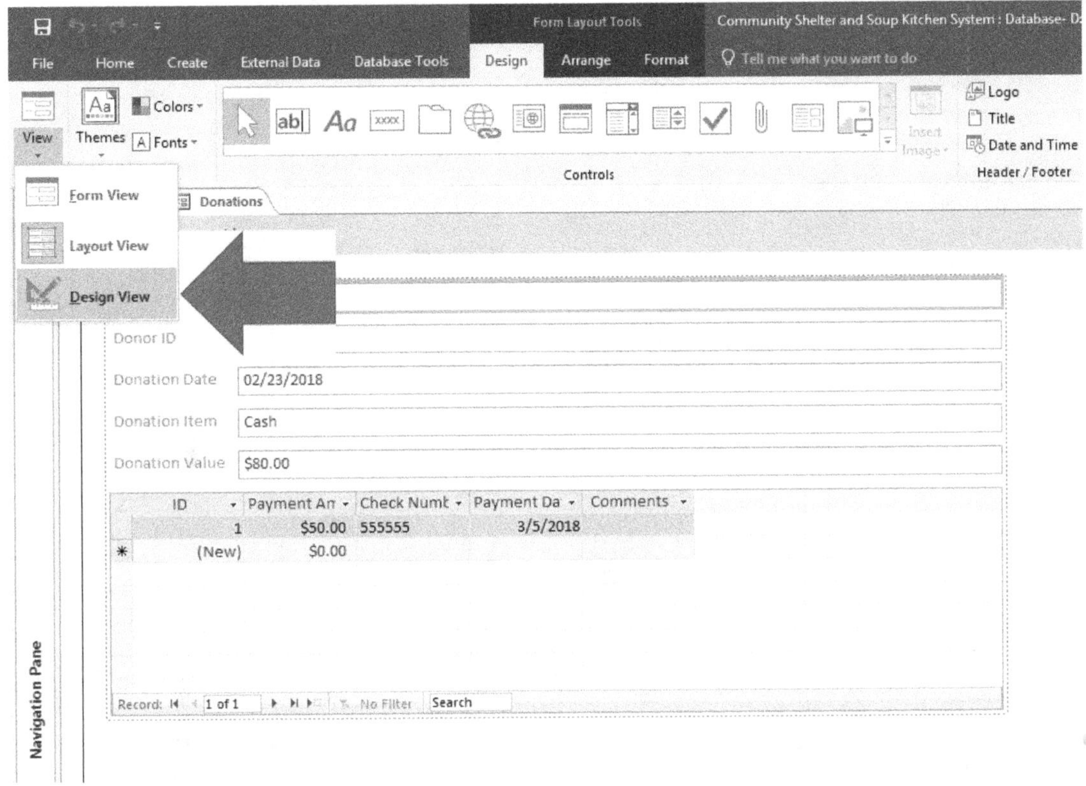

FIGURE 3.6   Created form of donations

Step 3: That will take you to the menu in figure 3.7. Fields, their labels, the sub-form, and all other items in the form are called **controls**. Resize the fields/**controls** just as you would resize a clipart by using the **double-arrow** (figure 3.7, arrow A) and make sure that they are reduced to approximately 6.5 inches.

Step 4: Click once (do **NOT** click and drag; click only once) the **AB** (**text box**) icon (figure 3.7, arrow B). It is found in what is called the **design toolbox** in the **Form Design Tools** ribbon in the **form design view** (figure 3.7, arrow C).

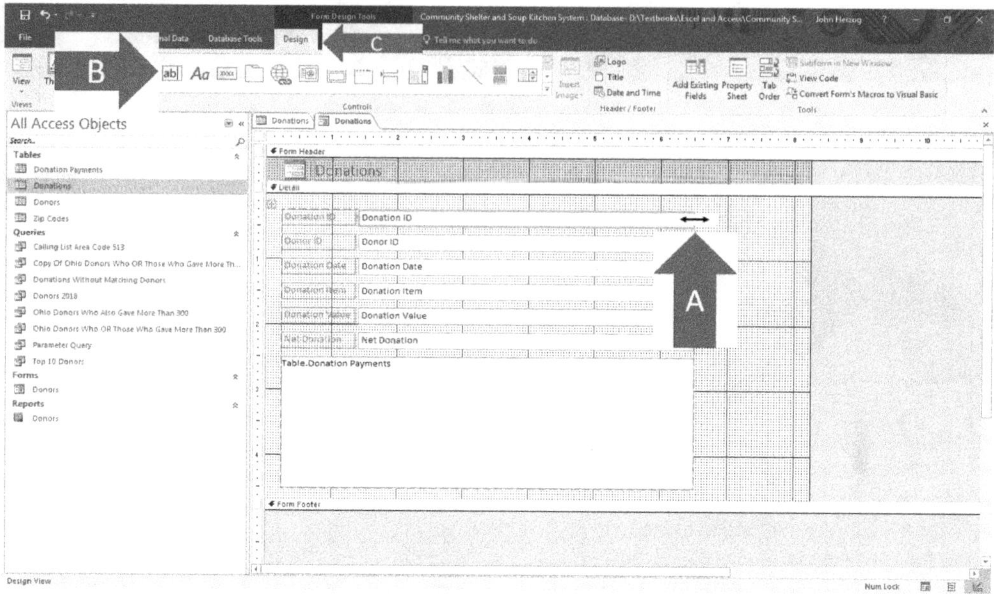

FIGURE 3.7   Resizing a form control

Step 5: Click to the right of the *donation value* field at approximately the *8-inch point (horizontally)* (figure 3.8, arrow). It will place the **text box** at that point, with its label to the left of where you clicked, and the calculated field itself will be to the right of where you clicked.

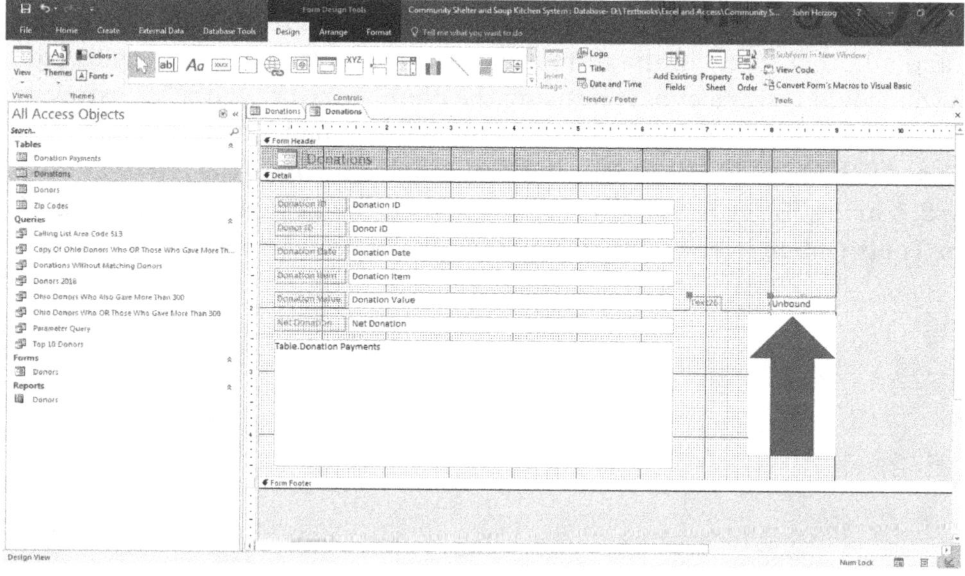

FIGURE 3.8   Adding an unbound text box in a form

Step 6: A **text box** is something in which you can type formulas, just as you did in the table design of *donations*. Inside the box it will read **unbound**. Click on the word **unbound**. When you do, the word **unbound** will vanish and it is then that you can type your formula. Type the formula that reads *=[Donation Value]*.25* inside the **text box** as seen in figure 3.9. You can do it one of two ways: You can type it directly in the **text box** (figure 3.9, arrow A), but the box is small, and it is difficult to see the entire formula. Therefore, making sure you still have the **text box** selected, you may want to click the **Property Sheet** (figure 3.9, arrow B) and you can then type the formula in the **Control Source** box (figure 3.9, arrow C) in the **property sheet**.

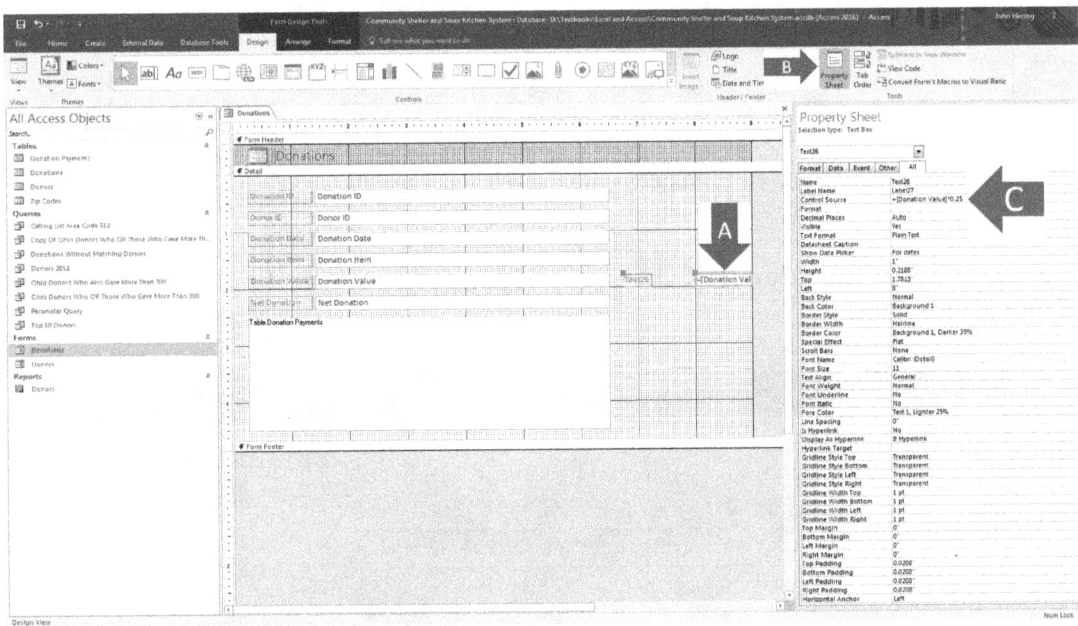

FIGURE 3.9   Adding a calculation to a text box in a form

NOTE: If you want to have a better view of what you are typing inside the **Control Source box**, you can also press and hold the **Shift key** and then tap the **F2 key**. That will give you the **Zoom box** (figure 3.10). Unfortunately, you can only use this feature in the **Control Source** box of the **property sheet**. After typing the formula in the **Zoom** box, click **OK**.

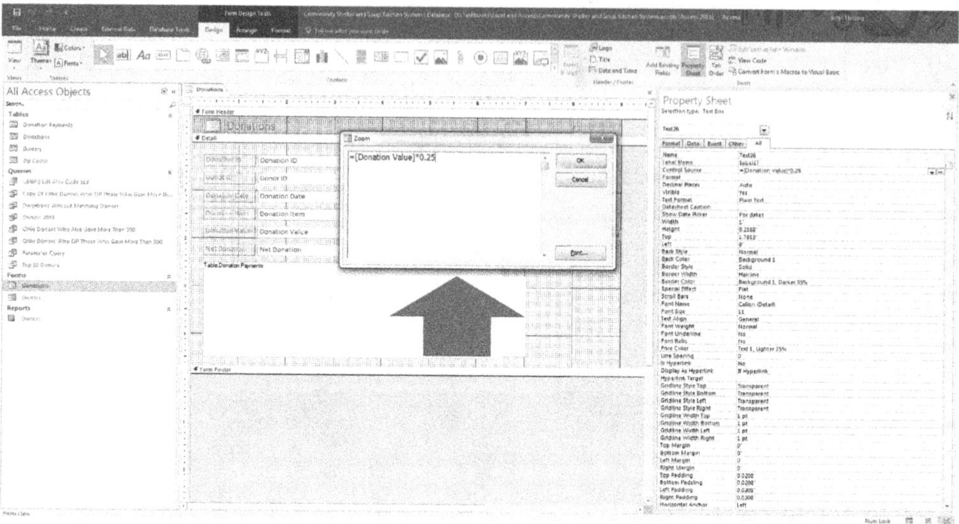

FIGURE 3.10   Adding a calculation to a text box in a form using the zoom

Step 7: With the **property sheet** still open, change the format to **currency** (figure 3.11, arrow).

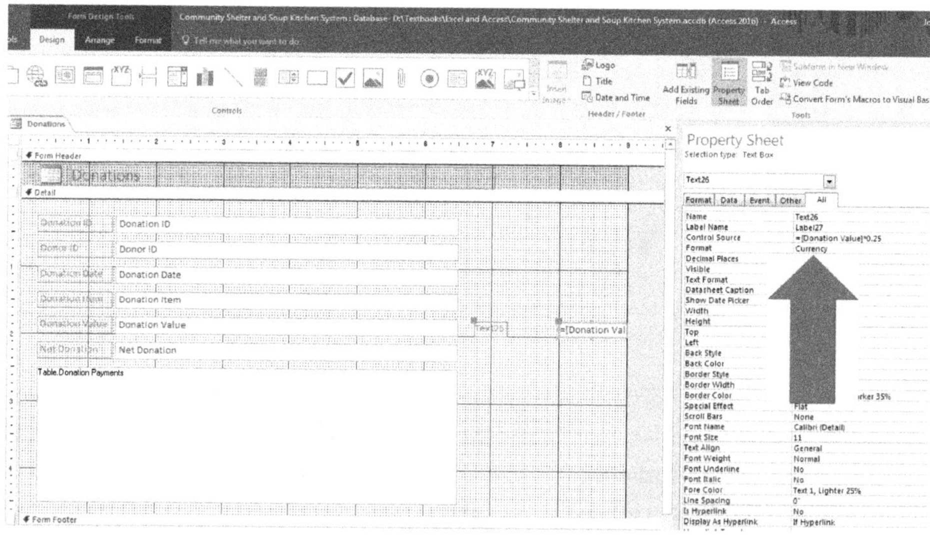

FIGURE 3.11   Changing the format of a calculation in a text box

Step 8: Click inside the label to the left of the **text box** and type over the word *Text20* so it will now read *Tax Reduction* (figure 3.12, arrow A).

NOTE: The label may read a different number other than *20*, but the label number is not relevant.

Step 9: Click the **Form View** button in the upper left of the **design view** screen (figure 3.12, arrow B).

NOTE: It is the **form view** that allows you to enter data as any end user of your database program would. **NEVER** attempt to put data into the fields in the **design view**.

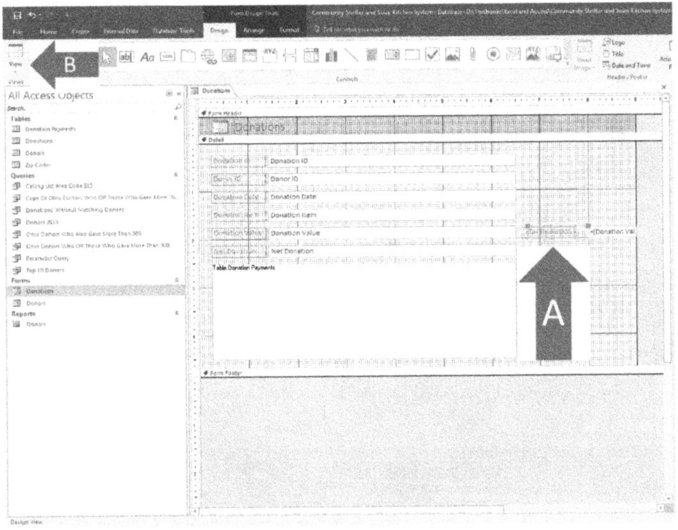

FIGURE 3.12   Changing the label of a calculation in a text box

You will then see in figure 3.13 that the formula/expression is displaying the calculation with the **currency** format.

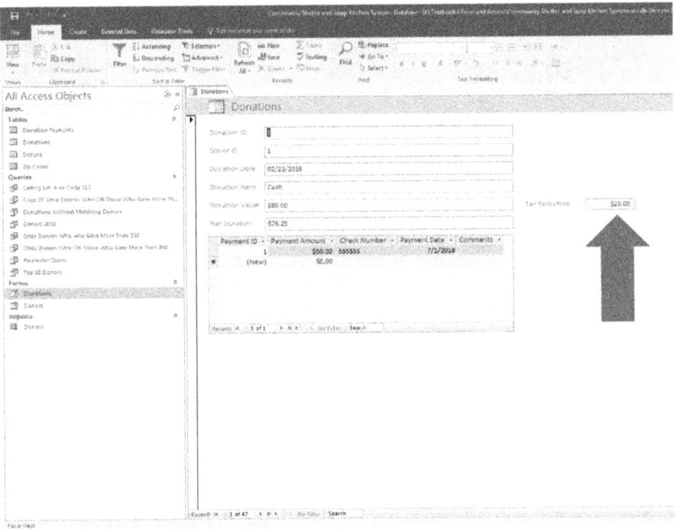

FIGURE 3.13   Results of a calculation in a text box

CREATING FORMULAS IN SELECT QUERIES

Sometimes you need to make calculations in **select queries**. For example, suppose you want to find out how much *donors* owe on their *donations* they have pledged in our *community shelter and soup kitchen system*. This would be considered another **record-by-record**, or **lateral** calculation. There are some users who would rather create calculations in **queries** than in the **table design**. In this case it would be necessary to do so in a **query**, because more than one **table** will be needed for this calculation. You can do so using the following steps.

Step 1: Start a new **query**. Include only the *donors* **table**, the *donations* **table**, and the *donation payments* **table**.

Step 2: Send the *donor contact last name, donor contact first name* fields from the *donors* **table** in to your **Query Design View grid**. Send the *donation date* and *donation value* fields from the *donations* **table** into the **Query Design View grid**. Send the *payment amount, payment date*, and *check number* fields into the **Query Design View grid**. Your **query design view** should look like figure 3.14.

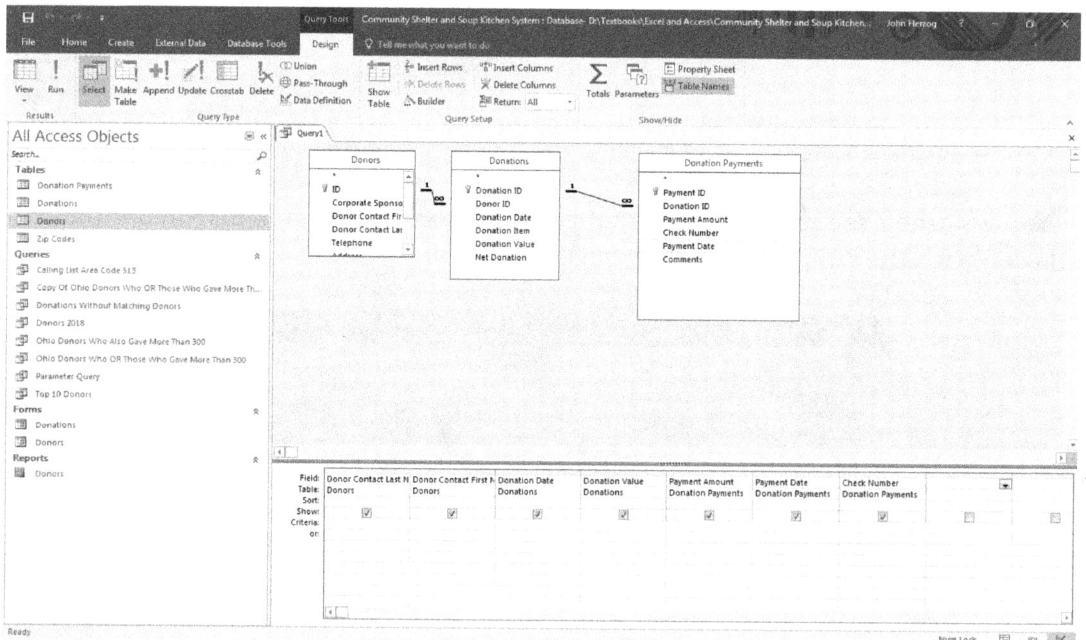

FIGURE 3.14　Tables and fields for a new query

Step 3: Click in the **TOP BOX OF THE FIRST BLANK COLUMN** of the **Query Design grid** (figure 3.15, arrow A).

Step 4: (Optional) To see better what you are typing, press and hold **Shift** and then the **F2** key to see the **Zoom box**.

Step 5: Type in the following formula (figure 3.15, arrow B):

Pledge Amount Due: [Donation Value]-[Payment Amount]

NOTE: The *pledge amount due* is followed by a **Colon (:)** (**NOT A SEMI-COLON**). The **colon** designates that you are creating a new field in this database, but in a **query** and not in a **table**. This also designates *pledge amount due* as the field name.

NOTE: The field name typed between the brackets [] must be exactly the same as the field name it is referencing. If there is a space in the field name, a space must also be placed in the expression.

NOTE: As mentioned earlier, just as you can do in **tables**, you can also add, multiply, and divide fields by other fields, or fields with other numbers in **queries**. For example, if you added a field to this **query** to create a *due date* (assuming that payments are due 30 days after the *donation date*) you could create an **expression** in that **calculated field** in a blank column that reads

Due Date: [Donation Date]+30

Step 6: Press **Enter** or click **OK** in order to close the Zoom box.

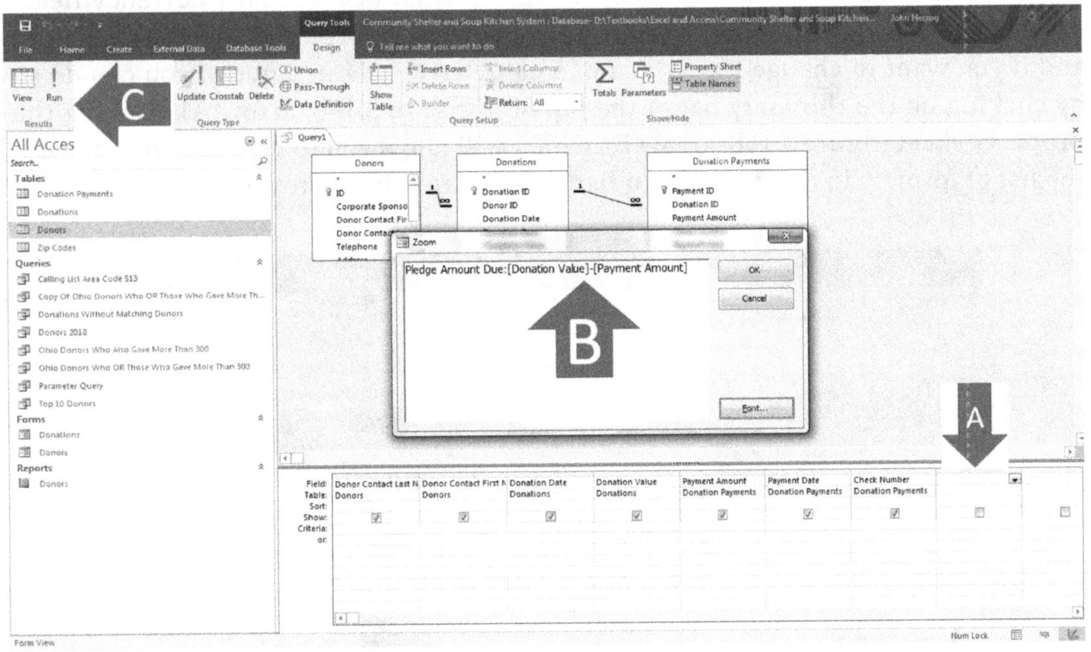

FIGURE 3.15 Formula in a query design

Step 7: Click the **Run** button in the **Design View** ribbon (figure 3.15, arrow C). The results will be what you see in figure 3.16a.

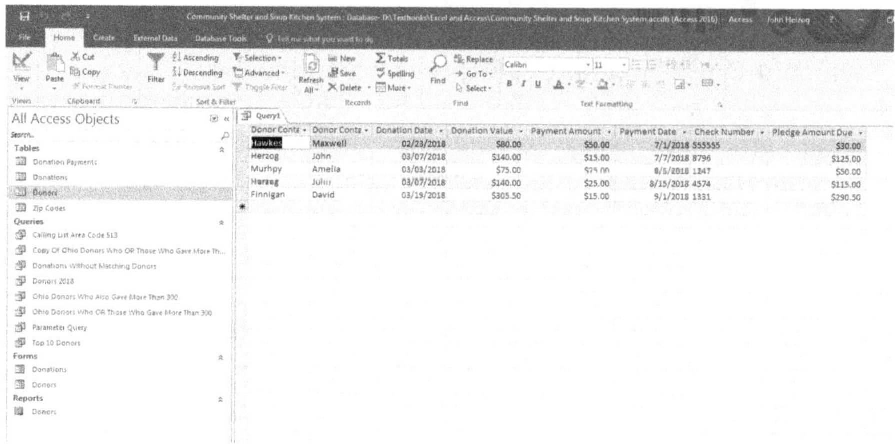

FIGURE 3.16a   Formula in a query results

NOTE: In this case you will get a currency format because only currency fields were in the calculating formula. However, if you get a format that you didn't want and if you want to change the format of a calculated field in a query, you can do so by clicking on the thin gray bar at the top of it (figure 3.16b, arrow A), clicking the **property sheet** (figure 3.16b, arrow B), which will allow you to choose a format from the list of formats in the **drop-down list** box (figure 3.16b, arrow C).

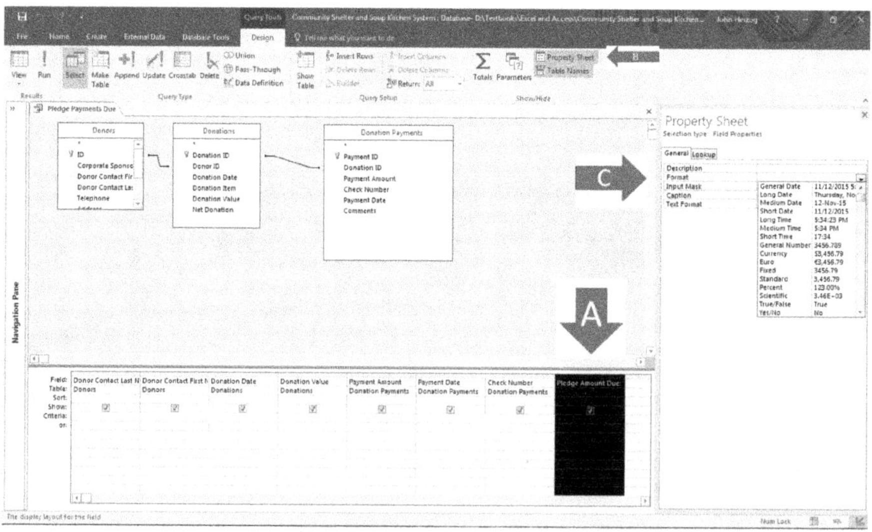

FIGURE 3.16b   Changing format of a query calculation

Step 8: Save the **query** as *Pledge Payments Due.*

TO GET CALCULATIONS OF TOTALS/AVERAGES, ETC. OF MANY RECORDS (AGGREGATE) IN A SELECT QUERY

Sometimes you may need to get totals (or averages etc.) in a **query**. For example, suppose you want to know how much has been donated from *donors* who live in each given *zip code*. You can do so by doing the following steps:

Step 1: Start a new **query** and include only the *donors* and *donations* **tables** (figure 3.17).

Step 2: Send the *zip code* field from the *donors* **table** into the **Query Design View grid**.

Step 3: Send the *donation value* field from the *donations* table into the **Query Design View grid**.

Step 4: Click the **Totals** icon (**Σ**) in the **Query Design** ribbon (figure 3.17, arrow A). Notice that when you do, a **total row** will appear in the **Query Design View grid** (figure 3.17, arrow B).

Step 5: In the **drop-down list box** of the **total row**, beneath the *donation value* field, choose **Sum** (figure 3.17, arrow C).

NOTE: You can also get averages (**AVG**), minimums (**MIN**), and other statistics about a particular field by using this procedure.

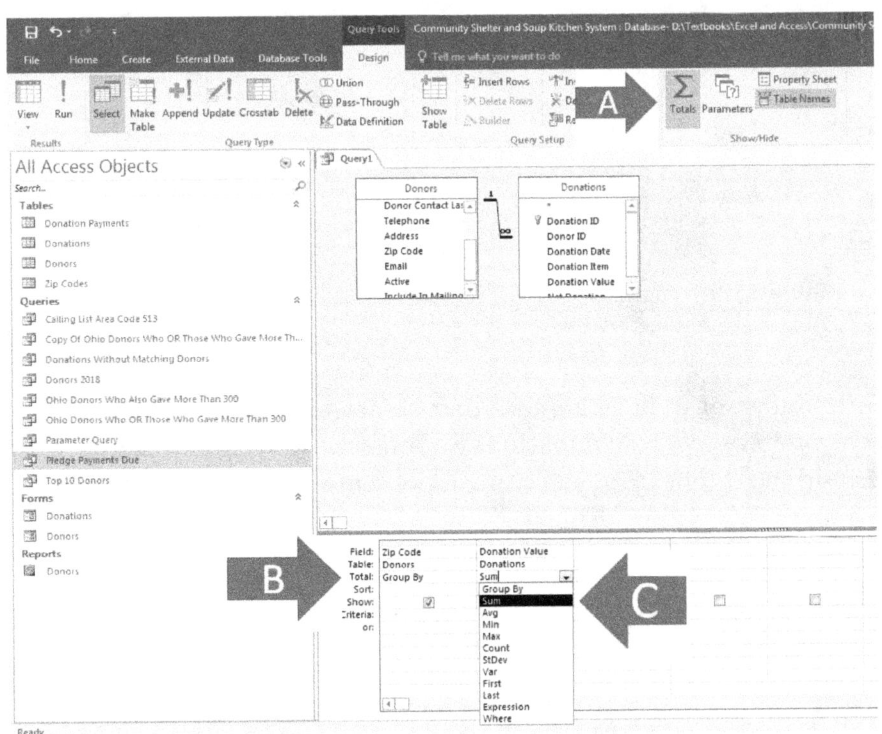

FIGURE 3.17 Using a total row in a query design

Step 6: Run the **query** as you have before. The results will be what you see in figure 3.18a. Name the query *Totals by Zip Code*.

NOTE: If you do **NOT** put the *zip code* field in the query, you will simply get a grand total of all *donations*. When the *zip code* field **IS** included, however the system will **group** all of the *zip codes* and thus put them together. It will then get a total (or average or whatever you use) by each of those *zip codes*. That's why it will default by saying **group by** in the **total row** when you click the **Totals (∑)** icon. This will hold true with **ANY** scenario, whether it be *zip codes*, *cities*, or any category you desire.

Zip Code	SumOfDona
44101	$1,524.65
45202	$4,737.40
45501	$1,339.25
49014	$1,271.85
49085	$820.00
49423	$85.00
49504	$529.00

FIGURE 3.18a   Results of a total row in a query design

NOTE: If you want to have a query that shows only the **DIFFERENT** *zip codes* from which you have received donations (in other words if you want to see all of the area codes with no duplicates), you can simply remove the **donation value** field and rerun the **query**. This will be discussed again later in this chapter.

CREATING CHARTS

You can create many different types of **charts** in Access, such as **bar charts**, **pie charts**, **column charts** and many more. They can be either two or three-dimensional. They can be created in either a **form** or **report**, using virtually the same process.

In this example, we will create a **pie chart** in a **report** of the *donations by zip code query* that we just created.

NOTE: Pie charts can be displayed in a **3D format**, but their data can only have two elements, (such as in this case): *zip code* and *donation value*. While other **charts** permit more than two sets of data (perhaps *zip code, donation value, donation item,* etc.), the only way you can show more than two elements in a **pie chart** is by creating a separate chart for each element.

To create a **pie chart**, use the following steps:

Step 1: Click the **Create** ribbon (figure 3.18b, arrow A).

Step 2: Click **Report Design** (figure 3.18b, arrow B).

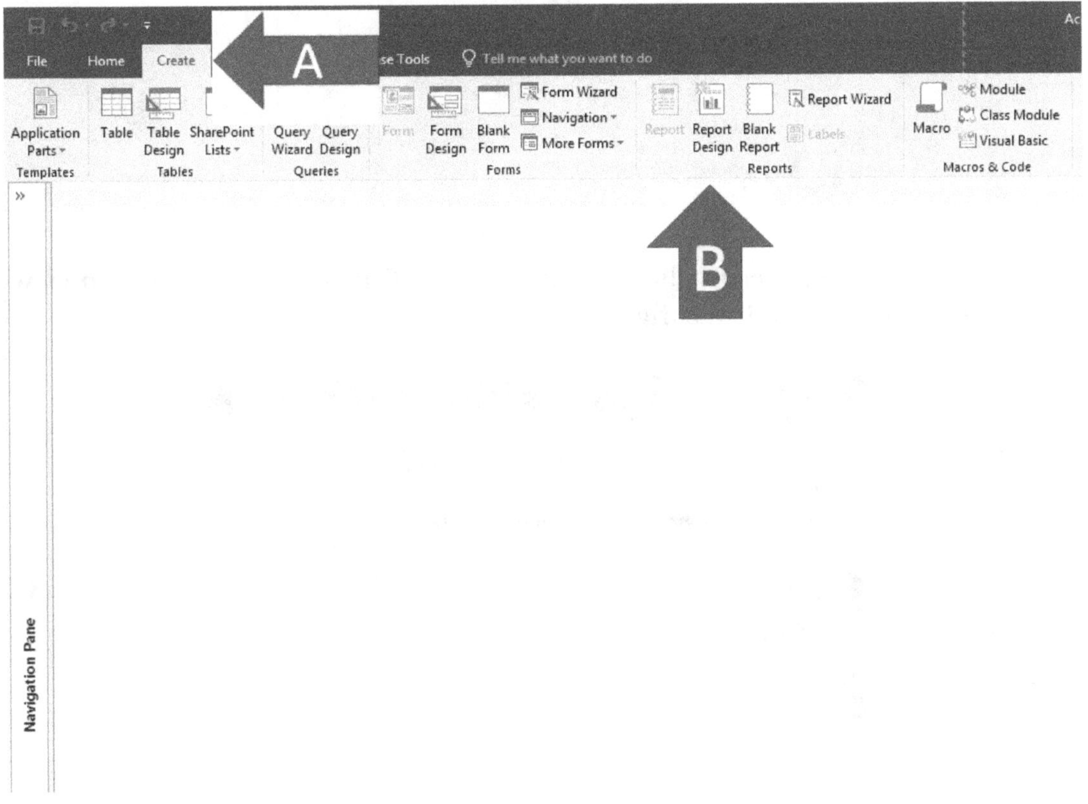

FIGURE 3.18b   Creating a graph

Step 3: That will take you to what you see in figure 3.18c. Open the **design toolbox** in the Design tab of the **Report Design Tools** tab to make sure that the **Use Control Wizard** setting is **ON**. If it is, it will be shaded and with a border (figure 3.18c, arrow).

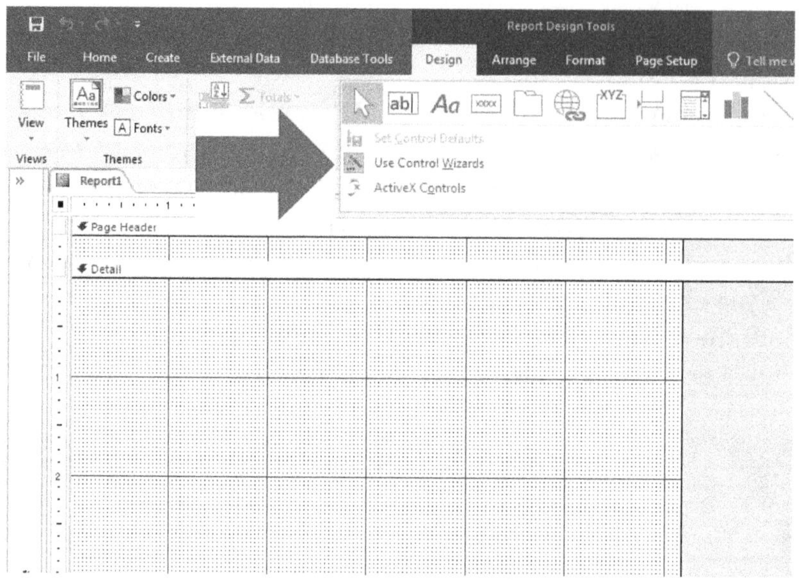

FIGURE 3.18c   Creating a graph

Step 4: Click once (do **NOT** click and drag) on the **Chart** icon in the **design view** of the **Report Design Tools** tab (figure 3.18d, arrow A).

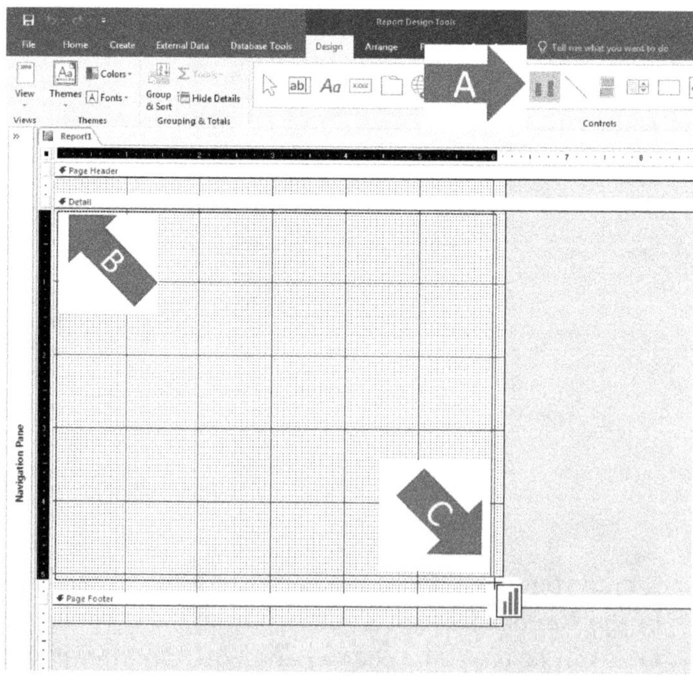

FIGURE 3.18d   Creating a graph

Step 5: Click and drag to draw a box, starting from the upper left of the **detail** area of the **report** (figure 3.18d, arrow B) to the lower right of it (figure 3.18d, arrow C). The box should be approximately *6 inches* wide by *5 inches* tall in this example. After letting go of the mouse, the **Chart Wizard** dialogue box will be displayed (figure 3.18e).

Step 6: Click the **Option button** that will display **queries** (figure 3.18e, arrow A). It will show all **queries** in this database.

Step 7: In this example, choose the **query** named *Totals by Zip Code* that we created earlier (figure 3.18e, arrow B).

Step 8: Click **Next** (figure 3.18e, arrow C).

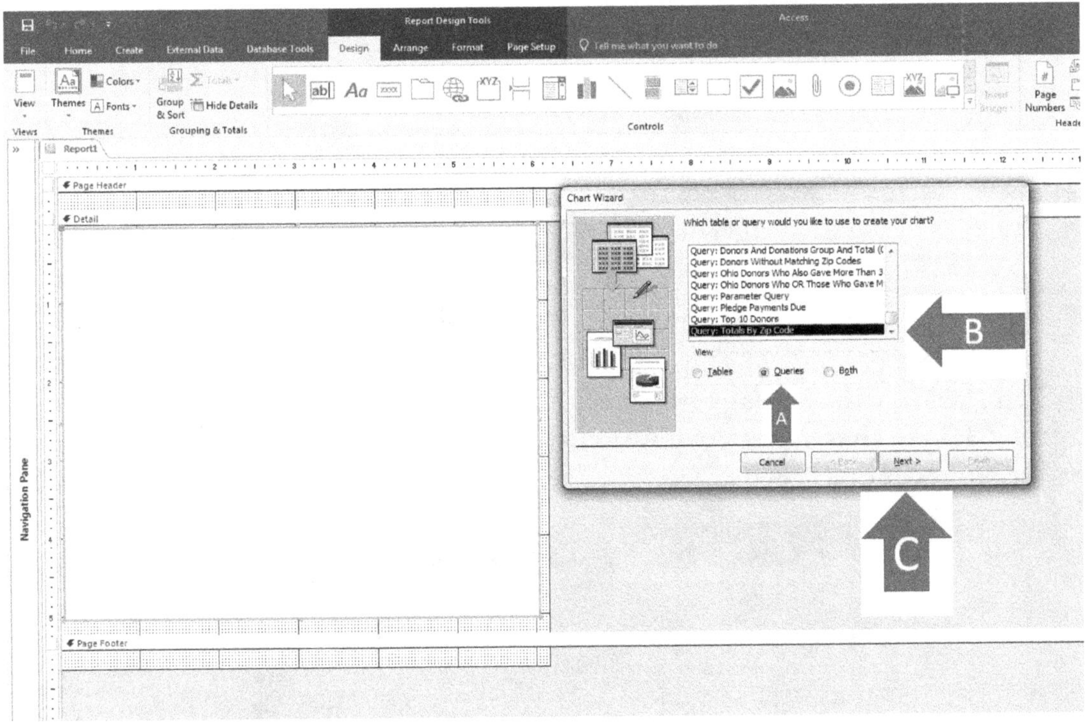

FIGURE 3.18e   Creating a graph

Step 9: That will take you to the screen you see in figure 3.18f. Select the fields you want in the **chart**. In this example we want both of the fields that are shown in the **Available Fields** window on the left to be sent to the **Fields for Chart** window on the right. Therefore, click the **double arrow** (figure 3.18f, arrow A).

Step 10: Click **Next** (figure 3.18f, arrow B).

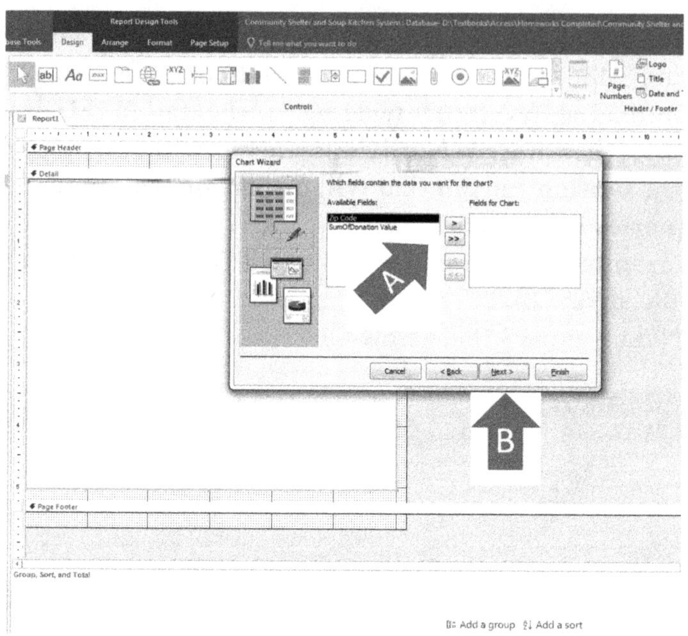

FIGURE 3.18f Creating a graph

Step 11: That will take you to what you see in figure 3.18g. In this example, we will create a **3D pie chart**, thus you must click it (figure 3.18g, arrow A) and then click **Next** (figure 3.18g, arrow B).

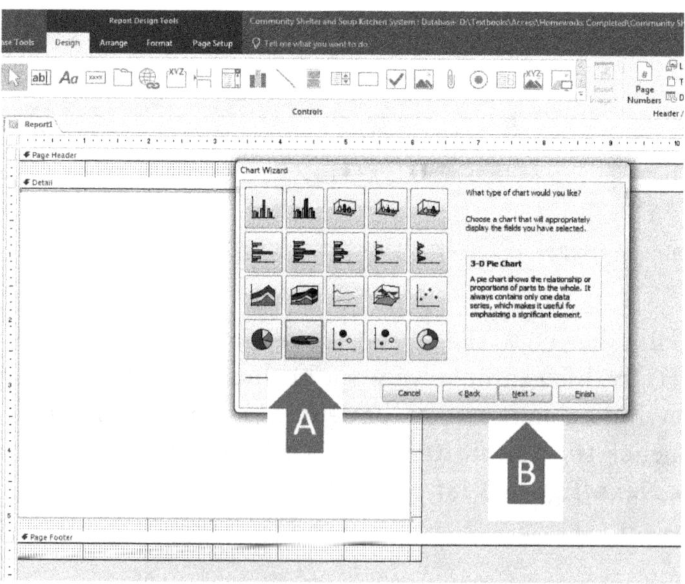

FIGURE 3.18g Creating a graph

Step 12: That will take you to the screen you see in figure 3.18h. If you want to leave the **pie chart** as is, where the *donation values* will be displayed in the **pie** and the *zip codes* will be displayed in a **legend** (which gives a breakdown of the chart values), you would then click **Next** (figure 3.18h, arrow B). On the other hand, if you would like the chart to reflect averages, minimums, or the like, instead of the raw numbers that would be used by default, you can double-click the *Sum of Donation Value* box to make such a change (figure 3.18h1, arrow). Doing so will show you the **Summarize Pop-Up** menu where you can make such a choice (figure 3.18h1).

NOTE: If you double-click the *zip codes* in this example, it will not allow you to view a **Summarize Pop-Up** menu, because text fields cannot have sums, averages, or any calculations.

Also, if you want to preview the chart, click **Preview Chart** (figure 3.18h, arrow C).

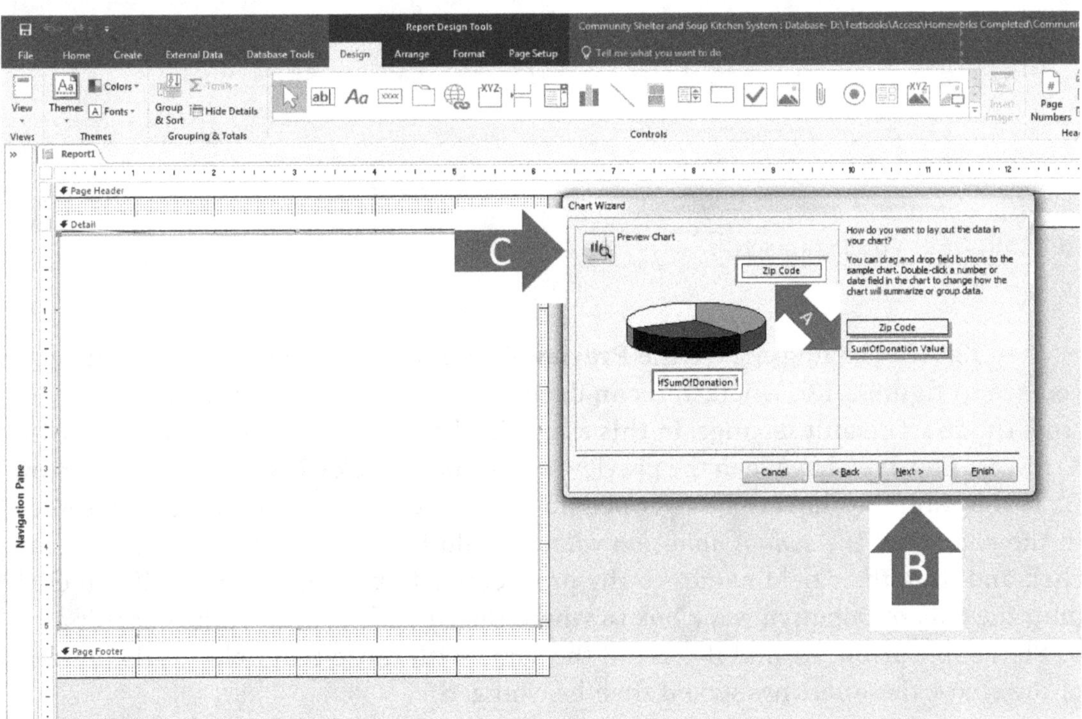

FIGURE 3.18h Creating a graph

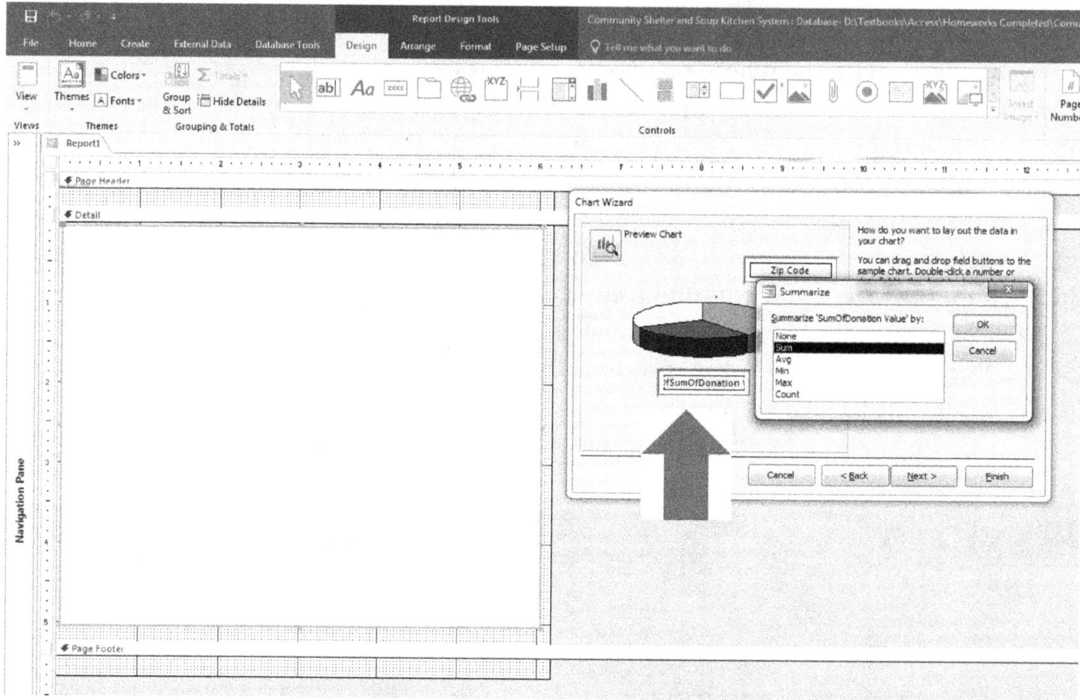

FIGURE 3.18h1   Creating a graph

Step 13: After having clicked the **Preview Chart** button, that will take you to what you see in figure 3.18i, where you can decide if you like the arrangement you have from the chart default settings. In this example, the *zip codes* are in the **legend** (3-18I, Arrow B). Once you have seen the preview of the chart, click **Close** (figure 3.18i, arrow C). If you want the **chart** to be swapped, to where the *zip codes* would be represented in the chart and the *sum of donation values* would be shown in the *legend*, you can click and drag their field names to the prospective boxes. For example, if you click drag the *Sum of Donation value* box to where the *zip code* is now in the chart (figure 3.18h, double arrow A), and if you did the same with those two fields at the bottom of the chart, these settings would then be changed.

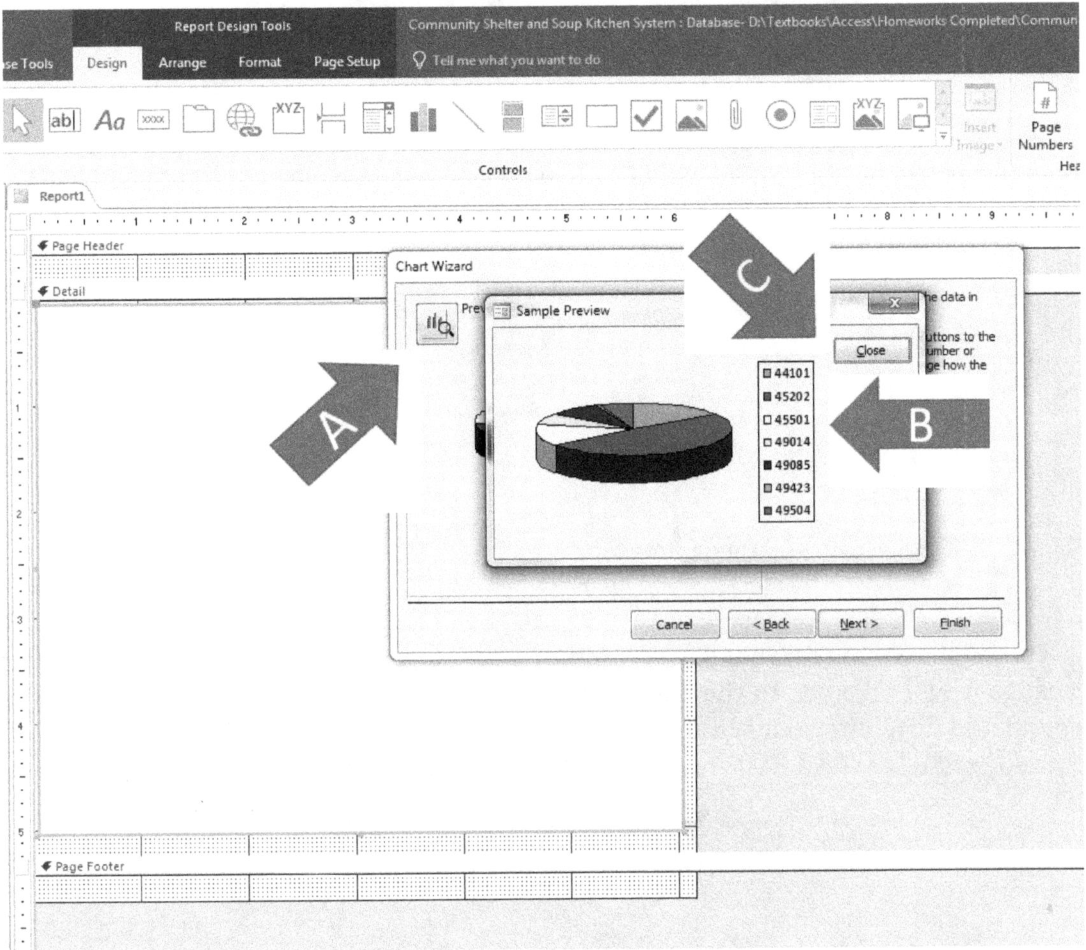

FIGURE 3.18i Creating a graph

Once you have made your changes, if any, click **Next** (figure 3.18h, arrow B). In this example, we will keep the default settings.

Step 14: That will take you to what you see in figure 3.18j. By default, Access will name the chart after the **query** you used for its data. If you wish to change it, type over the name Access used (figure 3.18j, arrow A). In this example, we will use the default name for the chart (*Totals by Zip Code*).

Step 15: Click **Finish** (figure 3.18j, arrow B).

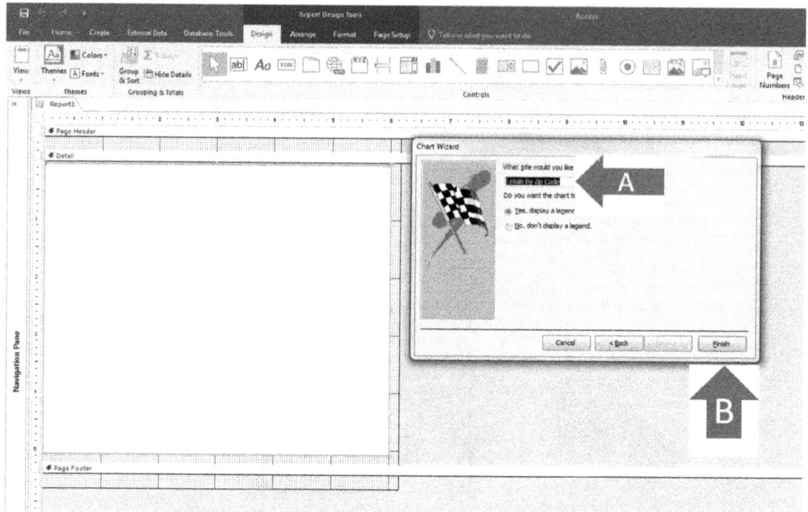

FIGURE 3.18j Creating a graph

That will take you to what you see in figure 3.18k. It may be a bit confusing to you, because it will take you to the **design view**, which will show a completely different **legend** and different values in the **data points** represented by the pie slices.

Step 16: To see the **TRUE** results, click **Print Preview** (figure 3.18k, arrow).

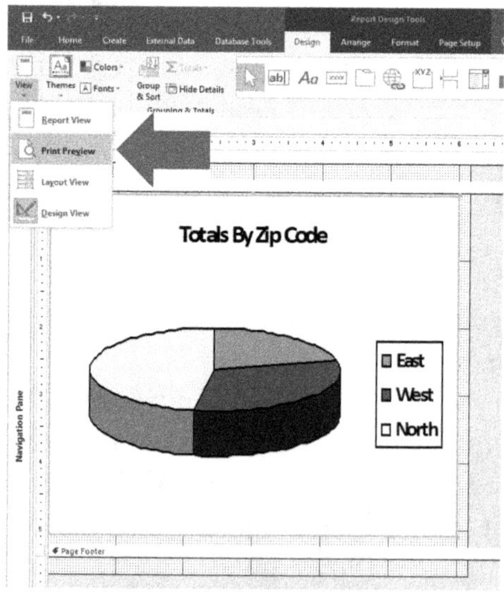

FIGURE 3.18k Creating a graph

Step 17: That will take you to what you see in figure 3.18l. If it's suitable to you, you can save it by using the means shown earlier in this book for saving an object If you wish to make changes to the chart's design, click **Close Print Preview** (figure 3.18l, arrow).

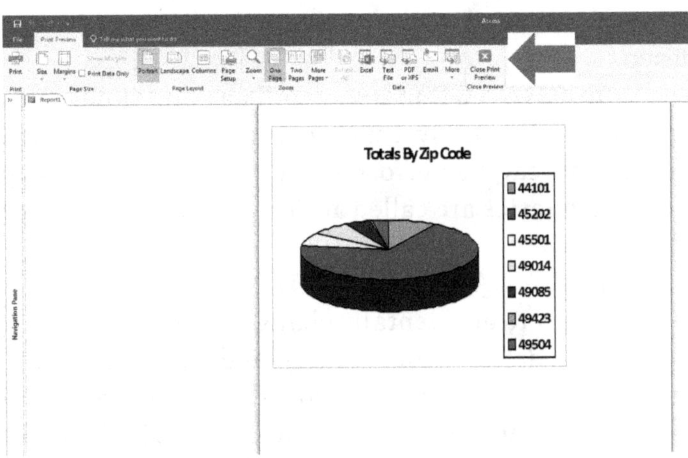

FIGURE 3.18l   Creating a graph

Step 18: That will take you back to the **design view** of the chart (that you saw in figure 3.18k). At that point you can resize it as you would resize any **control** in a **form** or **report**. You can also double-click the chart in the **design view** to get many, many other options for making changes. If you do, your screen will look like figure 3.18m. Across the top of the screen there are many menus you can explore to change things such as the **chart type** (figure 3.18m, arrow).

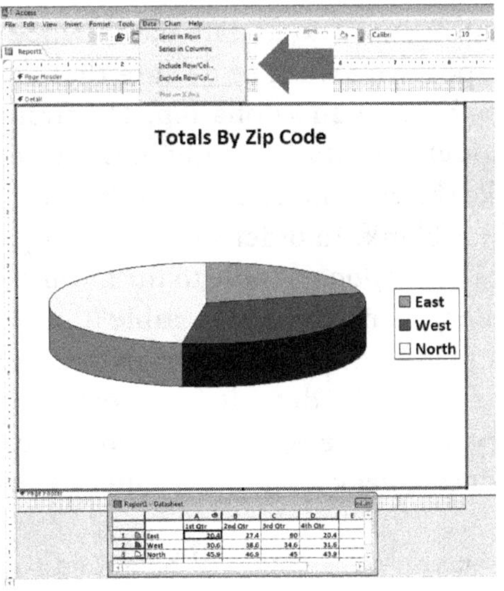

FIGURE 3.18m   Creating a graph

CREATING ACTION QUERIES

So far you have seen **select queries** that select (or display) data from different **tables**, showing only the fields you need, and showing only the records that fit the criteria that you have chosen.

Suppose you want to **CHANGE** many records at one time, rather than having to enter or change them individually, especially if you have to do these things with thousands of records. This can be performed quickly by creating a query that takes action. That is why such **queries** are called **action queries**: they don't **DISPLAY** data, they **CHANGE** data.

YOU MUST BE VERY CAREFUL WHEN YOU CREATE AN ACTION QUERY! It can be very easy to accidentally change thousands of records that should not be changed, or they can delete records that should not be removed! When that happens, the changes often cannot be corrected, and deleted records cannot be retrieved, unless a backup file of the database is utilized. **Action queries** include **update queries** that change thousands of records with one click of the mouse. Other types of **action queries** include **delete queries** that remove many records of certain criteria. There are also **append queries** that send many new records into a table instantly. Finally, there are **make table queries** that will make **select queries** export their data to a new **table**. They are often used so earlier entered data cannot be altered or to create a **combo box** (which will be discussed later).

CREATING ACTION QUERIES: UPDATE QUERIES

Our first **action query** will be used to update data; hence they are named **update queries**. For example, suppose you want to declare that *donors* in our database are active givers **ONLY** if they have given a donation. The first thing we need to do is to remove (or set to a **null value**) the data currently in the **active** field before we update it. By doing so we can make the current data reflect the most recent donations. A **null value** means that the field is blank. In order to create an **update query** to set all data currently in the *active* field of the *donors* **table** to **null**, you can do the following steps:

Step 1: Create a new **query** using the *donors* table (figure 3.19a).

Step 2: Send the *Active* field to the **Query Design View grid**.

Step 3: As mentioned earlier, the default for **queries** is to assume that you will be creating a **select query**. But since we are now going to create an **update query**, you need to click the **Update Query** icon in the **Query Design** ribbon (figure 3.19a, arrow A).

Step 4: Notice that when you do, a row will be added to the **Query Design View grid** that says **Update To** (figure 3.19a, Arrow B). Type the word *null* in **the Update To box** (figure 3.19a, arrow B).

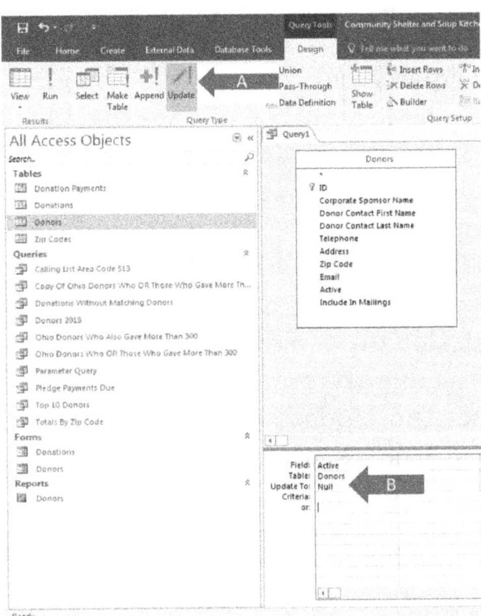

FIGURE 3.19a   Update query action confirmation

Step 5: Click the **Run** button in the **Design View** ribbon. No records will be displayed. It will only begin to take the action you have requested of it. You will see a **pop-up menu** that asks you to confirm whether you indeed want to update these records (figure 3.19b).

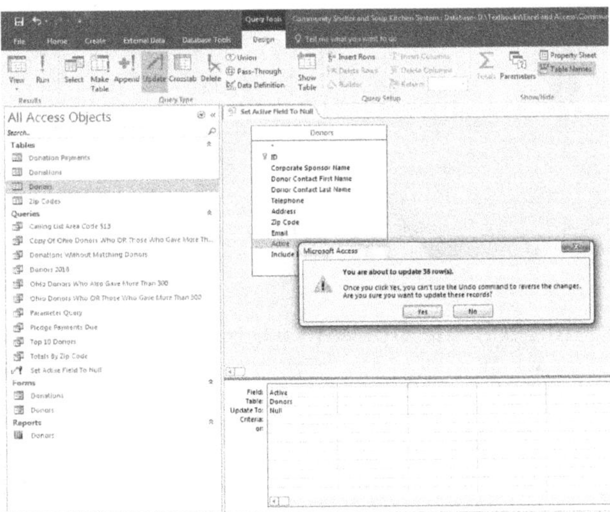

FIGURE 3.19b   Pop-up menu

Step 6: When it does, click **Yes**.

Step 7: Save the **update query** and name it *Set Active Field to Null*.

Step 8: Open the **donors** table now. You will see that all of the **active** field data in the *donors* table is blank.

Now let's have the query update the **active** field to *yes*, only in the records of each donor that have a matching record in the *donations* table. The logic here is simple: If a *donation* record links with the *donor*, then they must have given a *donation*, thus, they are *active* donors. You can do this in the following steps:

Step 1: Return to the **design view** of the *set active field to* **null** query that you just created. Add the *donations* table as a table from which we are to draw data.

Step 2: Make sure that the **Update Query** icon in the **Query Design** ribbon (figure 3.20a, arrow A) is still checked.

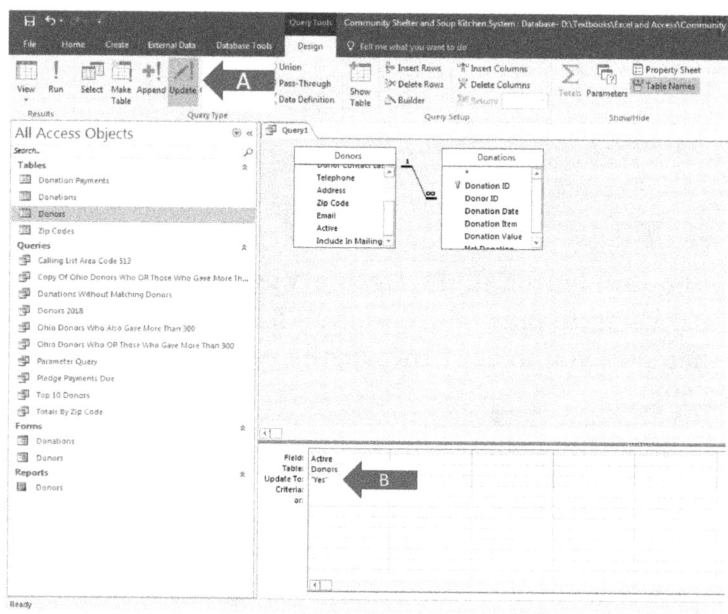

FIGURE 3.20a   Setting field values to yes in an update query

Step 3: Type the word *Yes* in the **Update To** box. When you do, it will put quotation marks around it because it is a **short-text field** and we have not yet converted this field to a **yes/no** field.

Step 4: Click the **Run** button in the **Design View** ribbon. Again, no records will be displayed. It will only begin to take the action you have requested of it. Once again, you will see a **pop-up menu** that asks you to confirm that you indeed want to update these records.

Step 5: When it does, click **Yes**.

Step 6: Using the process discussed earlier in this textbook, us the **Save As** feature in the **file menu** of Access to save the **update query** as *Declare Donors as Active*.

Take a look at the *donors* table now and see that only *donors'* records that matched with donations in the *donations* table have been declared as *active*.

NOTE: It would be wise at this point to create a **NEW update query**, without the *donations* table to declare that all other *donors* whose *active* fields are **NOT** marked as *yes* should have the word *no* placed into the *active* field of their records. In other words, declare that all other *donors* whose *active* fields now have a criteria of **null**, should have the word *no* placed into the *active* field of their records. To do so, you can repeat this process we have just completed, but with a different criteria and update value as shown in figure 3.20b.

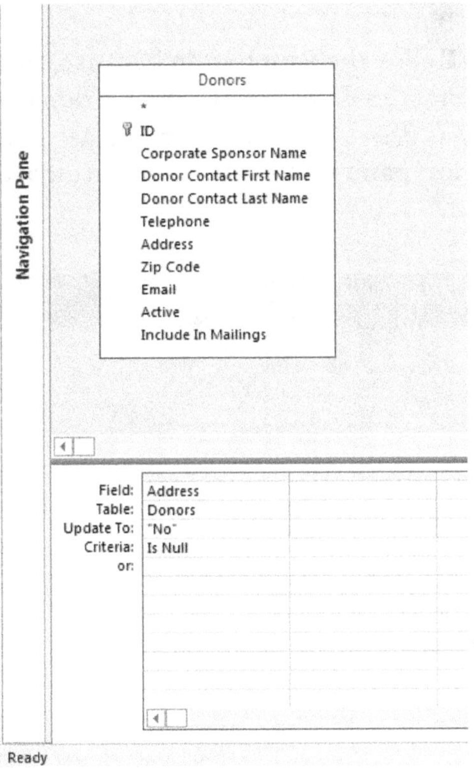

FIGURE 3.20b Criteria and update value

NOTE: When you look at the **navigation pane**, you will see that the **update query** icons are different than the **select query** icons. They have a pencil which indicates that running the query will change data. If you open the **query** as you would a **select query**, again it will not show you data. Opening the **update query** is the same as running it: Data will be changed.

CREATING ACTION QUERIES: DELETE QUERIES

Suppose now that you want to remove all of the *inactive donors*. It is often good to remove records if it is certain that they are of no use. Doing so can make your database run faster. Once again, be careful before you do! Doing this wrong could cause you to lose a great deal of important and irreplaceable data.

You can delete unnecessary data by creating a **delete query,** using the following steps:

Step 1: Create a new **query** using only the *donors* table (figure 3.21).

Step 2: Make sure that the **Delete Query** icon (figure 3.21, arrow A) is chosen in order to designate what kind of **action query** you are creating.

Step 3: Send the *active* field to the **Query Design View grid** (figure 3.21, arrow B).

Step 4: Enter the criterion for this **query** in the **Criteria box**, below the *active* field as *no* (figure 3.21, arrow B).

NOTE: Please do **NOT** click the **Run** button because, this query is only being created to show you how to create a **delete query,** but we don't want any records deleted.

Step 5: Save the query as *delete inactive donors*. Please notice again that the **query** as shown in the **navigation pane** will have an icon attached to it, indicating that it is a **delete query.**

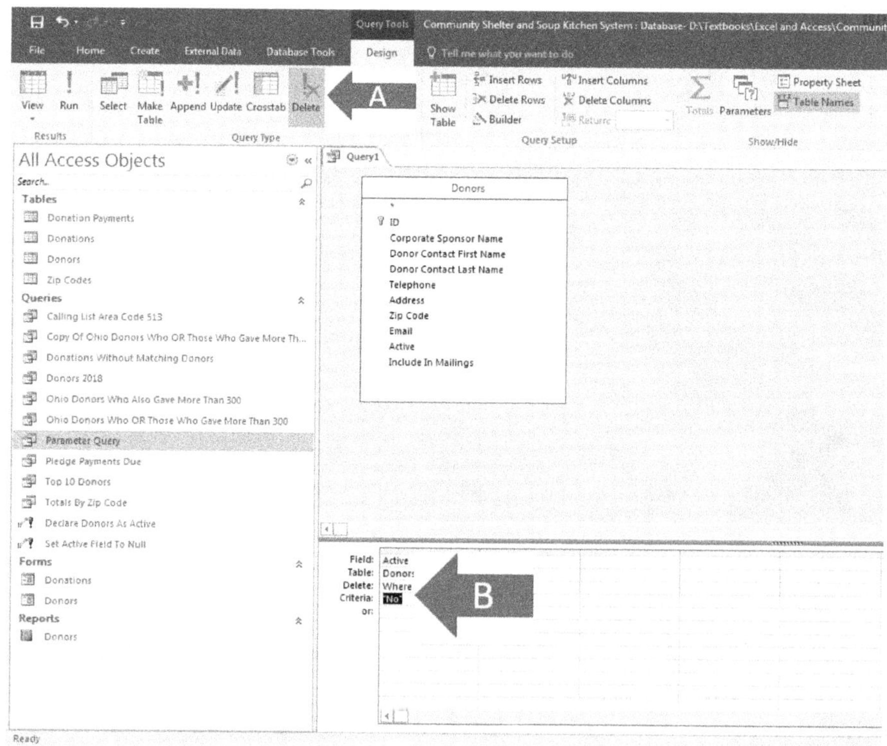

FIGURE 3.21   Delete query design

CREATING ACTION QUERIES: APPEND QUERIES

Suppose you want to bring new records in from a website or from another source and add them (or append them) to your *donors* table on a regular basis. When you import data (as mentioned in chapter 1), there is always an option to allow you to do that. However, in many cases such data will likely need to be qualified before it is added to your tables. Therefore, an **append query** can be created to send them into your table, rather than having to type them in.

In order to prepare for this discussion, using the methods described in chapter 1, import the spreadsheet named *Donors from Download.xlsx* that is associated with this textbook into your *Community Shelter and Soup Kitchen System* database with the following requirements:

a. Check the box that reads First Row Contains Column Headings.
b. Make sure that **NO** field names or field types are changed.
c. Make sure that the **No Primary key** option is chosen when asked about a **primary key.**
d. Name **table** *Donors from Download.*

To create an **append query** that will send those records into the *donors* table, do the following steps:

Step 1: Create a new **query** using only the *donors from download* **table** you just imported into the database as shown in figure 3.22. By placing the *donors from download* **table** in the top of **the query design**, you are declaring it the **source table**. That is because it is the source of the records you are about to append.

Step 2: Make sure **ALL** of its fields are sent into the **Query Design View grid**.

NOTE: There is a quick way to get **ALL** of the fields into the grid **WITHOUT** having to double click of them individually. If a table contained the limit of 255 fields, it would take forever to double-click each one of them to get them into the grid. Instead, double-click the name of the table at the top of the **table box** that contains it. When you do, the fields inside it will go dark (figure 3.22, arrow).

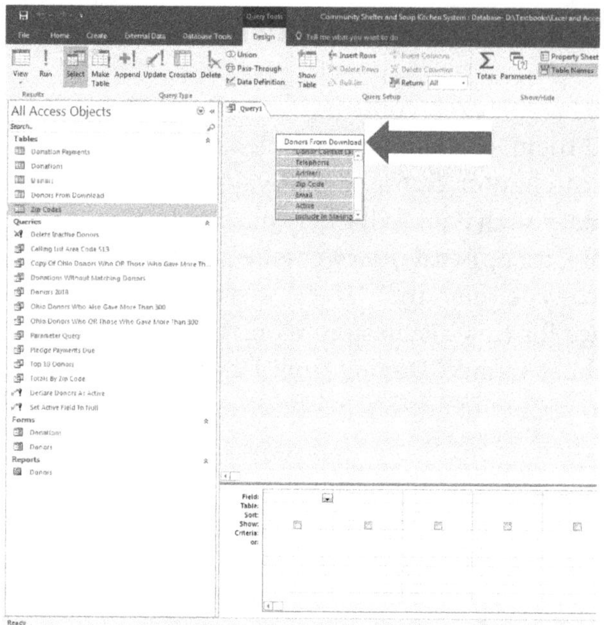

FIGURE 3.22   Append query design (Entering all fields into the grid)

Click and drag **ANY ONE** of those fields down to the **Query Design View grid** and they will all be inserted instantly (figure 3.23, arrow A).

Step 3: Declare the query type as an **append query** by choosing that option in the **Query Design** ribbon (figure 3.23, arrow B).

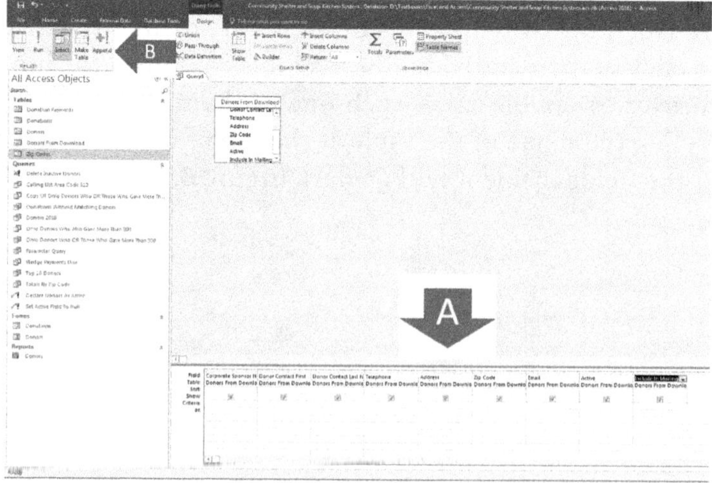

FIGURE 3.23   Append query design

Step 4: When you do this, the **Append Pop-Up menu** will appear (figure 3.24, arrow), asking you which table these records are to be sent (or appended). This would be called the **destination table**. In this example, choose the *donors* **table** as the **destination table** (figure 3.24, arrow).

Step 5: Click **OK**.

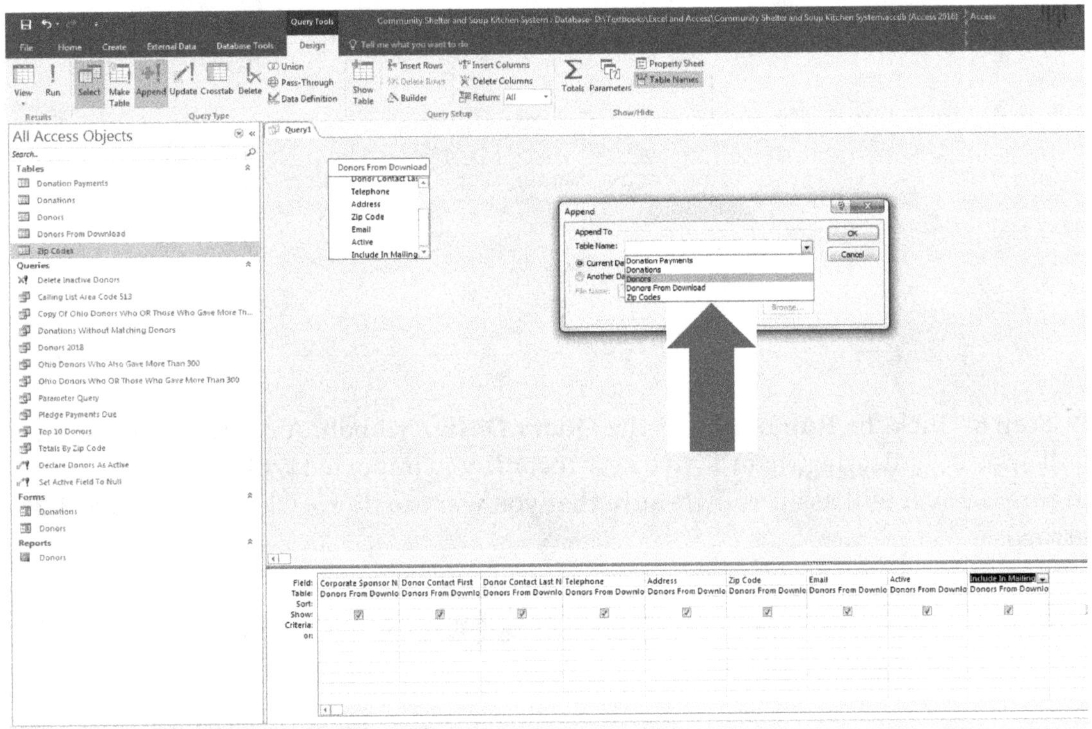

FIGURE 3.24   Append query design (Declaring the destination table)

Step 5: When you do this, the **Append To** line will then be displayed in the **Query Design grid** (figure 3.25, arrow). At the same time, **Access** will then determine that the data from *the corporate sponsor name* field of the *donors from download* table will be sent to the *corporate sponsor name* field of the *donors* table (figure 3.25, arrow). It will then also determine that the data from the *donor contact first name* field of the *donors from download* table will be sent to the *donor contact first name* field of the *donors* table. It will do this for all of the fields.

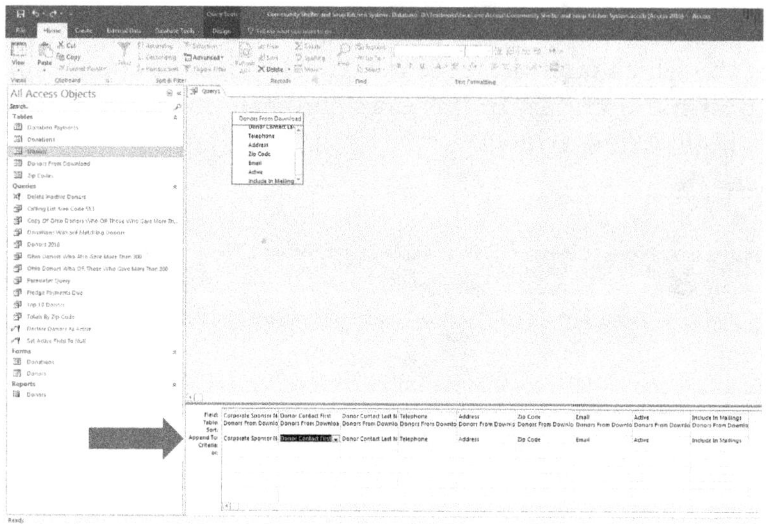

FIGURE 3.25   Append query design

Step 6: Click the **Run** button in the **Query Design** ribbon. A confirmation screen will appear, indicating that the five new records you imported will now be appended (figure 3.26). It will ask if you are sure that you want to do so. Click **Yes** (figure 3.26, arrow).

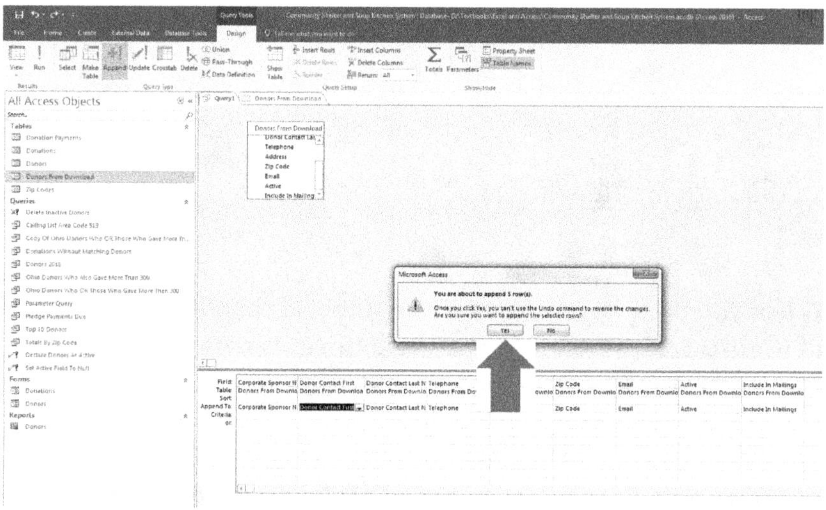

FIGURE 3.26   Append query confirmation

Step 7: If you open the *donors* table you will see that the new records have been added.

Step 8: Name the query *Append New Donors*. Please notice once again that the query as shown in the **navigation pane** will have an icon attached to it, indicating that it is an **append query**.

NOTE: If you get errors in attempting to append records, it could be because the **field data types** between the **source table** and the destination table don't match, or there may be duplicates. If that ever happens in any one of your Access databases, you would then need to correct these problems before attempting to append the records again.

CREATING ACTION QUERIES: MAKE TABLE QUERY

For various reasons, the results in a given **query** need to be sent into a new **table**. One reason for this is that some **reports** often become slow when they are reading many different **queries** with tens of thousands of records of several different **tables**, while such **queries** are often reading data that will not likely change from week to week or month to month. You can send such data to a **table**, which will make the **reports** run much faster.

Sometimes organizations need to take data from existing tables and use it for another purpose. For example, suppose you wanted to make a table showing the different types of *donated items* you have received in the *Community Shelter and Soup Kitchen System* database. You can create a **make table query** that would take all of the entries from the *donations* **table** to create a *donation item* **table**.

In order to do this, we would first have to make sure that **ALL** duplicates of the *donation items* would be removed and then a **table** of the data would be created. To do this, take the following steps:

Step 1: Create a new **query** using only the *donations* **table** and using only the *donation item* field (figure 3.27).

Step 2: Click the **Totals** icon (Σ) in the **Query Design** ribbon (figure 3.27, arrow A).

Step 3: The **Totals line** will appear, which will put the word **Group By** beneath the *donation item* field (figure 3.27, arrow B).

NOTE: Earlier it was indicated that using the **Group By** feature in the **Totals line** would remove duplicates and that is why it is being used here.

Step 4: Click the **Make Table** icon in the **Query Design** ribbon (figure 3.27, arrow C).

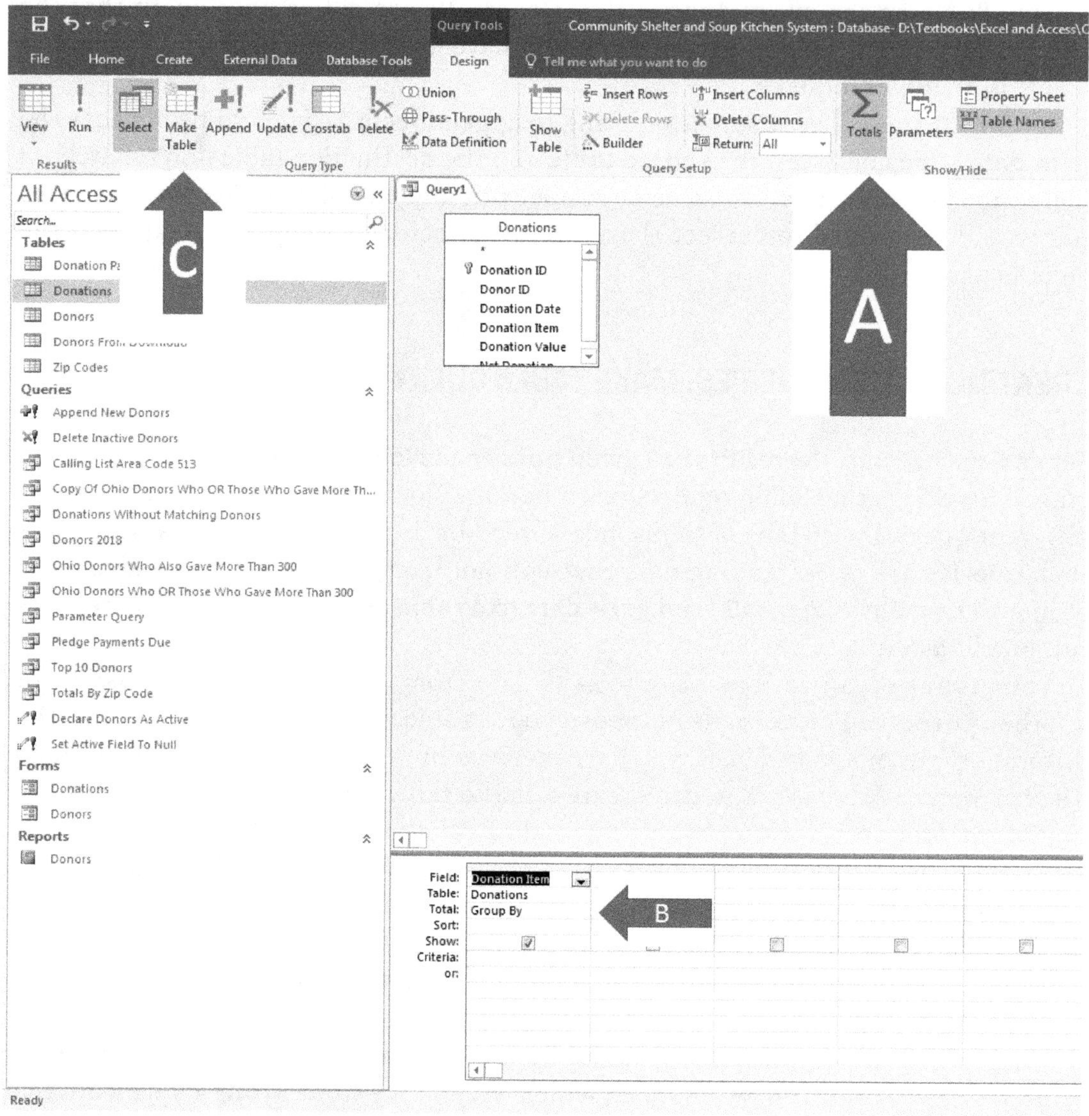

FIGURE 3.27 Make table query design

Step 5: The **Make Table pop-up menu** will appear asking you to name the **table** to which you want to send this data (figure 3.28a). Type freehand, the name of the new **table** which will be **Donation Items** (**Alternate**) in the **Table Name** box. We want this new **table** to be in this database, hence we will leave the **Current Database** option checked. Click **OK**.

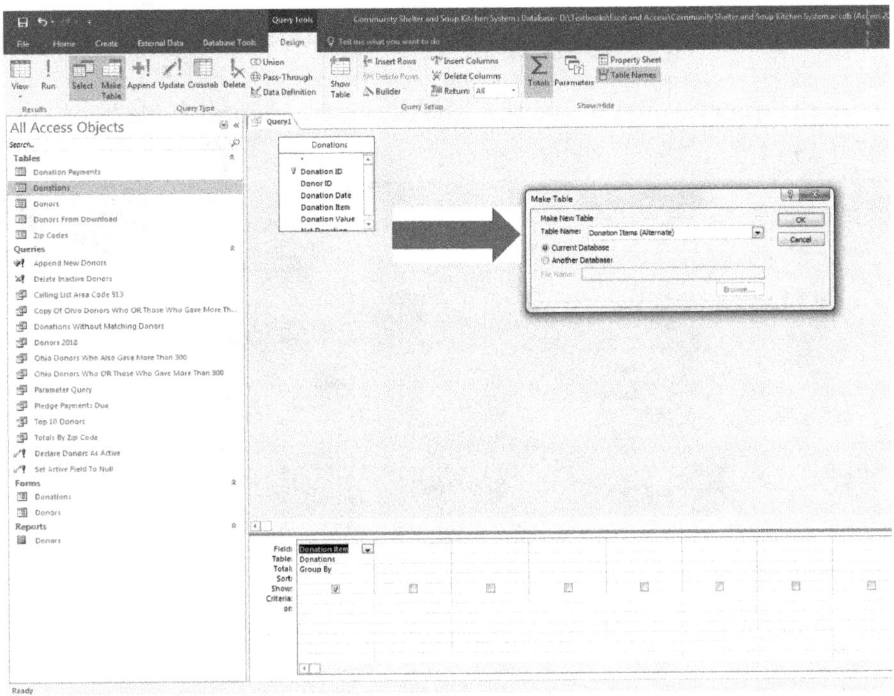

FIGURE 3.28a Make table query (declaring the new table name)

Step 6: Click the **Run** icon in the **Query Design** ribbon.

Step 7: The new **table** will be created. You can open it to see that all of the items are there and that there are no duplicates.

NOTE: In real life, sometimes you will have many variations from users of the spelling of the same items and the list may have to be modified.

Step 8: Save the query as *Donation Items Make Table*. Please notice once again that the query as shown in the **navigation pane** will have an icon attached to it, indicating that it is a **make table query**.

NOTE: On some occasions if you don't plan to reuse these **action queries**, they may not have to be saved, but in our case we will save them.

AN ALTERNATE WAY TO REMOVE DUPLICATES IN A QUERY

Step 1: Put the *item* field in the **Query Design View grid** (figure 3.28b, arrow A) (or whatever field it is of which you want to see no duplicates, as sometimes there are those who want **ALL** duplicate records to be removed).

Step 2: Click **ANYWHERE** in the **blank area** above the **Query Design grid** (figure 3.28b, arrow B).

Step 3: Click the button to open the **property sheet** (figure 3.28b, arrow C).

Step 4: In the **property sheet**, change the **unique value setting** to **yes** (figure 3.28b, arrow D).

If you run the **query**, you will see that the duplicate records have been removed.

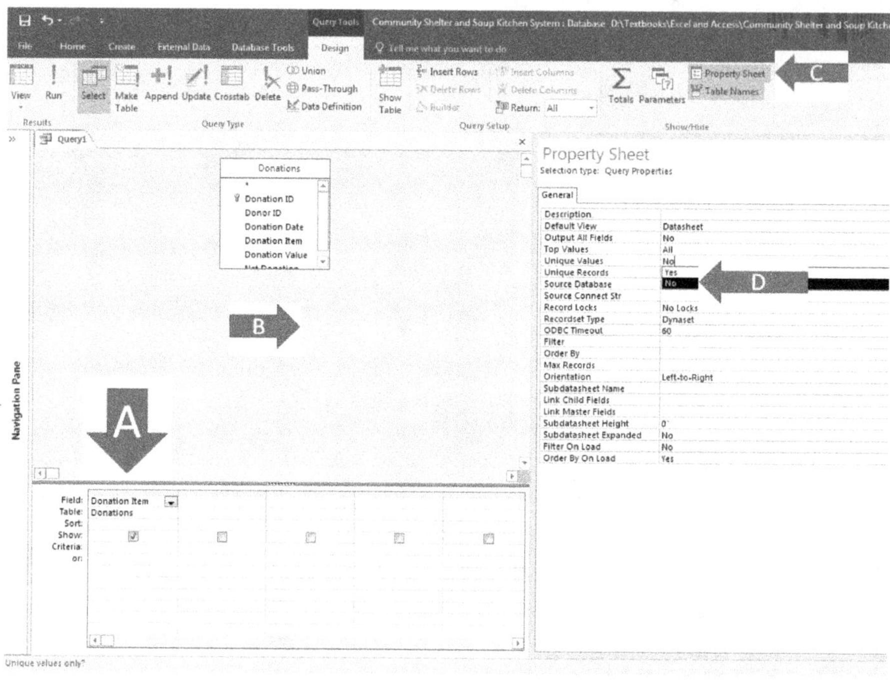

FIGURE 3.28b   Removing duplicates

CREATING A GROUP AND TOTAL REPORT

Sometimes you may want reports that will show subtotals. You may also want to show information without repeating other data. For example, suppose you created a **query** of the *donors* and their *donations*. If *donors* have more than one *donation*, their names would be repeated with each one of those *donations*. It might also be that you would want to sort the fields in various orders. A **group and total report** can make that happen. In most cases, you almost always need to make a **query** before you make a **report** to make sure it has only the fields and data you want. The **report** you made of the raw *donors* **table** in chapter 1 is a rare exception.

Therefore, before making the **group and total report**, make a **query** of all of the *donors* and their *donations*, using the *ID, donor contact last name,* and *donor contact first name* fields from the *donors* **table** and the *donation value and donation date* fields from the *donations* **table**. Close and save the query as *Donors and Donations Group and Total.*

In order to create the **group and total report** of that **query**, take the following steps:

Step 1: Click the **Create** ribbon (figure 3.29, arrow A).

Step 2: Click **Report Wizard** (figure 3.29, arrow B).

Step 3: When you do these things, the **Report Wizard pop-up menu** will appear (figure 3.29, arrow C).

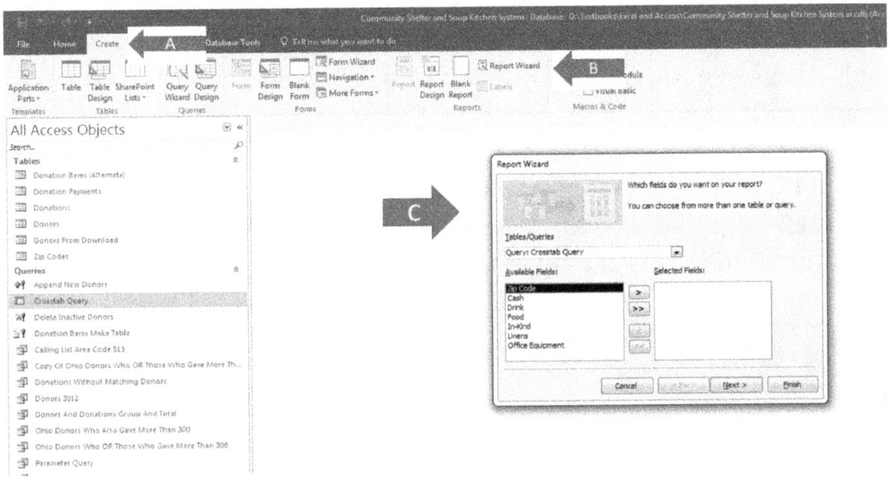

FIGURE 3.29   Report wizard (group and total report)

Step 4: If the **table** or **query** you want to use as the source of this **report** is not the one selected, you must then do so by clicking it in the **Table/Queries drop-down list box** (figure 3.30, arrow). Of course, in this case, you need to select the *Donors and Donations Group and Total* **query** you just created as shown.

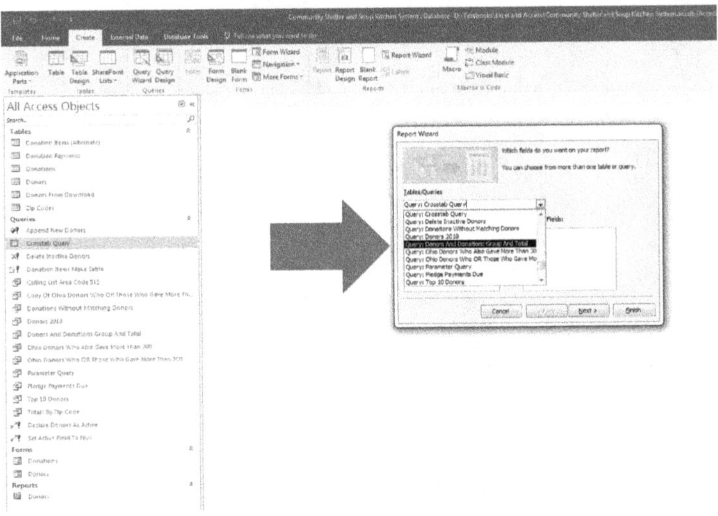

FIGURE 3.30   Report wizard (group and total report)

Step 5: That will take you to what you see in figure 3.31. You must now send the fields from the **Available Fields** window (figure 3.31, arrow A), which shows all of the fields of the **query**, to the **Selected Fields** window (figure 3.31, Arrow C). Notice the arrows located to the right of the **Available Fields** window (figure 3.31, arrow B). The **single arrow** is designed to send the field that is currently selected into the **Selected Fields** window (figure 3.31, arrow C). By default, the first field is selected and would be sent into the **Selected Fields** window if you clicked the **single arrow**. Each time you click the **single arrow**, it will always send the selected field in the **Available Fields** window to the **Selected Fields** window. That way you can control the order in which the fields will be displayed in the **report**. On the other hand, if you want to send **ALL** of the fields to the **Selected Fields** window at one time (in the order in which they now appear), you can click the **double arrow** (figure 3.31, arrow B). That is what we want to do in this example: Click the **double arrow** (figure 3.31, arrow B).

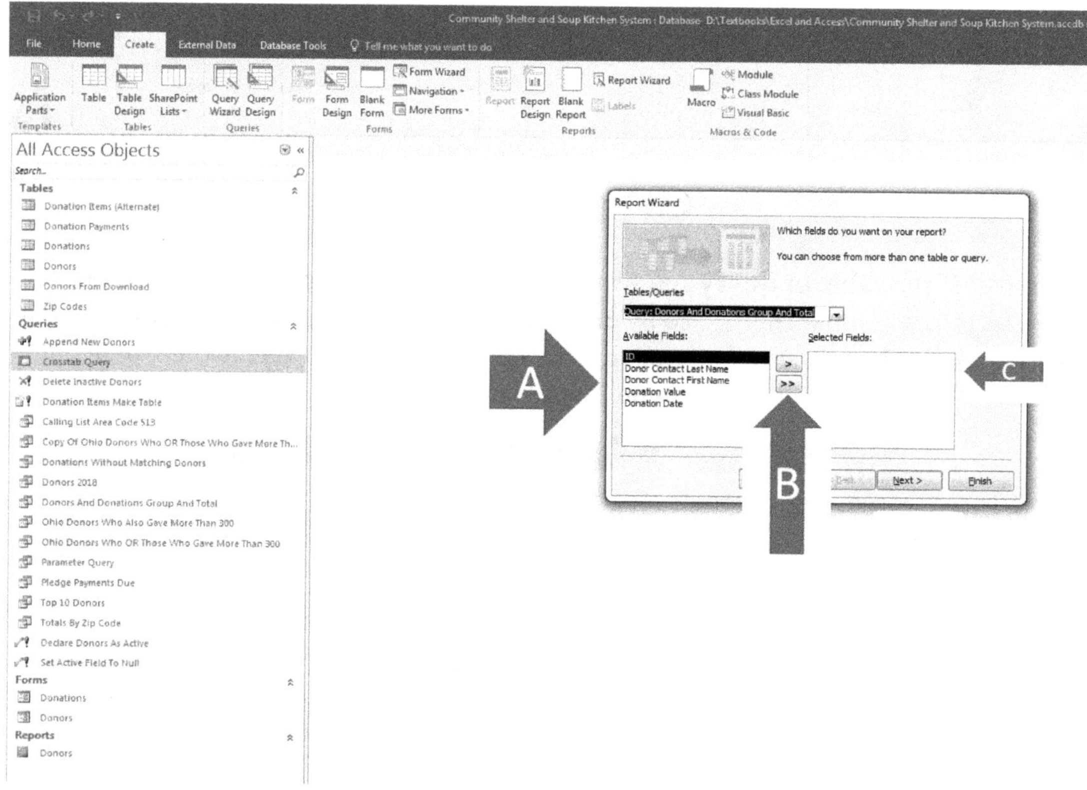

FIGURE 3.31   Report wizard (group and total report)

Step 6: That will take you to what you see in figure 3.32. Notice that all of the fields are now in the **Selected Fields** window (figure 3.32, arrow A).

Step 7: Click **Next** (figure 3.32, arrow B).

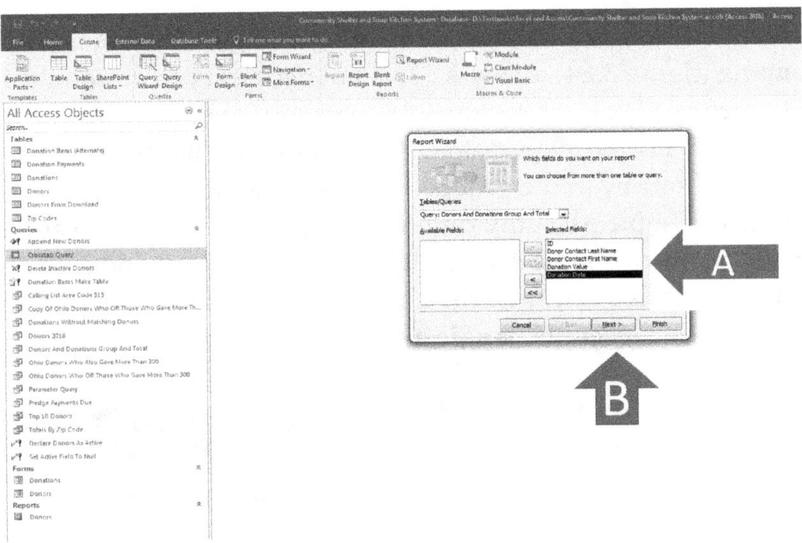

FIGURE 3.32   Report wizard (group and total report)

Step 8: That will take you to the menu seen in figure 3.33. This menu will to allow you to confirm that you do in fact want to **view** (or **group**) this report by *donors* (figure 3.33, arrow A) which will almost always be the **parent table**. In figure 3.33, Arrow B is pointing to the window that shows the *donors* **table** fields in the top, which means that the report will group by those fields. Click **Next** (figure 3.33, arrow C).

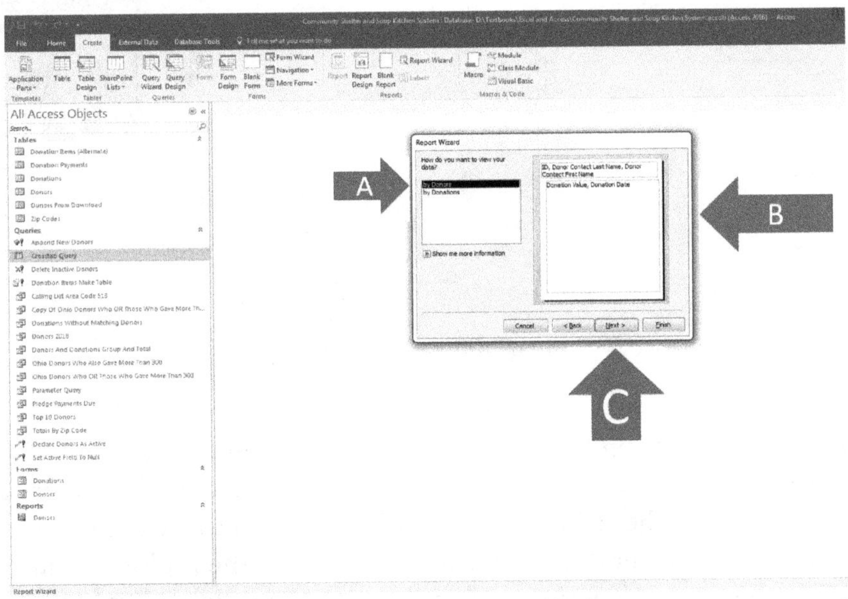

FIGURE 3.33   Report wizard (group and total report)

Step 9: That will take you to what you see in figure 3.34. This menu will allow you to add a **grouping** (figure 3.34, arrow A). It is not necessary to do so in this example; thus, you only need to click Next (figure 3.34, arrow B). The things that **ARE** grouped will be sorted by the **primary key** field (and in this case, that would be the *ID* field). Changing what you sort by in the grouped area will be discussed in a later chapter.

NOTE: In this example, the only time you would have needed to add a grouping would be if you had the *city* of the *donors* included in this **query**, and if you had wanted to show groups and totals by *city* and then by *donor*. To do so, you would have selected the *city* field and then clicked the **single arrow** to its right.

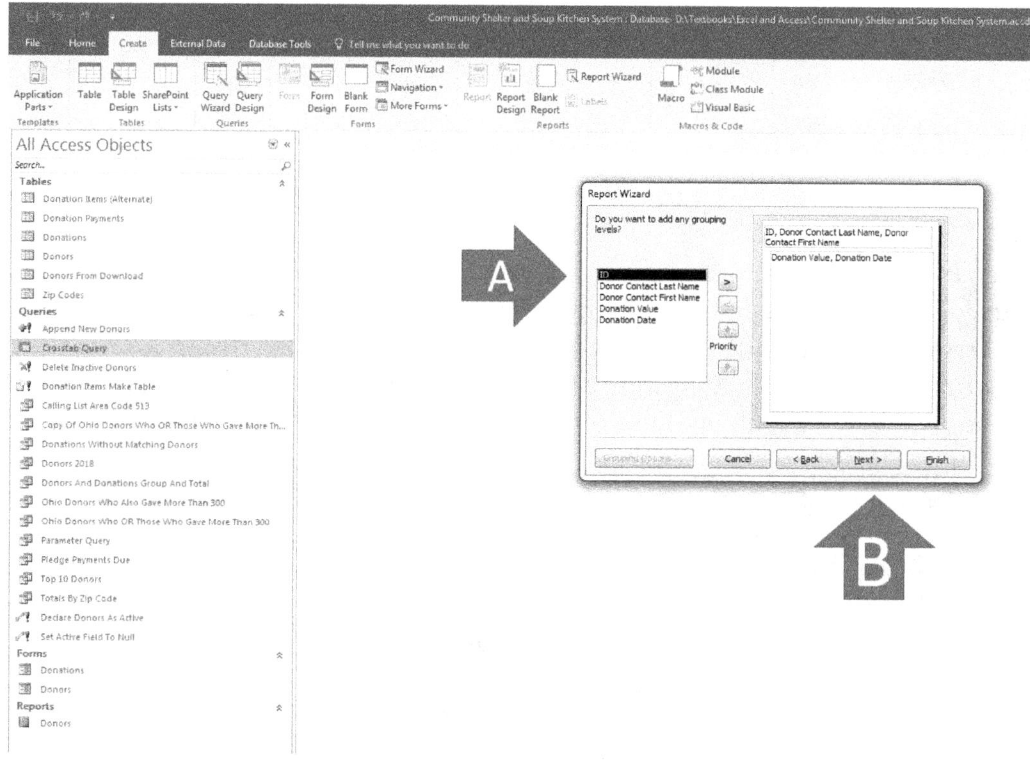

FIGURE 3.34   Report wizard (group and total report)

Step 10: Clicking the **Next** button in the previous menu will take you to what you see in figure 3.35. Here, you can sort what data is **NOT** grouped, and what is **NOT** grouped usually will be the fields from the **child table**. In this example, let's sort by the *donation value* in **descending order** (which will put the highest *donation values* first). To do so, choose the *donation value* field in the **drop-down list box** (figure 3.35, arrow A) and then click the **Ascending** button until it reads **Descending** (figure 3.35, arrow B).

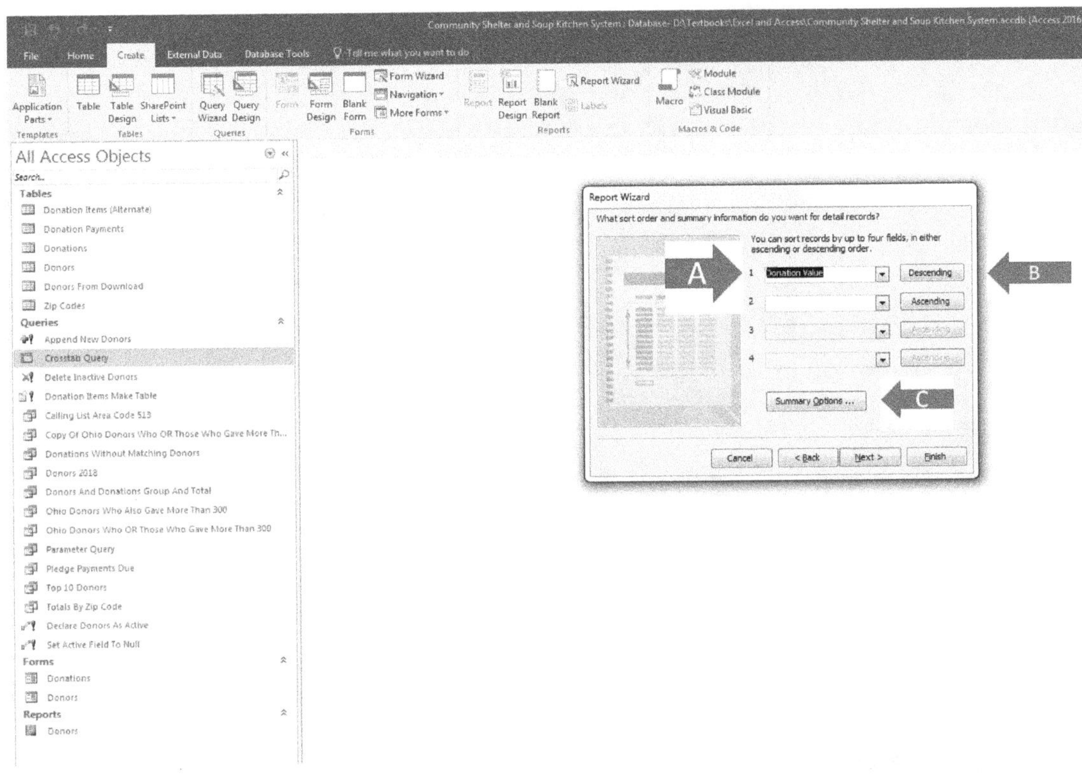

FIGURE 3.35 Report wizard (group and total report)

Step 11: IF YOU WANT TO GET SUBTOTALS, YOU MUST NOT MISS THIS STEP! SKIPPING THIS STEP WILL MAKE IT MUCH MORE DIFFICULT TO GET SUBTOTALS LATER! YOU MUST NOW CLICK SUMMARY OPTIONS (figure 3.35, arrow C).

Step 12: When you click **Summary Options**, that will make the **Summary Options pop-up menu** appear (figure 3.36). If you want to get a **sum** (or a **total**) at each change in *donors*, you will need to make sure that the **Sum** checkbox is checked (figure 3.36, arrow A). Notice that the option for getting averages, minimums, and maximums are also available. In this case, the **sum** is all that is needed. There are no other fields available for calculating, because the remaining fields in the *donors and donations group and total* **query** are not **currency** or **number** fields. Leave the **Show** options at their default settings in this case (figure 3.36, arrow B). You are now set to show the **totals** of the *donation values* at the change in each *donor* (the **detail**) and the **grand totals** (the **summary**) of this report.

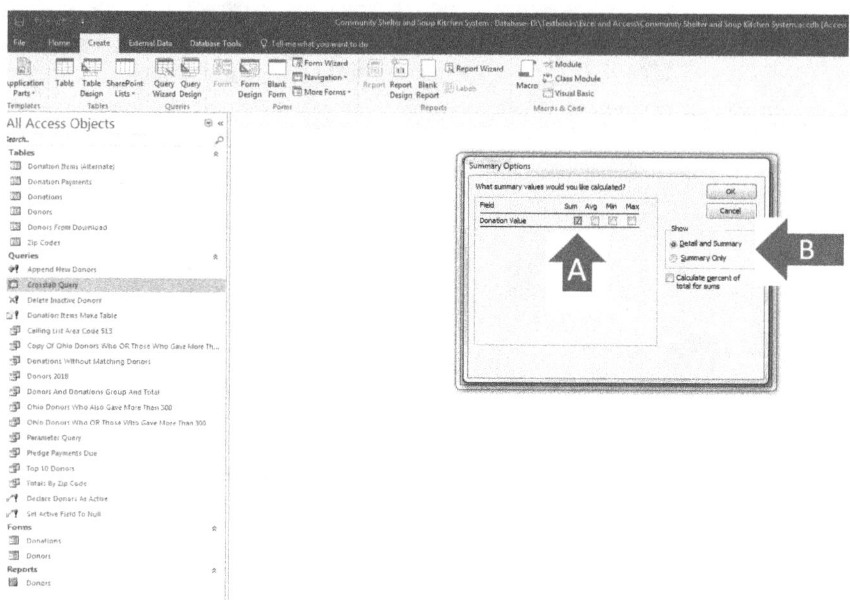

FIGURE 3.36 Report wizard (group and total report)

Step 13: Click **OK**.

Step 14: That will return you to the previous menu (figure 3.37). Click **Next** (figure 3.37, arrow).

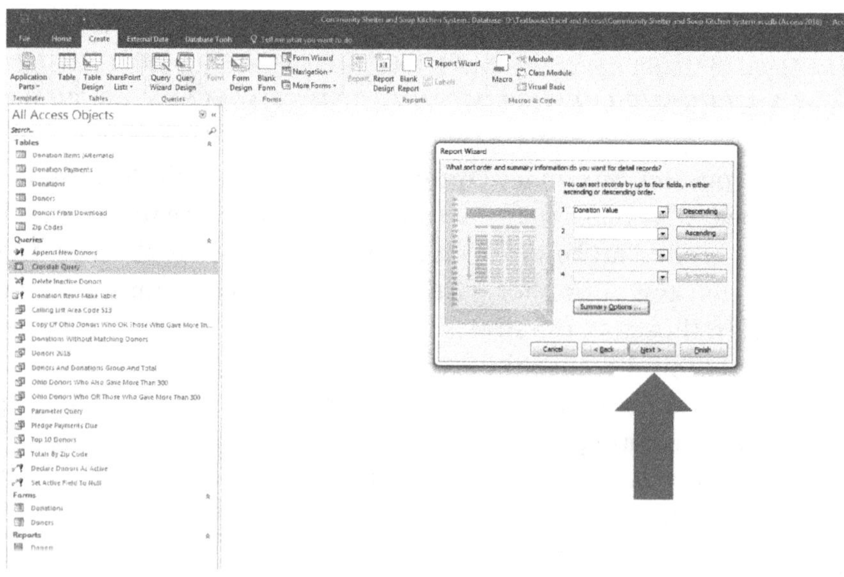

FIGURE 3.37 Report wizard (group and total report)

Step 15: That will take you to the menu in figure 3.38. It allows you to choose a **layout** and **orientation**. You can experiment with them as you like, but in this example, click the **Block Layout** (figure 3.38, arrow A) with the **landscape** (horizontal) **orientation** (figure 3.38, arrow B).

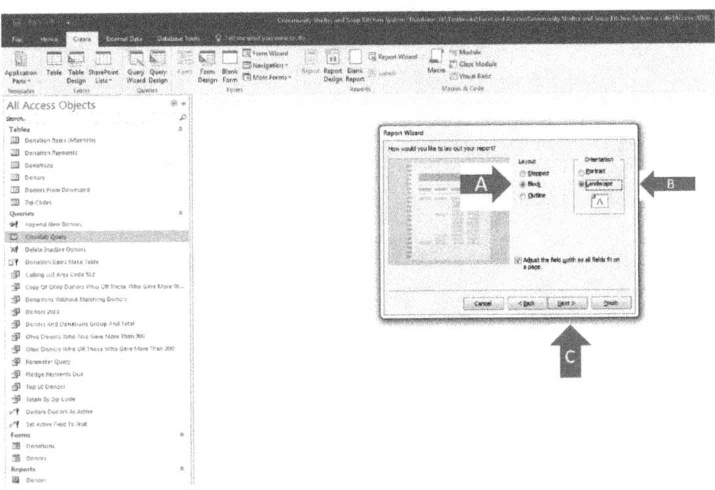

FIGURE 3.38   Report wizard (group and total report)

Step 16: Click **Next** (figure 3.38, arrow C).

Step 17: That will take you to the menu shown in figure 3.39, which allow you to name the report where it asks **What title would you like for your report?** (figure 3.39, arrow A). In the box, name the report *Donors and Donations Group and Total*. Keep all of the other defaults as they are now.

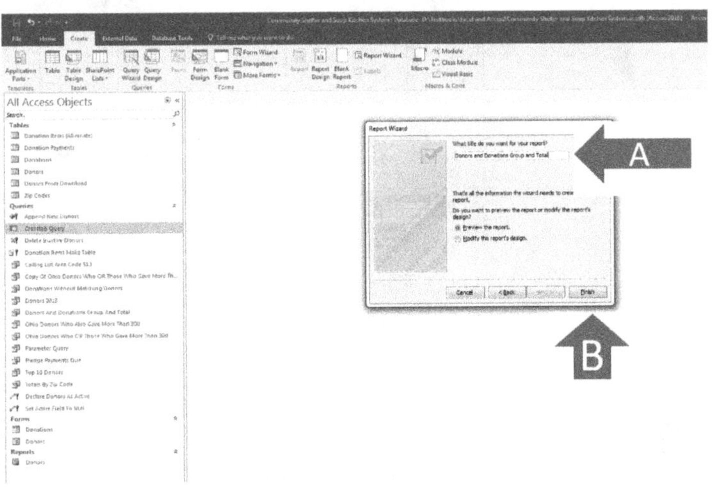

FIGURE 3.39   Report wizard (group and total report)

Step 18: Click **Finish** (figure 3.39, arrow B).
That will take you to your finished product shown in figure 3.40a.

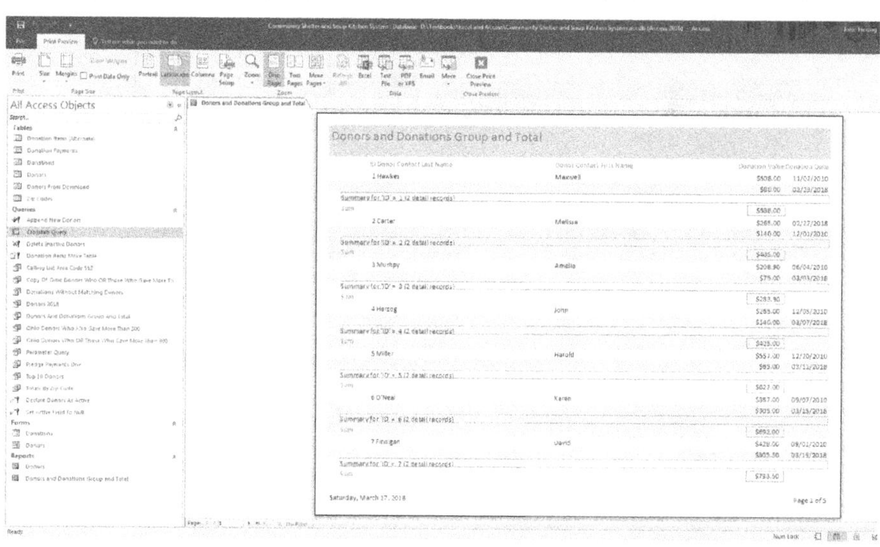

FIGURE 3.40a   Group and total report

Open the **design view** of the report by right-clicking anywhere on the report and then choosing **Design View** in the **Shortcut/Context Sensitive menu** (figure 3.40b, arrow).

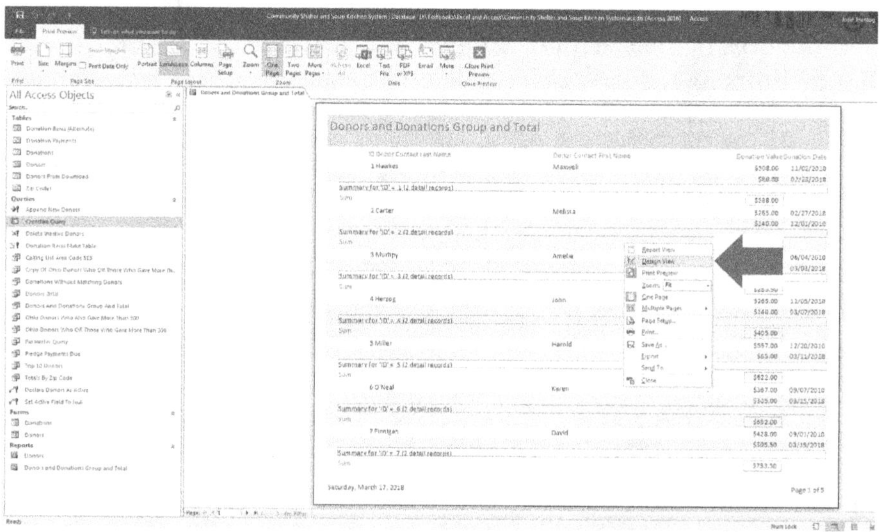

FIGURE 3.40b   Group and total report completed

That will take you to what you see in figure 3.41. There you will see that the calculated field, which gives the **sum** of the *donation values* (figure 3.41, arrow). The expression in the **text box** reads *=SUM([Donation Value])*. It is just like that of the **SUM** function used in Excel, except that its argument is referencing a field name and not cells. You can always use such an expression in any **report** or **form**. You must remember to put the square brackets around the field name that you are referencing in such an expression, and to spell it just as the field name is spelled. **Calculated fields** can be created by using a **text box,** and they work the same way here as they do in forms. More on that in **chapter 4.**

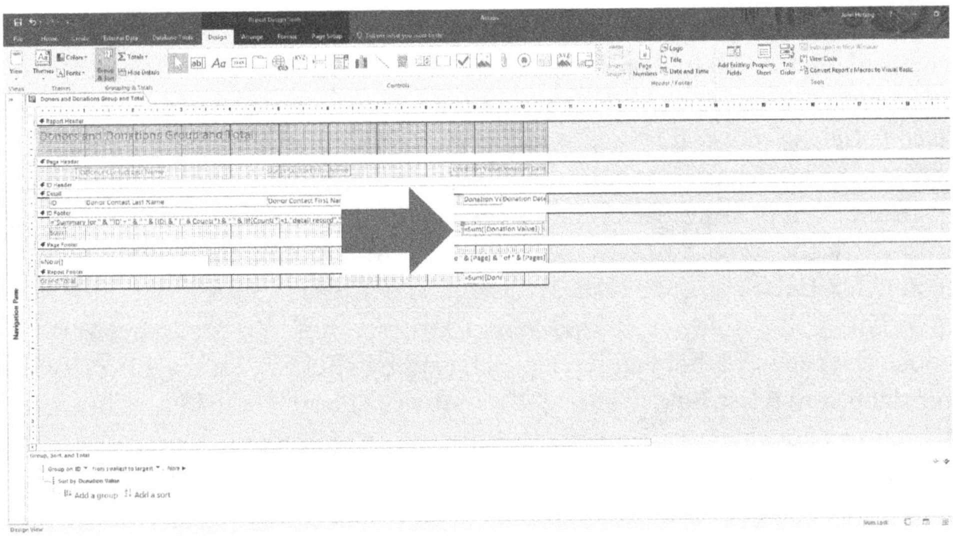

FIGURE 3.41   Report calculation

NOTE: There are times that the **group and total** reports don't display records in the order you want. That is because they usually group best by the **primary key** field. In most cases, the **primary key** field is an arbitrary number created by an **autonumber** field as was discussed earlier. If you group records by their names it can be a problem if you have many people with the same names. To avoid a lot of difficulty, therefore it is best to add a field by which you would want to group the data that includes the **primary key**, as well as the last and first names of a person (or some other element of the parent record by which you would want to sort). In this database, you could merge the *donor contact last name, donor contact first name*, and *ID* fields into one field, **IN THAT ORDER**. To do so you can add a calculated field in the first column of the **Query Design View grid** with the expression

Key Field: [Donor Contact Last Name]&[Donor Contact First Name]&[ID]

This is using what is called the **concatenate** language and will be discussed in greater depth in chapter 8.

CREATING LABELS

Sometimes people want to create labels of the records held in a database. They may be used to place on boxes for shipping in the case of inventory databases, or they can display names and addresses for doing mass mailings. There are many different ways they are used. The most common use of labels is for sending mailings to people whose data is stored in a database, such as customers, vendors, church members, students, donors, and many more. It is important to remember the brand name and the type of labels you have purchased for this endeavor. They are usually designated by a **product number.**

In our example, we will first create a **query** as our source of data for the labels. The **query** we will create will have no criteria and we will use the *donor contact first name, donor contact last name,* and *address* fields from the *donors* **table,** and the *city, state,* and *zip code* fields from the *zip codes* **table.** Name the **query** *Mailing Labels.*

To create the labels, use the following steps:

Step 1: Select (in the **navigation pane**) the *mailing labels* **query** that you just created.

Step 2: Click the **Create** ribbon (figure 3.42, arrow A)

Step 3: Click **Labels** (figure 3.42, arrow B).

Step 4: That will display the **Label Wizard** (figure 3.42). In the **Label Wizard** you can choose the type of label you have purchased by clicking on the **Filter By Manufacturer drop-down list box.** (figure 3.42, arrow C). There are different brand names from which to choose as you can see in figure 3.43. The default brand that Access uses is often the **Avery** labels. When you choose **Avery,** their product types will be displayed in the **Product Number** window (figure 3.42, arrow D). There are many. If the **product number** of the labels you purchased doesn't match the **product numbers** that you see listed in the menu, you can always customize the label dimensions by clicking **Customize** (figure 3.42, arrow F). In this example, we will assume that you have purchased **Avery** labels with a **product number** of *5160.* As you can see, that is the one that is already chosen. It will give you 1 inch by 2 5/8 inches in each label and there will be three columns of labels on each page (figure 3.42, arrow D). Those labels are designed to be printed on pages (for **sheet feed**), rather than the **continuous** labels, as you can see (figure 3.42, arrow G).

Step 5: Once you have chosen the label type to be used, click **Next** (figure 3.42, arrow E).

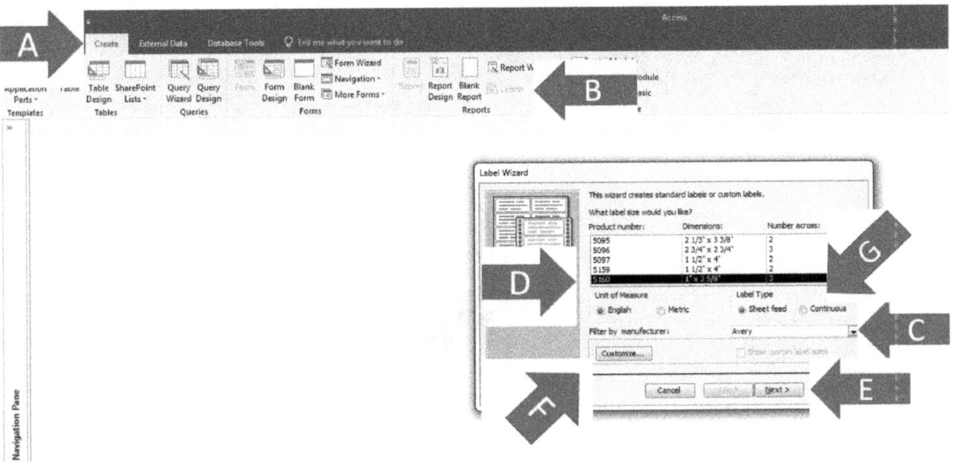

FIGURE 3.42 Report calculation

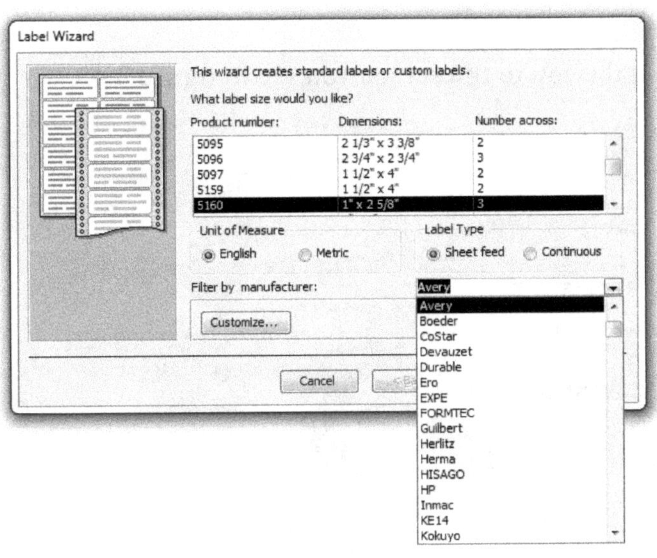

FIGURE 3.43 Report calculation

Step 6: That will take you to what you see in figure 3.44. It is here that you can change the font for your labels. In this example, we will use all of the settings that you see in figure 3.44. Click **Next** (figure 3.44, arrow).

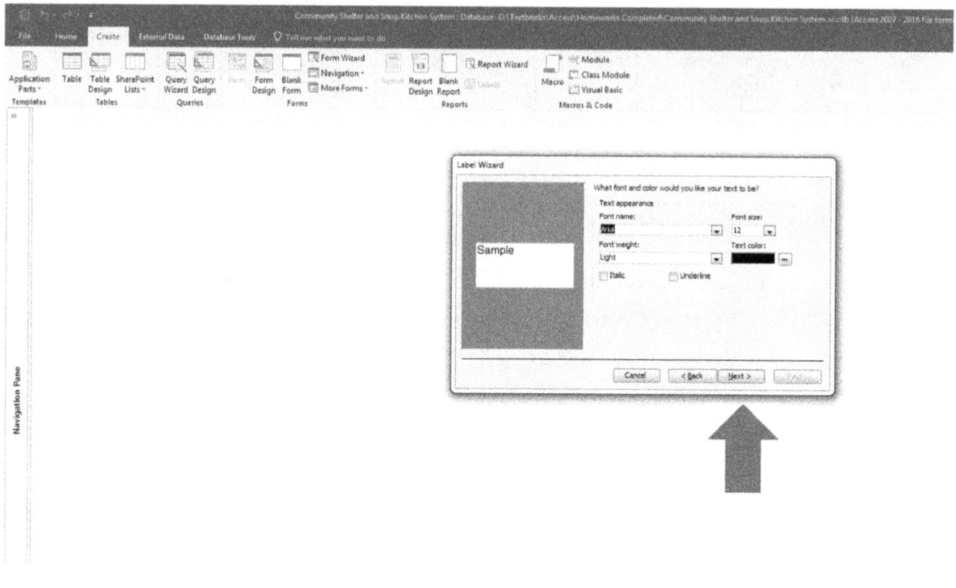

FIGURE 3.44   Report calculation

Step 7: That will take you to the screen you see in figure 3.45.

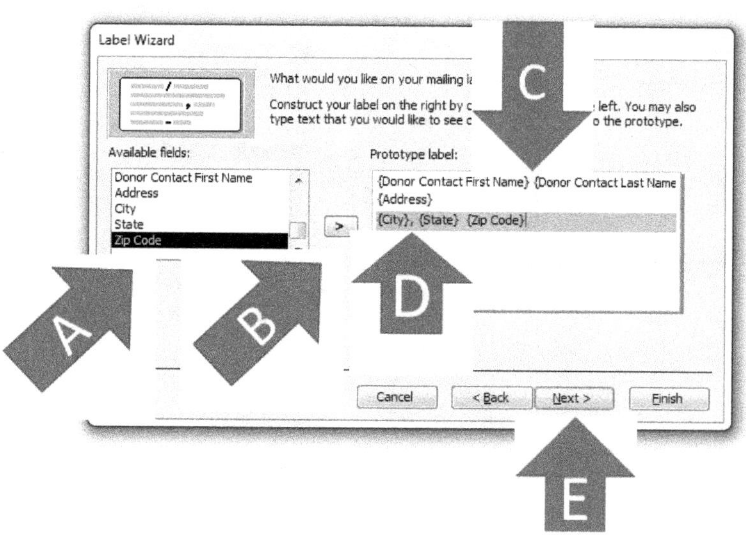

FIGURE 3.45   Report calculation

At this point, you must choose the fields from the **query** that you want for the labels that are shown in the **Available Fields** window (figure 3.45, arrow A). As you select one of them by clicking on it, you must then click the tiny arrow to the right (figure 3.45, arrow B) to send it to the **Prototype Label** window (figure 3.45, arrow C). You must choose each additional needed field and send them in the order that you want them to appear on the label. Notice in figure 3.45, that the *donor contact first name* was chosen first. After sending it to the **Prototype Label** window, the **space bar** was pressed, thus adding a space before the *donor contact last name* (figure 3.45 at arrow C), just as you would place a space between a first and last name if you were typing the labels freehand. To get to the second line of the label, press the **Enter** key on your keyboard, also as you would if you were typing an address freehand. After sending the *address* field to the **Prototype Label** window by clicking the tiny arrow again, press **Enter** again, to go to the next line in order to insert the *city, state,* and *zip code* fields. After you send the *city* field to the **Prototype Label** window, press the **comma** (,) button on your keyboard to insert a **comma** (figure 3.45, arrow D) also as you would if you were typing this address freehand. Also, be sure to add two spaces between the *state* and the *zip code* fields as well before you send them to the **Prototype Label** window.

Step 8: Click **Next** (figure 3.45, arrow E).

Step 9: That will take you to what you see in figure 3.46. If you want to sort the labels, you can use the same method of choosing fields discussed earlier, by clicking on the first field you want to sort in the **Available Fields** window (figure 3.46, arrow A), and then clicking the **single arrow** (figure 3.46, arrow B) to send it over to the **Sort by** window (figure 3.46, arrow C). In this example, the process was repeated to get the *donor contact last name, donor contact first name,* and *address* fields into the **Sort by** window to make sure that the labels are sorted in that order.

Step 10: Click **Next** (figure 3.46, arrow D)

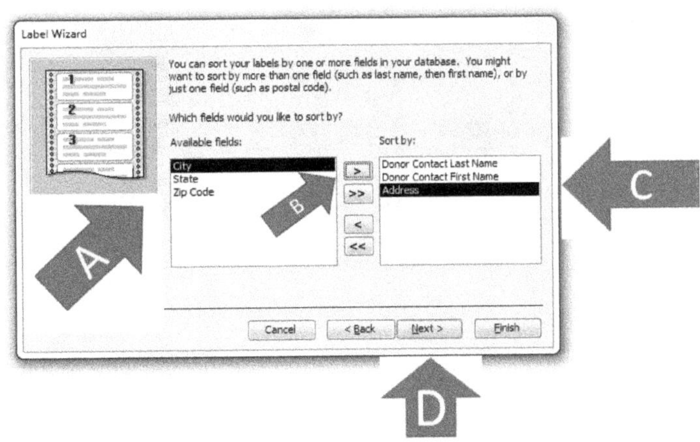

FIGURE 3.46 Report calculation

Step 11: That will take you to the screen you see in figure 3.47. At this point Access will ask you what you want to name the object. It will always start a **labels report** name with the word *labels*. Therefore, it is best that you remove the first *labels* in the name and instead name the object *Mailing Labels* rather than *Labels Mailing Labels* (figure 3.47, arrow A). Simply click to the left of the first *labels* and then press and hold the **Delete** key until the first *labels* is gone.

Step 12: Click **Finish** (figure 3.47, arrow B). The results will be what you see in figure 3.48.

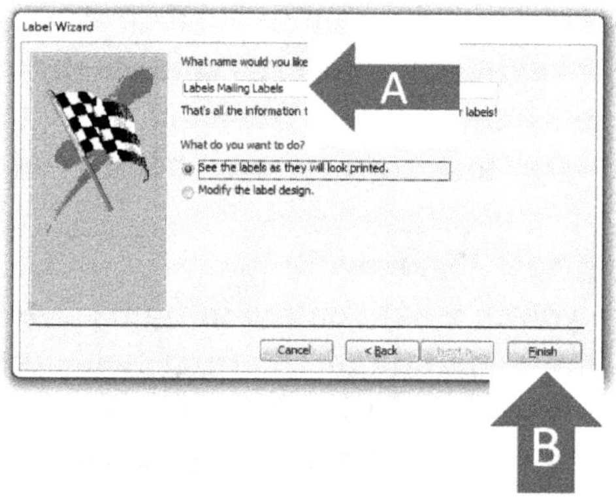

FIGURE 3.47   Report calculation

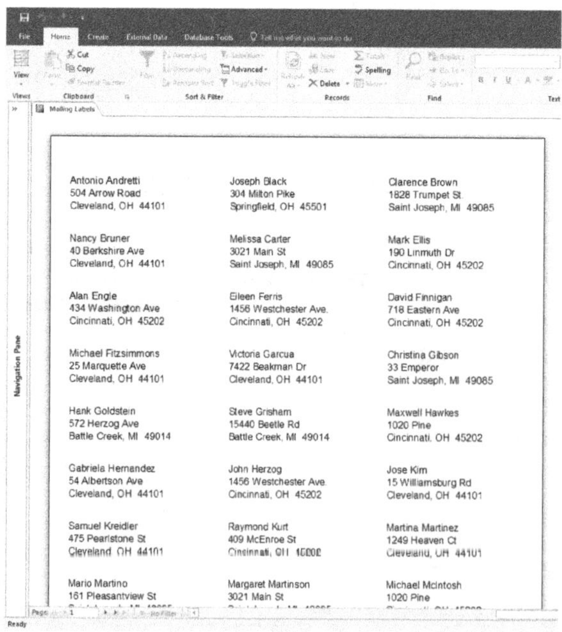

FIGURE 3.48   Report calculation

Chapter 3: Assignment (Summary of Tasks)

1. Open the *Community Shelter and Soup Kitchen* system.
2. Change the *donations* table by adding a field in the **design view** named *Net Donation* that will automatically subtract 3.75 from each of the *donation values*. Make sure the field is a **currency** format.
3. Add a **total row** at the bottom of the *donations* table in the *datasheet view* that will add up all of the *donation values*.
4. Create a **simple form** of the *donations* table and name it *Donations*. Make an addition to it with the following specifications:
   a. In the **form design view**, shrink the width of all of the **field controls** inside the **sub-form** to half their widths (to approximately *6.5 inches*)
   b. Add a **text box** to the right of the *donation value* **field control** that will multiply the *donation value* times *.25*.
   c. Make sure it has a **currency** format.
   d. Change the caption of its label to read *Tax Reduction*.
5. Create a new **select query** using the following specifications:
   a. Use only the *donors* table, the *donations* table, and the *donation payments* table.
   b. Use the donor contact first name and donor contact last name fields from the *donors* table.
   c. Use the *donation date* and *donation value* fields from the *donations* table.
   d. Use the *payment amount, payment date,* and *check number* fields from the *donation payments* table.
   e. Add a field after the *check number* field named *Pledge Amount Due* that is found by subtracting the *payment amount* field from the *donation value* field.
   f. Save the query as *Pledge Payments Due*.
6. Create a new **query** with the following specifications:
   a. Use the *zip code* field from the *donors* table.
   b. Use the *donation value* field from the *donations* table.
   c. Using the **Totals** icon in the **Design View**, select the **Group By** option of the *zip code* field and get a **SUM** of the *donation value* field.
   d. Save the query as *Totals by Zip Code*.
7. Make a **3D pie chart** of the *totals by zip code* query you just created, using all of the default settings, with the *zip codes* showing in the **legend**.
8. Create an **update query** that will remove all of the data in the *active* field in the *donors* table. Name the query *Set Active Field to Null*.
9. Create an **update query** that will use the *donors* and the *donations* tables. Make sure the **update query** will change the data in the *active* field to read *yes* in

the *donors* table only if those *donors* have a related *donation*. Name the query *Set Active Field to Yes*.

10. Create an **update query** that will change all of the *active* fields in the *donors* table to read *No* if they are **null**. Name the query *Set Active Field to No*.

11. Import the spreadsheet named *Donors from Download.xlsx* that is associated with this textbook as a table, and name it *Donors from Download*.
    a. Allow Access to use the column headings of that spreadsheet as the field names.
    b. Do **NOT** change any of the field types or field names.
    c. Create an **append query** using the *donors from download* table that will send all of its fields' data to the *donors* table.
    d. Name the query *Append Donors*.

12. Create a **delete query** that would remove all *donors* where the *active* field is equal to *no*.

   NOTE: Please do **NOT** click the **Run** button. Save the query as *Delete Inactive Donors*.

13. Create a **make table query** using the *donations* table, making sure that every **DIFFERENT** kind of *donation item* is sent to the newly created table only once (using the **Group By** setting in the query). Name the new table as *Donation Item (Alternate)*.

14. Create a new **select query** using the following specifications:
    a. Use ID, donor contact last name, and donor contact first name fields from the donors table.
    b. Use the *donation value and donation date* fields from the *donations* table.
    c. Save the query as *Donors and Donations Group and Total*.

15. Create a **group and total report** of the *donors and donations group and total* query that you just created using the following specifications:
    a. Use all of the fields.
    b. Make sure that the report will view the data by *donors* and do not add any additional groupings beyond that of the *donors* table.
    c. Using the **summary options**, get a **SUM** of the *donation values* field.
    d. Sort the **detail records** from the **child** table by *donation value* in **descending** order.
    e. Use the block layout and the landscape orientation settings.
    f. Name the report *Donors and Donations Group and Total*.

16. Create a new **select query** with no criteria using the *donor contact first name*, *donor contact last name*, and *address* fields from the *donors* table, and the *city*, *state*, and *zip code* fields from the *zip codes* table. Name the **query** *Mailing Labels*.
    a. Make labels of all of the records in the query, using the default settings and using the **Avery product number** of 5160.
    b. Save the report as *Mailing Labels*.

Chapter 3: Assignment (Alternate) (Summary of Tasks)

1. Open the village bookstore database system.
2. Change the *book titles* table by adding a field in the **design view** named *Stocking Fee* that will automatically multiply the *vendor cost* field times *.02* for each *book title*. Make sure the field is a **currency** format.
3. Change the *inventory orders* table by adding a **total row** at the bottom of it in the *datasheet view* that will calculate the **SUM** of the *quantity ordered* field.
4. Create a **simple form** of the *inventory orders* table and name it *Inventory Orders*. Make an addition to it with the following specifications:
   a. In the form *design view*, shrink the width of all of the **field controls** inside the **form** to half their widths (to approximately *6.5 inches*)
   b. Add a **text box** to the right of the *quantity ordered* **field control** that will multiply the *quantity ordered* times *.15*.
   c. Make sure it has a **currency** format.
   d. Change the caption of its label to read *Shipping Cost*.
5. Create a new **select query** using the following specifications:
   a. Use the *vendors* table, the *inventory orders* table, and the *shipments received* table.
   b. Use the vendor contact first name and the vendor contact last name from the vendors table.
   c. Use the date of order and quantity ordered fields from the inventory orders table.
   d. Use the quantity received, received date, and freight bill number fields from the shipments received table.
   e. Add a field after the *freight bill number* field named *Back Order Quantity* that is found by subtracting the *quantity received* field from the *quantity ordered* field.
   f. Save the query as *Back Orders*.
6. Create a new **query** with the following specifications:
   a. Use the quantity ordered and the book title fields from the *inventory orders* table.
   b. Using the **Totals icon** in the **design view**, select the **Group By** option of the *book title* field and get a **SUM** of the *quantity ordered* field.
   c. Save the query as Quantities Ordered by Book Title.
7. Make a **3D pie chart** of the *quantities ordered by book title* query you just created, using all of the default settings, with the *book titles* showing in the **legend**.
8. Create an **update query** that will remove all of the data in the *active* field in the *book title* table. Name the query *Set Active Field to Null*.
9. Create an **update query** that will use the *inventory orders* and *book titles* tables. Make sure the **update query** will change the data in the *active* field to read

yes in the *book titles* table only if those *book titles* that have a related *inventory order*. Name the query *Set Active Field to Yes*.

10. Create an **update query** that will change all of the *active* fields in the *book titles* table to read *No* if they are **NULL**. Name the query *Set Active Field to No*.

11. Import the spreadsheet named *Append Vend.xlsx* that is associated with this textbook and name it *Append Vend* as a table.
    a. Allow Access to use the column headings of that spreadsheet as the field names.
    b. Do **NOT** change any of the field types or field names.
    c. Create an **append query** using the *append vend* table that will send all of its fields' data to the *vendors* table.
    d. Name the query *Append Vendors*.

12. Create a **delete query** that would remove all *book titles* where the *active* field is equal to *no*. **NOTE:** Please do **NOT** click the **Run** button. Save the query as *Delete Inactive Book Titles*.

13. Create a **make table query** using the *vendors* table, making sure that every **DIFFERENT** *credit limit* bracket is sent to the table only once (using the **Group By** setting in the query) and using the *credit limit* field. Name the new table as *Credit Limit Brackets (Alternate)*.

14. Create a new **select query** using the following specifications:
    a. Use *ID* and *vendor name* fields from the *vendors* table.
    b. Use the *quantity ordered* and *date of order* fields from the *inventory orders* table.
    c. Save the query as *Vendors and Orders Group and Total*.

15. Create a **group and total report** of the *vendors and orders group and total* query that you just created using the following specifications:
    a. Use all of the fields.
    b. Make sure that the report will view the data by *vendors*, and do not add any additional groupings beyond that of the *vendors* table.
    c. Using the **summary options**, get a **SUM** of the *quantity ordered* field.
    d. Sort the **detail records** from the **child** table by *date of order* in **descending** order.
    e. Use the block layout and the landscape orientation settings.
    f. Name the report *Vendors and Orders Group and Total*.

16. Create a new **select query** with no criteria using the *vendor contact first name, vendor contact last name,* and *vendor address* fields from the *vendors* table, and the *city, state,* and *zip code* fields from the *zip codes* table. Name the **query** *Mailing Labels*.
    a. Make labels of all of the records in the query, using the default settings and using the **Avery product number** of 5160.
    b. Save the report as *Mailing Labels*.

Making Easier Forms to Enter Data

IN THIS CHAPTER you will learn how to do the following:

1. Create a main and subform
2. Create an aggregate calculation in a sub-form
3. Create **combo boxes**
4. Create an **aggregate calculation** in a form
5. Create an **input mask**
6. Set tab order
7. Set a **password** to an Access database

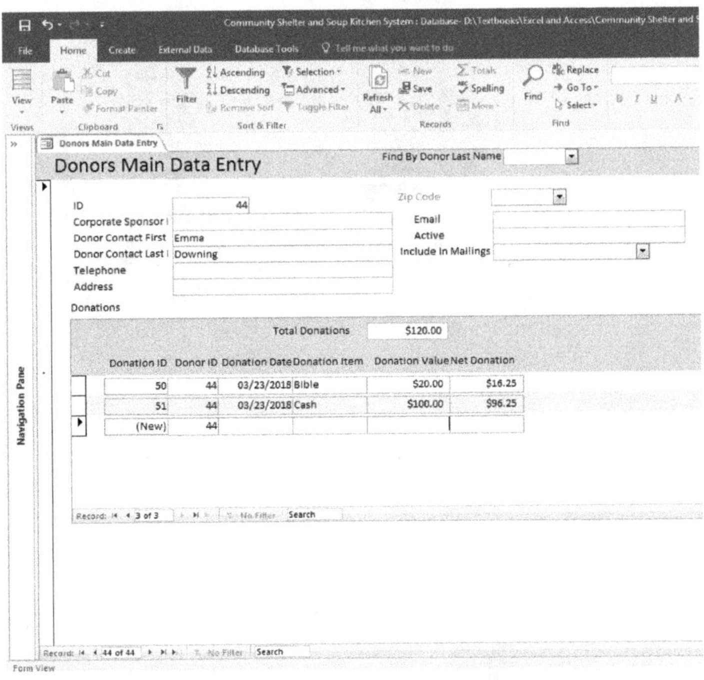

FIGURE 4.1 Sample of a main and subform

CHAPTER 4 TASK: CREATE A MAIN FORM AND A SUB-FORM WITH COMBO BOXES (DROP-DOWN LIST BOXES) IN THE COMMUNITY SHELTER AND SOUP KITCHEN SYSTEM

When you want to enter the **primary key** of a *donor* in the *donations* table to associate that *donor* with a given *donation*, it becomes difficult to do so if you don't know the *ID* number assigned to that *donor*. It is also difficult if you want to enter a new *donation* when the *donor* is new and is not yet in the *donor* **table**. That's why a **main** and **sub-form** are needed as shown in figure 4.1. They would allow you to enter a *donor* at the top of the menu and then in the lower portion of the menu. Hence, you will be able to enter any *donations* each *donor* may have given. This makes it possible to enter data into multiple tables in the same screen. **THIS MAKES DATA ENTRY SPEED UP DRAMATICALLY!**

When you create a **main form** and a **sub-form**, the **main form** will almost always read its data from the **parent** table, which in this case is the *donors* table. The **sub-form** will almost always read its data from the **child** table, which in this case is the *donations* table. This can be accomplished by taking the following steps:

Step 1: Click the Create ribbon (figure 4.2, arrow A).

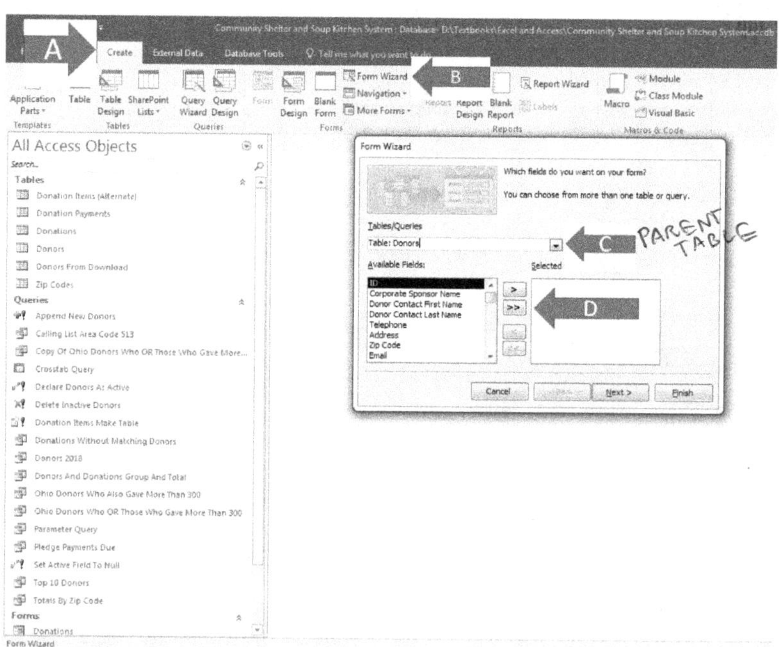

FIGURE 4.2   Creating a main and subform using the form wizard

Step 2: Click **Form Wizard** (figure 4.2, arrow B). That will display the **Form Wizard pop-up menu.**

DO NOT LEAVE THAT POP-UP MENU UNTIL STEPS 3 THROUGH 6 ARE COMPLETED. LEAVING THE MENU BEFORE THAT WILL PREVENT YOU FROM CREATING A SUBFORM!!

Step 3: In the **drop-down list box** labeled **tables/queries,** you will choose the table needed for the **main form,** which will be the **parent** table (which is the *donors* table in this example) (figure 4.2, arrow C). Fortunately, the *donors* table may already be chosen for you, but if the one you want isn't already selected in the **tables/queries** box, you can always click on the **drop-down list box** (figure 4.2, arrow C) and choose the table or query you **DO** want.

Step 4: Click the **double arrow** (figure 4.2, arrow D). That will move all of the fields from the *donors* table in the **Available Fields** window, to the **Selected Fields** window. It is by this step that you choose the fields you want in the **main form. DO NOT LEAVE THIS SCREEN IN THIS STEP!**

Step 5: Click on the **Tables/Queries** box **drop-down-list** (figure 4.3, arrow A) and choose the *donations* table.

Step 6: Click the **double arrow** (figure 4.3, arrow B). That will move all of the fields from the *donations* table in the **Available Fields** window to the **Selected Fields** window. It is by this step that you choose the fields you want in the **sub-form.**

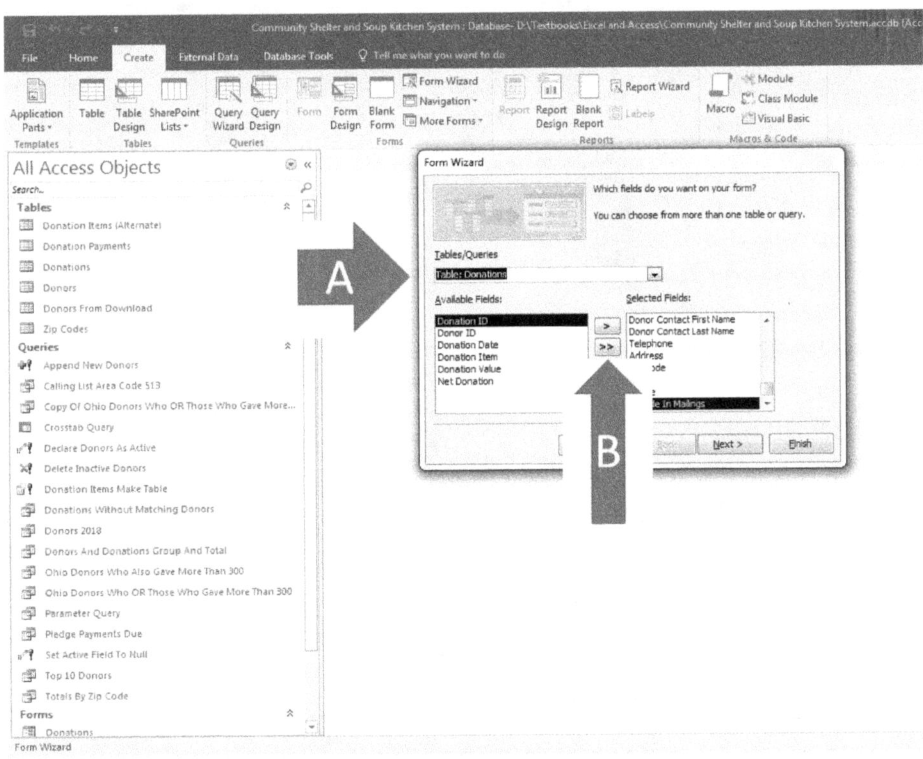

FIGURE 4.3 Creating a main and subform using the form wizard

REMEMBER, DO NOT CLICK NEXT AND LEAVE THIS SCREEN UNTIL ALL OF THESE STEPS, 3 THROUGH 6, ARE COMPLETE. DOING SO WILL MAKE IT SO YOU WILL NOT HAVE A SUB-FORM!!

Step 7: The menu will now look like what you see in figure 4.4. Click **Next** (figure 4.4, arrow).

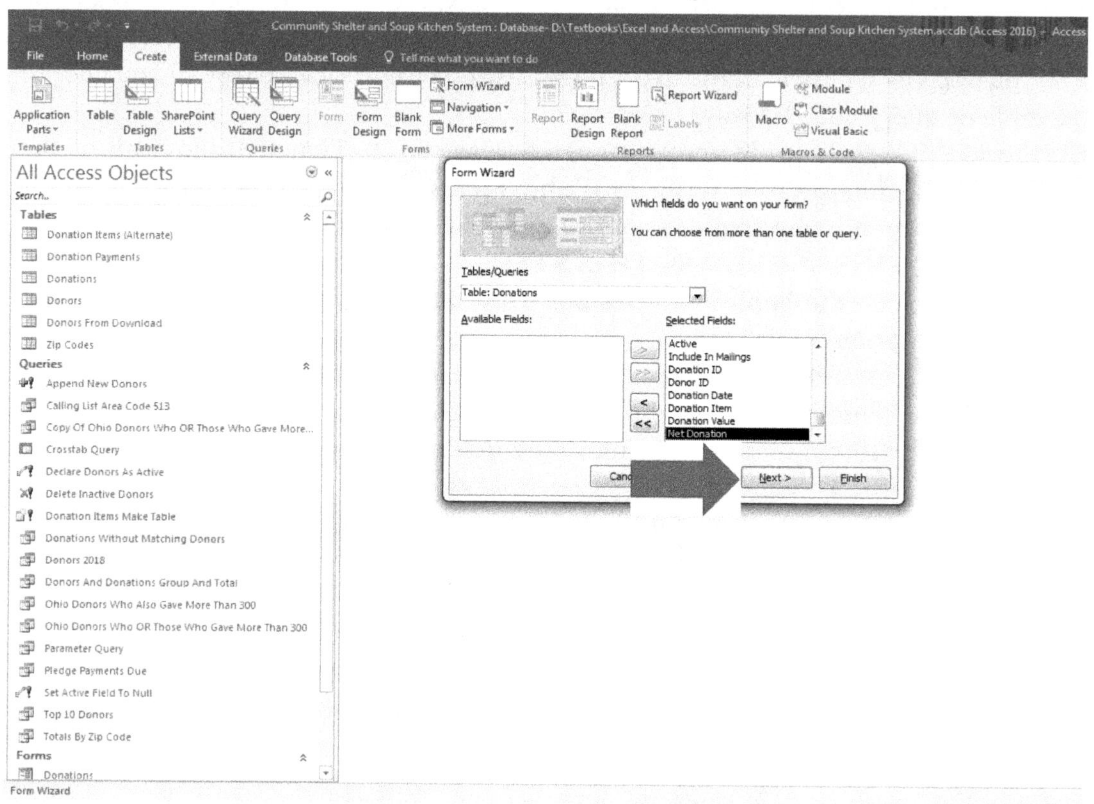

FIGURE 4.4 Creating a main and subform using the form wizard

Step 8: That will take you to the screen shown in figure 4.5. This screen will only confirm with you that you have chosen the fields you have chosen for both the **main** form and the **sub-form**. Notice that the **parent** table fields are on the top of the gray box (figure 4.5, arrow A), which will indicate that those are the fields that will be shown in the **main form**. It also shows the **child** table fields that are in the box embedded within the gray box (figure 4.5, arrow B), which will indicate that those are the fields that will be shown in the **sub-form**.

Step 9: Click **Next** (figure 4.5, arrow C).

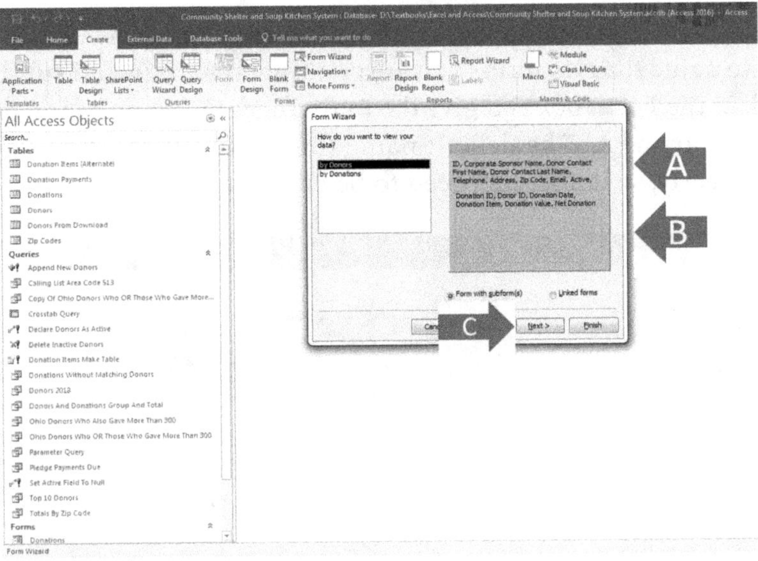

FIGURE 4.5   Creating a main and subform using the form wizard

Step 10: That will take you to the screen you see in figure 4.6. For reasons you will see later, you will almost **ALWAYS** want to click the **Tabular Layout** for your **subform**. The **datasheet layout** is **HIGHLY** restrictive, but it is the default; thus, you will need to click the **Tabular Layout Option button** (figure 4.6, arrow A).

Step 11: Click **Next** (figure 4.6, arrow B).

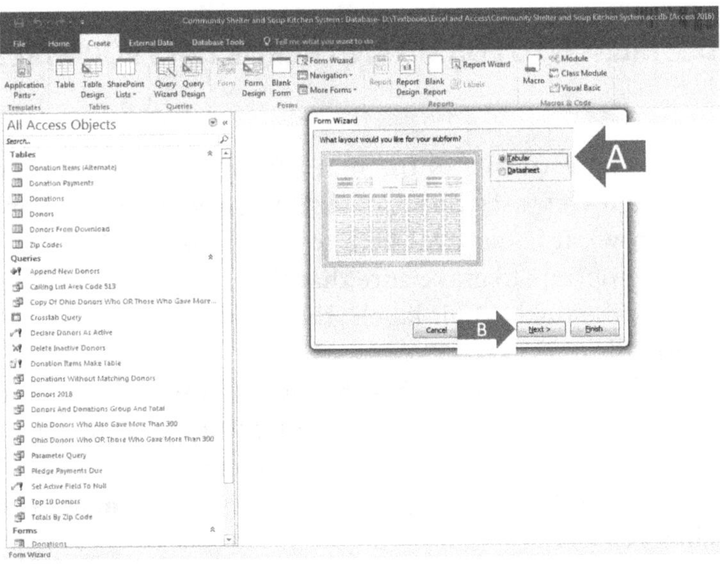

FIGURE 4.6   Creating a main and subform using the form wizard

Step 12: That will take you to the screen you see in figure 4.7. At this point you will be asked to name the two forms you are creating: the **main** and **sub-form**. In this example, click in the **Form** box and give the **main form** the name of *Donors Main Data Entry* (figure 4.7, arrow A) The default name for the **sub-form**, *Donations sub-form*, will suffice perfectly, so it does not need to be changed.

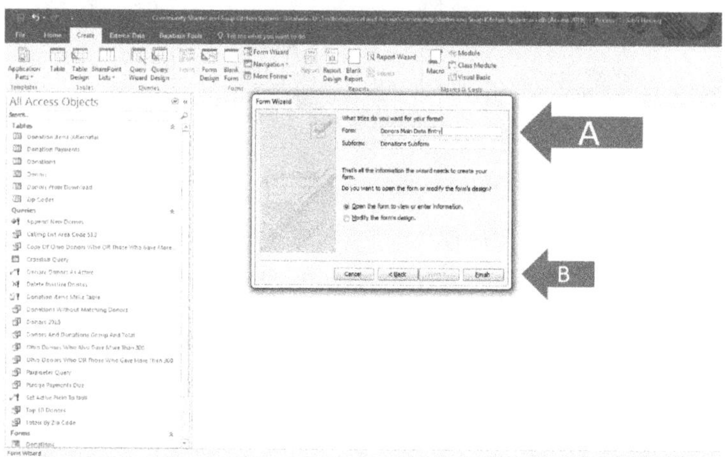

FIGURE 4.7   Creating a main and subform using the form wizard

Step 13: Click **Finish** (figure 4.7, arrow B). You will see a result comparable to what you see in figure 4.8.

MAKING ADJUSTMENTS TO FORMS AND REPORTS IN FORM/REPORT DESIGN VIEW

Now let's make a few adjustments to the **main form** and the **sub-form** by opening them up in their **design views**. To do so, click **Design View** in the upper left of your menu (figure 4.8, arrow). It is our goal here to make the widths of the fields in the form a little more appropriate to make sure that the forms will be more user friendly. Please remember that the fields, their labels, the sub-form, and all other items in the form are called **controls**.

NOTE: This may be one of those times that you may want to press the **F11** key at the top of your keyboard in order to hide the **navigation pane**, to get a better view of the **main** and **sub-form** and to better make adjustments.

NOTE: The procedures for adjusting **controls** within a **form** are IDENTICAL to doing the same in a **report**.

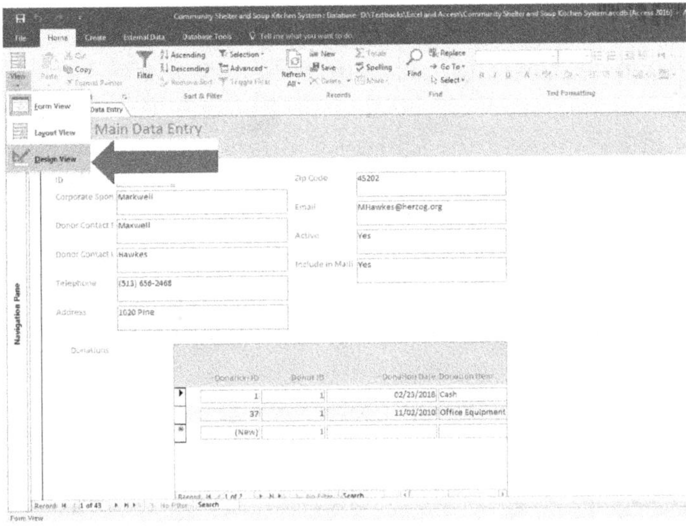

FIGURE 4.8   Main and subform changing design

You can resize and adjust the **sub-form** by taking the following steps:

Step 1: Click anywhere on the **sub-form** in the **design view** (figure 4.9, arrow). When you do, a thick orange border will appear around it.

Step 2: Press and hold your **Shift key** (preferably the one on the left side of your keyboard) and then press and hold the **Cursor Control arrow** → that points to the right. Continue to press and hold it until you can see the entire *net donation* field in the **sub-form** (figure 4.9, arrow).

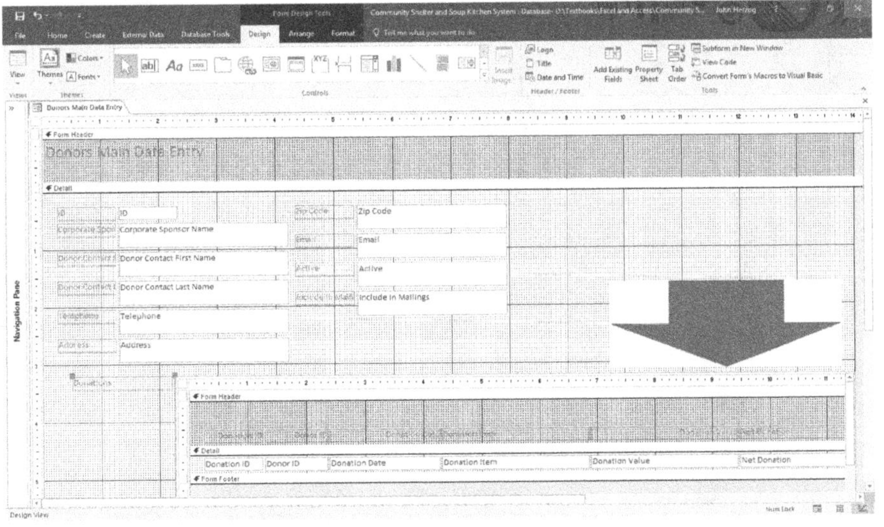

FIGURE 4.9   Main and subform changing design

NOTE: In order to change the size of any control you can do so by pressing and holding the **Shift key** and then tapping the **Cursor Control key that points Up ↑** (to make controls shorter), tapping the **Cursor Control key that points down ↓** to make the control taller, tapping the **Cursor Control key that points left ←** to make the control more narrow, and tapping the **Cursor Control key that points right →** to make the control wider.

NOTE: You can also resize a **control** by placing your mouse (if you are more confident using the mouse) at a corner and resizing it as you would a clipart.

Now, let's select all of the fields in the **main form** to make some adjustments to them. It is important to make the menu easy to read and making controls that same size. This can be done in a few steps without having to resize them individually.

TO MAKE MANY CONTROLS THE SAME SIZE IN FORM/REPORT DESIGN VIEW

Step 1: Select all of the fields in the **main form** by clicking and dragging and drawing a box around them (figure 4.10).

Step 2: Let go of your mouse. When you do, all of the **form controls** in the **main form** will be selected with orange borders (figure 4.11).

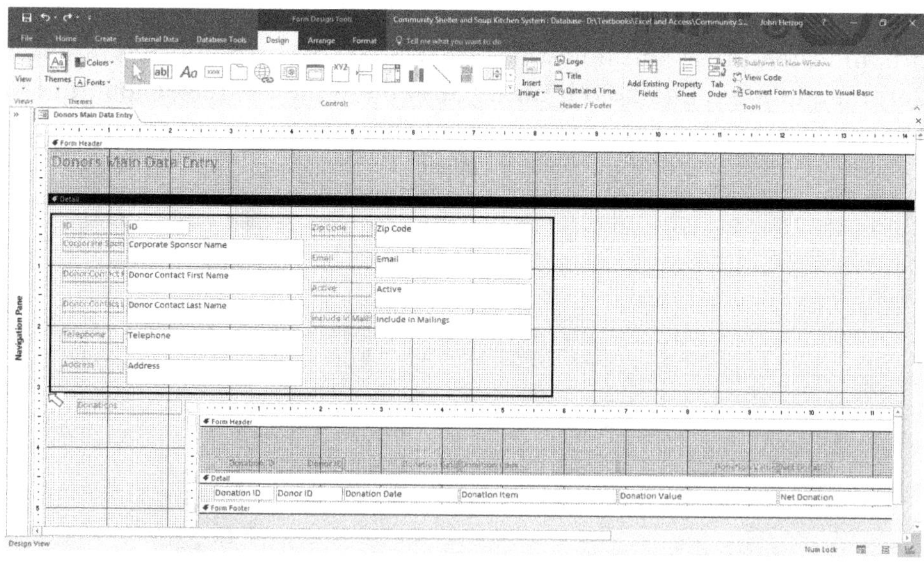

FIGURE 4.10   Selecting controls

Step 3: Click the **Arrange** ribbon in the **Form Design Tools** of the **form design** (figure 4.11, arrow A).

Step 4: Click **Size/Space** in the **Arrange** ribbon (figure 4.11, arrow B).

Step 5: Click **To Shortest** in the **Size/Space menu** (figure 4.11, arrow C). This will find the shortest, selected control in the form and make all of the selected controls the same height as that one.

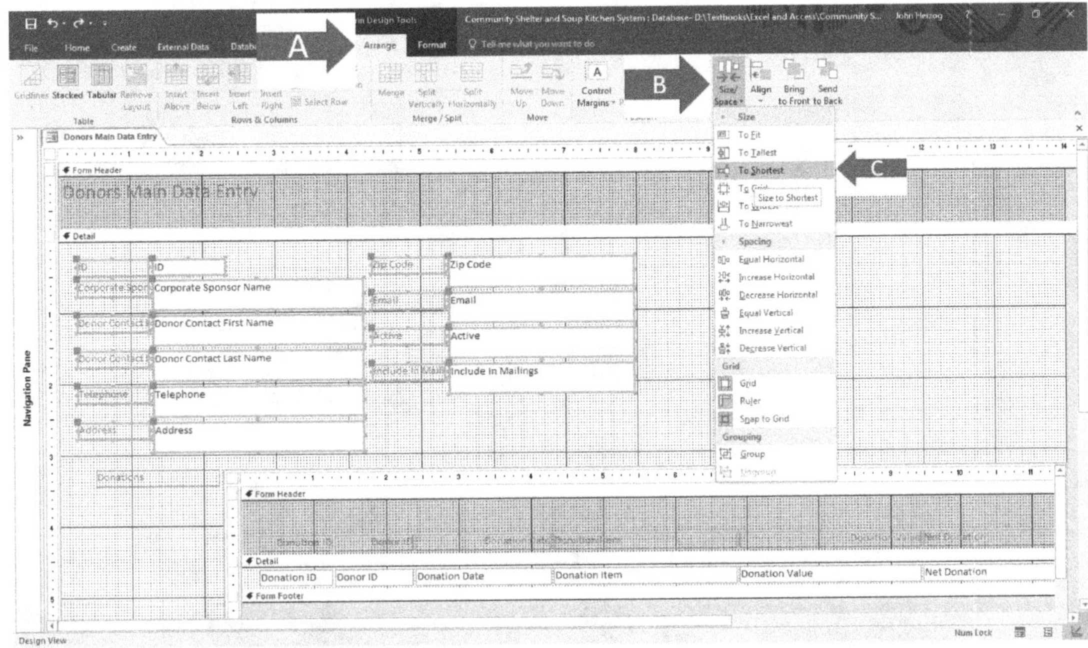

FIGURE 4.11   Changing control sizing

That will give you what you see in figure 4.12.

TO ALIGN CONTROLS IN FORM/REPORT DESIGN VIEW

In order to make the controls gravitate or align to the top of the **form** you can take the following steps:

Step 1: **KEEPING THE SAME CONTROLS SELECTED**, click the **Arrange** ribbon in the **Form Design Tools** (figure 4.12, arrow A)

Step 2: Click **Align** in the **Arrange** ribbon (figure 4.12, arrow B).

Step 3: Click **Top** in the **Align menu** (figure 4.12, arrow C). By doing this, you will have all of the controls align with the highest selected control in the form.

The result can be seen in figure 4.13.

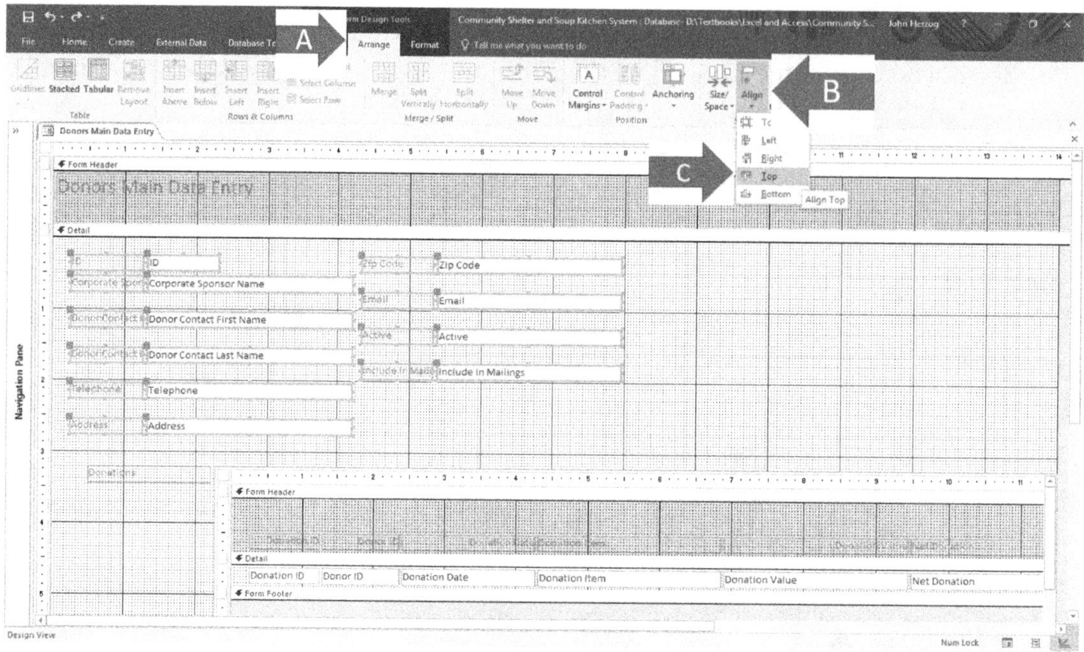

FIGURE 4.12 Changing control alignment

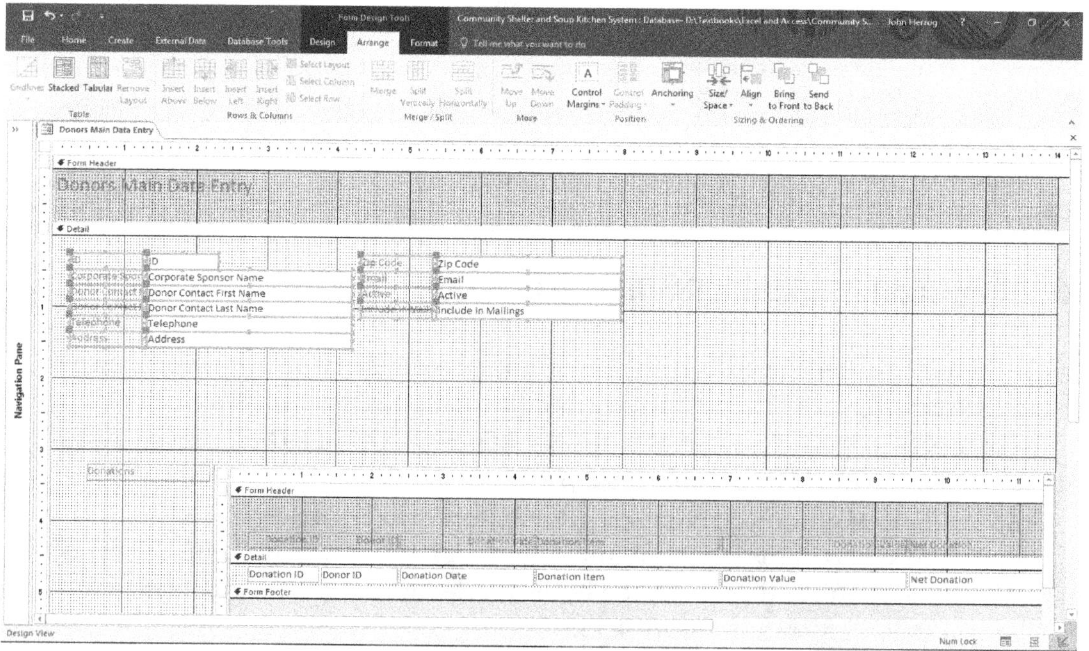

FIGURE 4.13 Changing control alignment

TO RESIZE A LABEL USING BEST FIT IN THE FORM/REPORT DESIGN VIEW

In the **design view** of a **form** or **report**, at the top of the screen there is a **header**. That is typically where you can place labels that are titles for what the **form** or **report** data represent. It is also used for search boxes in forms that will be discussed later. Below the **header** is the **detail area**. That is where the data is placed for data entry for forms. The same thing is true for the **report design**. The **header** is used for placing titles, and the **detail area** holds the actual data to be viewed. The **header** and **detail** areas are separated by what is often called the **header bar**.

At this point the **label** in the **header** of the *Donor Main Data Entry* form needs to be resized. As it is now, there is a lot of wasted space beneath it. Those who enter data do not like to scroll any more than they have to, but dead space such as this takes up more space in a form than is needed and thus causes the need for more scrolling. You can resize the **label** by performing a **best fit** function by using the following steps:

Step 1: Click on (or select) the label, and then place your mouse at a corner of the label in the **form header** with the caption that reads *Donor Main Data Entry* (figure 4.14a).

Step 2: Double-click your mouse. Doing so will resize the label to where it will display all of the text it holds. See the results in figure 4.14b. Notice how the orange selection border is now smaller and is only large enough to contain all of the label's text.

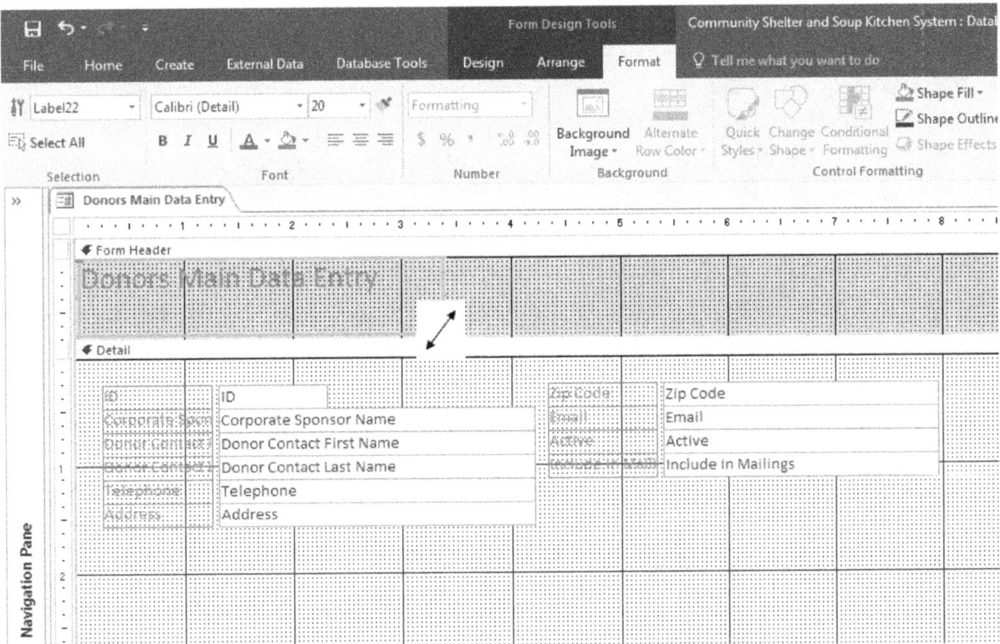

FIGURE 4.14a Best fit with form labels (before)

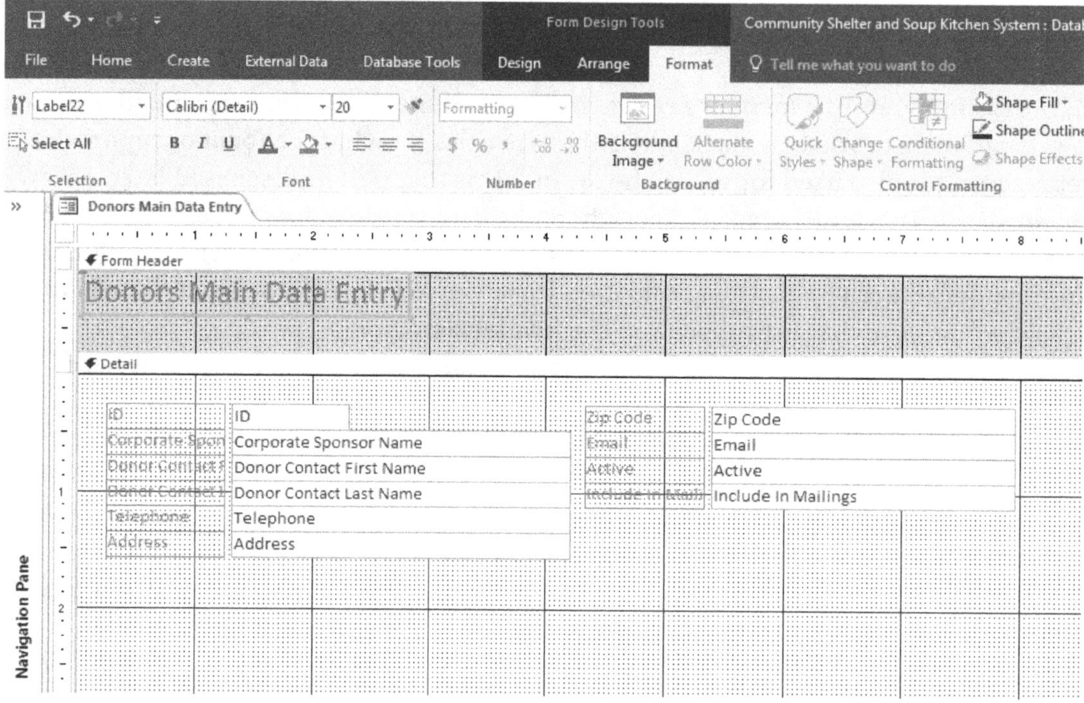

FIGURE 4.14b Best fit with form labels (after)

TO CHANGE THE FONTS IN LABELS AND CONTROLS IN THE FORM/REPORT DESIGN VIEW

The default font in Access is often very light and needs to be a darker color. To change any font, you can take the following steps:

Step 1: Select the given label or control. In this example we will select them all by pressing **Ctrl A** (which is the **Select All command**) (figure 4.15a).

Step 2: Click the Format ribbon in the Form Design Tools (figure 4.15a, arrow A).

Step 3: Click the **Black Text 1** font color (figure 4.15a, arrow B).

Step 4: The results are seen in figure 4.15b. You should see now that the text is darker.

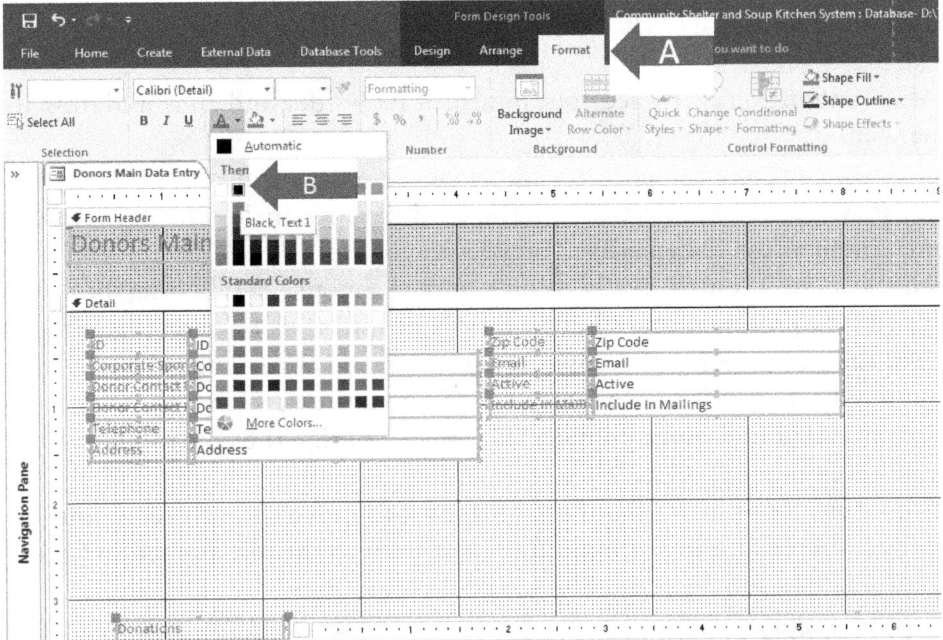

FIGURE 4.15a   Changing font colors in form design (before)

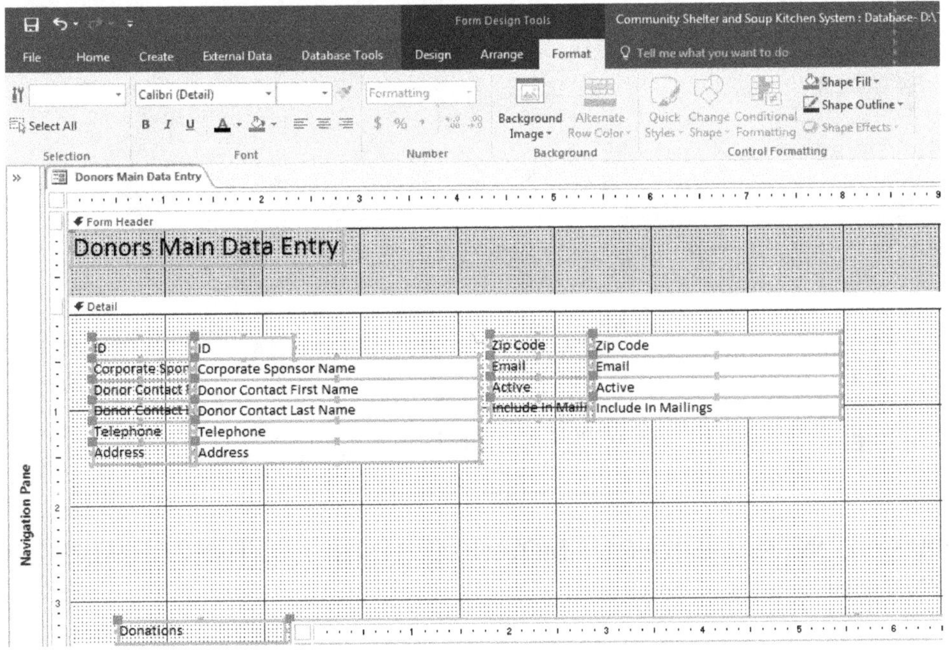

FIGURE 4.15b   Best fit with form labels (after)

TO MOVE A CONTROL IN THE FORM/REPORT DESIGN VIEW

To Move a **control** you can place your mouse at its border and click and drag it wherever you want it.

NOTE: Do **NOT** double-click inside a label or a data control as you try to move it. Doing so will cause Access to begin the editing mode inside them. If you do so by accident, simply press **Escape** (**Esc.**) on your keyboard.

When you move a control by dragging it with the mouse, it will often be difficult to move it to a very precise place. That's why it is best to click once on it and then press any one of the **cursor control arrows keys** (←↓↑→) on your keyboard in the direction you want it to be moved. You can also press and hold the **Ctrl key** as you tap the arrow keys. This is called the **nudge command** as it will **nudge** the control in a way that is slower and even more precise.

At this point, let us now take these measures to move the **sub-form** and its label closer to the other fields in the main form and then to the left as seen in figure 4.16.

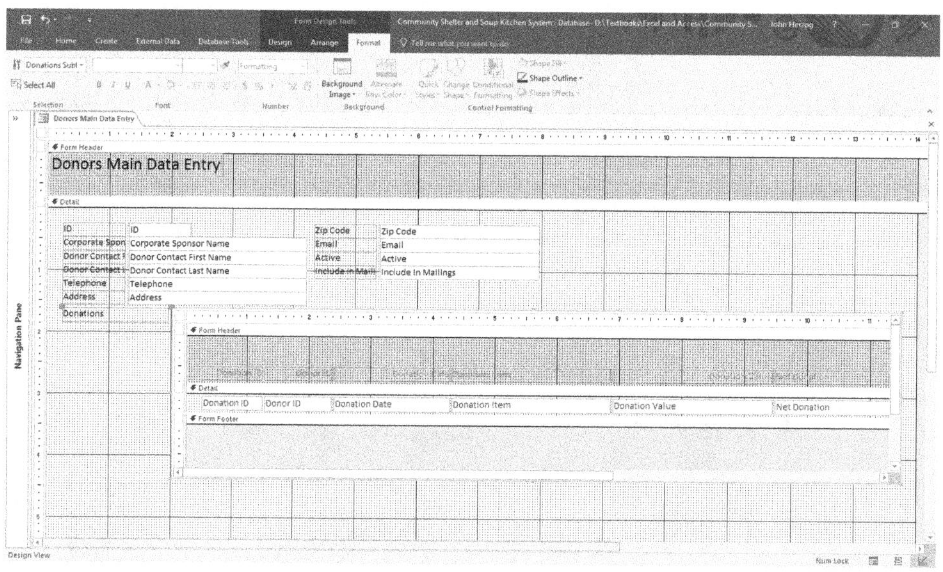

FIGURE 4.16　Moving controls without moving labels

TO MOVE A CONTROL WITHOUT MOVING ITS LABEL IN THE FORM/REPORT DESIGN VIEW

At this point you will want to move the **sub-form control** down and left so that it will be below its label. This will remove dead space to the right and make it easier for data entry. However, you may notice that whenever you move controls, their labels that are associated with them will always move with them. To move a control without moving its label you can do the following steps:

Step 1: Place your mouse in the upper left of the control and a small cross with arrows attached to the mouse arrow will appear (figure 4.17a).

Step 2: Click and drag the control where you want it (in this case down and left below its label as shown in Figure 4.17b).

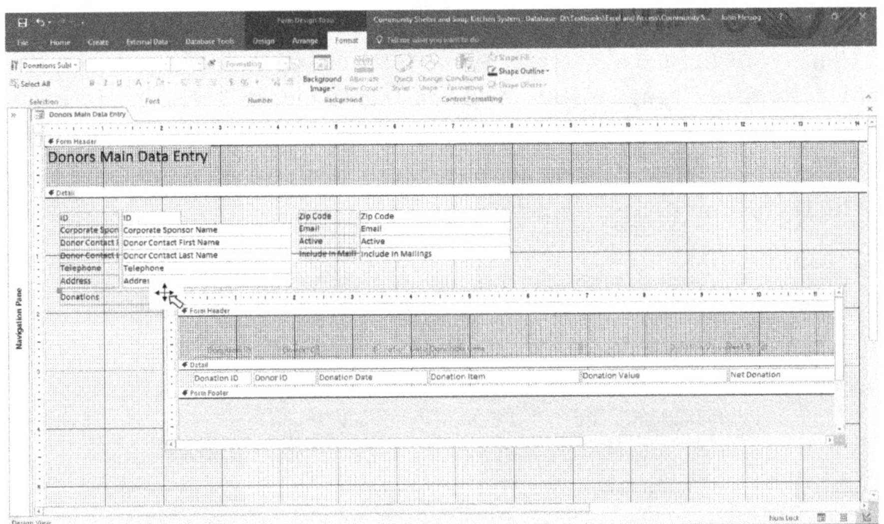

FIGURE 4.17a Moving controls without moving labels

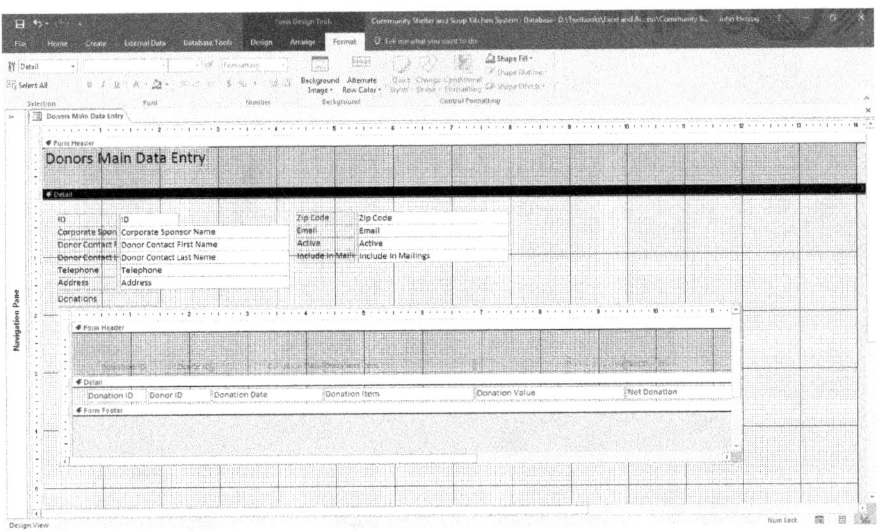

FIGURE 4.17b Moving controls without moving labels

SHORING UP WASTED SPACE IN HEADERS AND FOOTERS IN THE FORM/REPORT DESIGN VIEW

At this point of our creation of the **main** and **sub-form**, there is still a lot of wasted space. Remember, that is not good, because this will cause users to have to do a lot of scrolling in your forms and menus if they are too big for the screen resolution on their computers. Scrolling takes time and makes it difficult to enter data properly.

To remove the wasted space in the **header** do the following:

Step 1: Place your mouse at the top of the **header bar,** which separates the **detail area** from the **header** (figure 4.18a). You will see a small, black cross with arrows pointing up and down.

Step 2: Click and drag the cross at the **header bar** straight up as far as it can be dragged. The result can be seen in figure 4.18b.

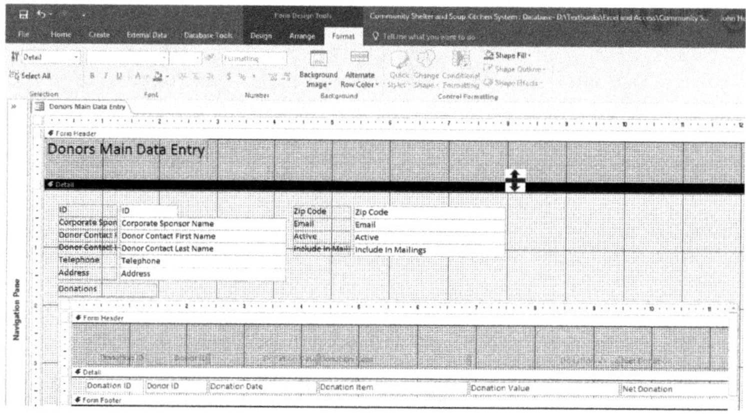

FIGURE 4.18a   Shoring up the header in form design

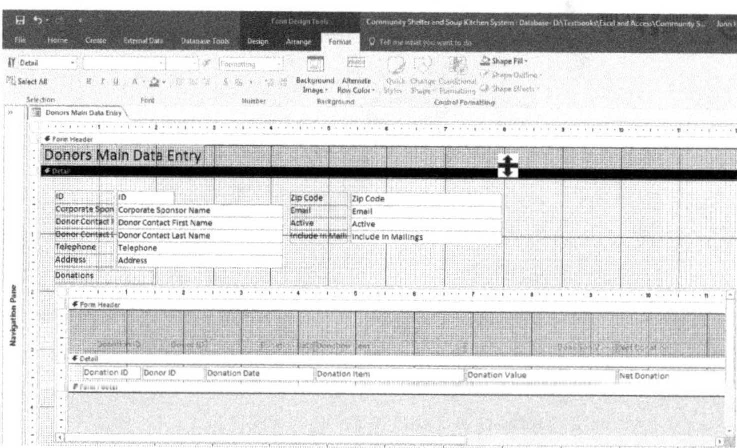

FIGURE 4.18b   Shoring up the header in form design

To shore up the wasted space on the right of the form, you can take the following steps:

Step 1: Place your mouse at the right edge of where the **form grid** stops (figure 4.19a, arrow). You will see a small, black cross with arrows pointing right and left.

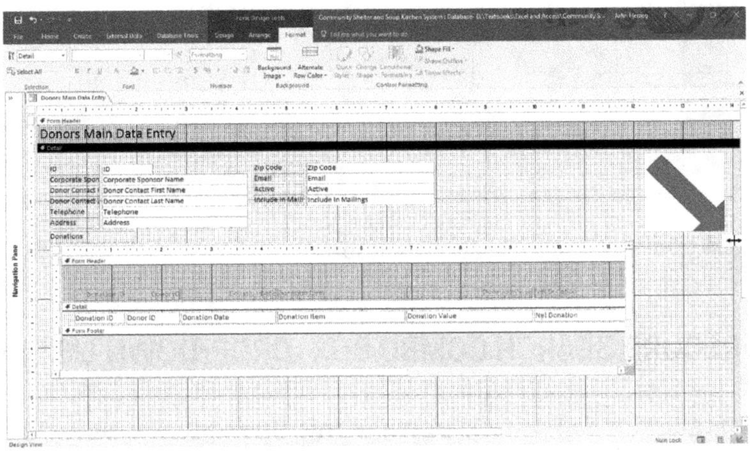

FIGURE 4.19a   Shoring up the form dead space

Step 2: Click and drag the arrow to the left as far as it will go. The result is seen in figure 4.19b.

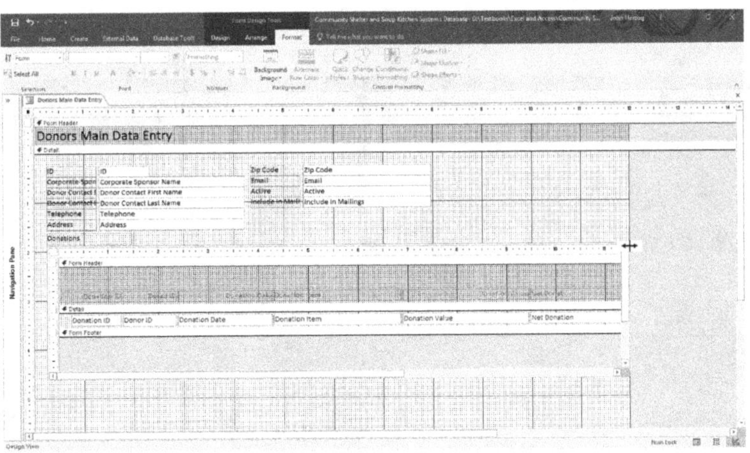

FIGURE 4.19b   Shoring up the form dead space

The same procedure can be done with the **form footer**. The **footer** and **detail** areas are separated by what is often called the **footer bar**. By placing your mouse above the

footer bar and then clicking and dragging the **footer bar** up as high as it will go, you will remove dead space that is below the footer. The result can be seen in figure 4.19c.

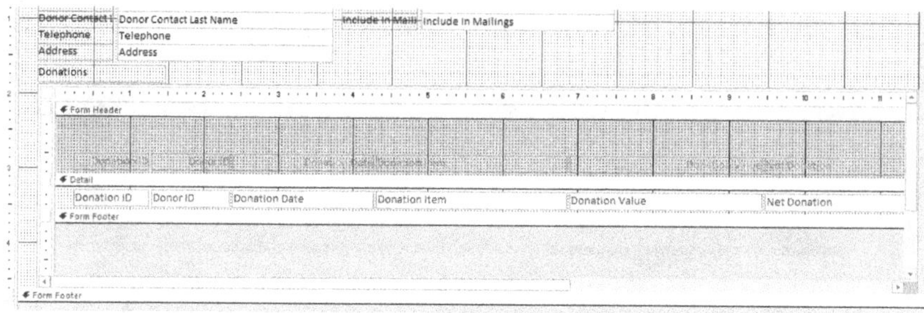

FIGURE 4.19c   Shoring up the form dead space

CREATING A RECORD SEARCH COMBO BOX (DROP-DOWN LIST BOX)

Now we need to take some measures to make this **form** more usable. Very commonly, in about every menu that you see on the internet, or when using just about any program, you will see **drop-down list boxes**. We have used them in this book many times. In Access they are called **combo boxes**. There are three types. The first is called a **record search combo box**. When you open the box, you can search for the record you need to see or edit. It can be created in a few easy steps as follows:

Step 1: In the form design view click the small scroll bar arrow on the Design Controls toolbox in the Form Design Tools (figure 4.20).

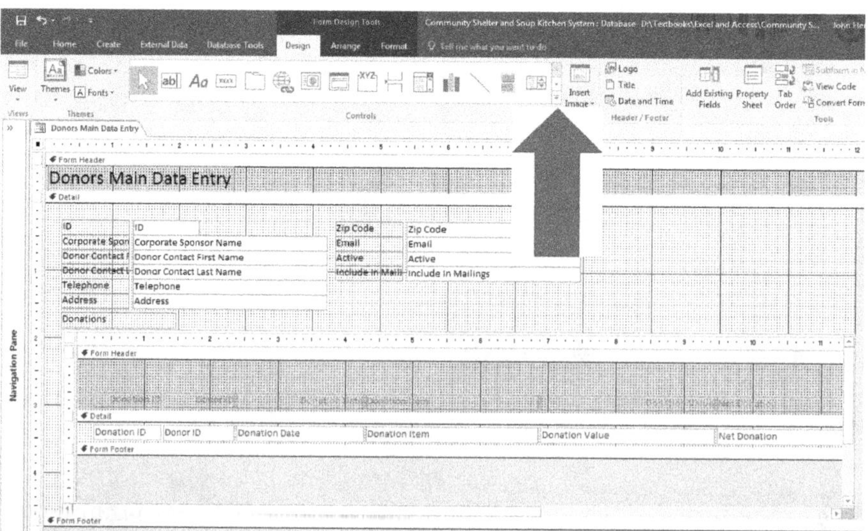

FIGURE 4.20   Creating a search combo box

Step 2: When you do it will open the menu. Make sure that the **Use Control Wizards** setting is **ON**. If it is, it will be recessed, and it will be a different color with a small border around it (figure 4.21, arrow A).

Step 3: Click once (do not click and drag) the **Combo Box** icon (figure 4.21, arrow B).

NOTE: Do **NOT** get the **Combo Box** mixed up with the **List Box**. They look similar, but their characteristics and functionality are very different.

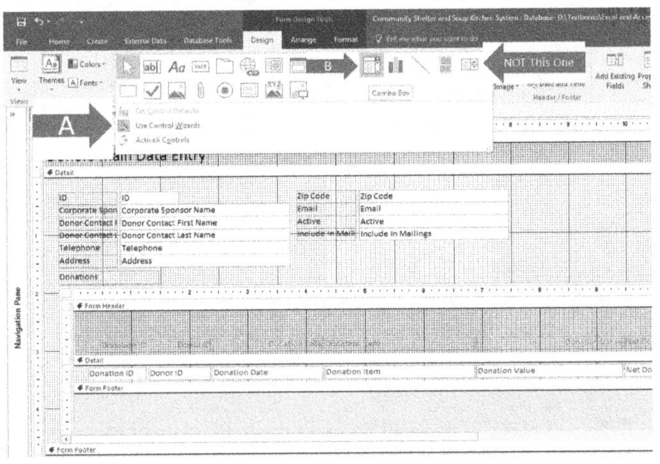

FIGURE 4.21 Creating a search combo box

Step 4: When you click on the **Combo Box** icon, it will attach a tiny **combo box** to your mouse, as you see in figure 4.22. Access will plant the **Combo Box** wherever you click the mouse. In this example, click (do not click and drag) on the grid line at the **6-inch mark** in the **form header** (figure 4.22, arrow).

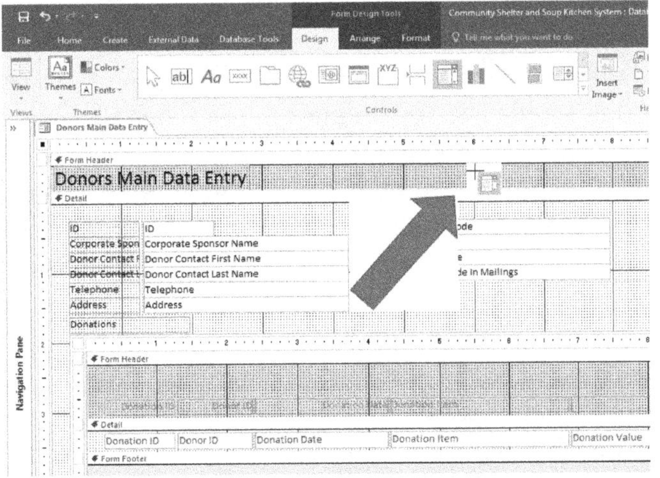

FIGURE 4.22 Creating a search combo box

Step 5: When you do, a **pop-up menu** will appear that is called the **Combo Box Wizard Menu** (as seen in figure 4.23). In that menu, you will want to choose the last of the **Option buttons** displayed there that is labeled **Find a record on my form based on the value I selected in my combo box** (figure 4.23, arrow A). Why? Because the **combo box** you are creating will be used to **FIND** a record and not to be used to change any data. That is why the box will be **unbound** as seen at arrow B in figure 4.23. Once you have chosen the **3rd Option button**, click **Next** (figure 4.23, arrow C).

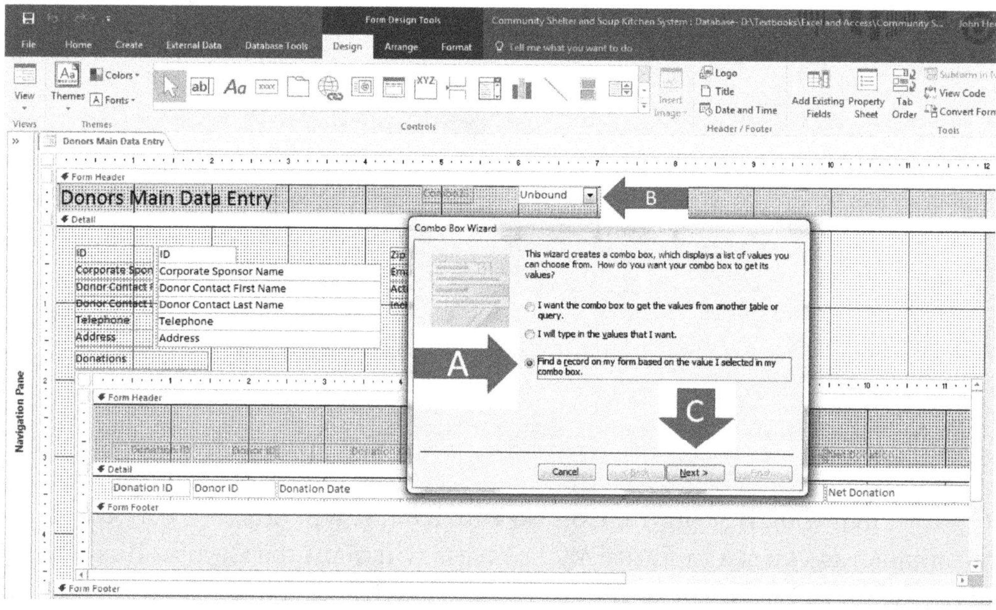

FIGURE 4.23   Creating a search combo box

Step 6: That will take you to the screen you see in figure 4.24. In this menu, you are asked in the **Available Fields** window to choose the fields that you want to put into the search box. These are the fields that the user will see. The **primary key** field must always be included and should always be the one you click and send first. In this case that would be the *ID* **field** (figure 4.24, arrow A) and it is already chosen. Remember, in order to click and send **ONLY ONE, GIVEN FIELD** from the **Available Fields** window to the **Selected Fields** window, you must click on the field you want to send and then click the **Single Arrow** button (figure 4.24, arrow B). From there you need to choose the fields that you feel are needed to make sure that the user can know if it is in fact the record he or she is looking for. For example, in this case you are looking for entered *donors*; therefore, you may want to send the *donor contact last name*, *donor contact first name*, and *address* fields. If you don't include the address in the **combo box**, the user may have a problem knowing which *donor* to click to find if there are two *donors* with the same name in the list. Therefore, you need to click on each of those fields in the order given, and click the **Single Arrow** button (figure 4.24, arrow B). That will shift them from the **Available Fields** window to the **Selected Fields** window (figure 4.24, arrow C).

Step 7. After all of the fields you want for the **combo box** are in the **Selected Fields** window, click **Next** (figure 4.24, arrow D).

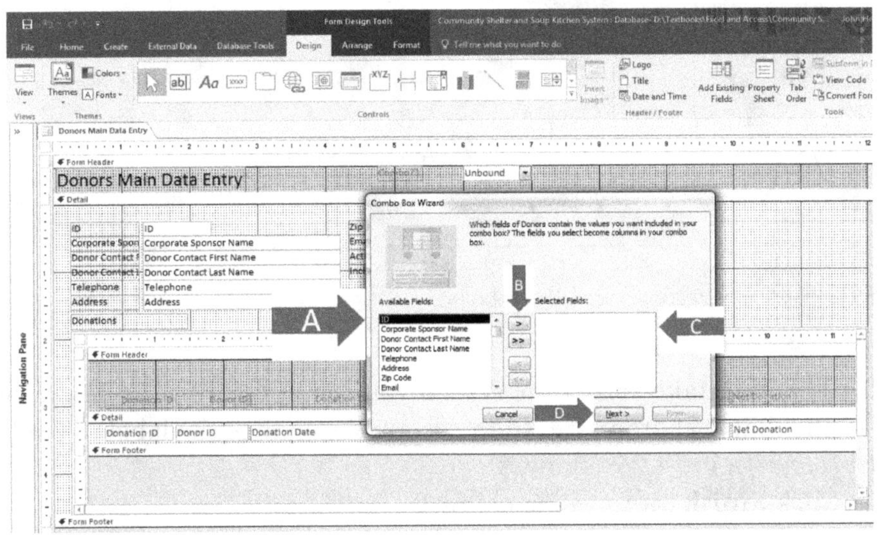

FIGURE 4.24   Creating a search combo box

Step 8: That will take you to the screen you see in figure 4.25. In this menu, you need only to determine if the **key column** (or **key field**) is to be hidden (figure 4.25, arrow A) and whether you wish to widen the columns. When you create a search box, you would almost always want to hide the **key column**, because **key fields** (or ID fields) that often hold **primary key** numbers that are generated by autonumbering field types will ask users to find a record by those ID numbers. Those ID numbers mean nothing to the user. They will want to find records by things such as names, telephone numbers, or Social Security numbers. Hiding the **key field** would then put the first field with recognizable data in view. In this example, by hiding the **key field**, the user would then be searching by the *donor contact last name, donor contact first name,* and then by the *address* field in that order.

If you want to widen the columns, you can do so by double-clicking the lines in the column headings that separate them, or clicking and dragging them, just as you would resize a column in a spreadsheet or in the datasheet view of a table or query. In this example, the column widths are wide enough and do not need to be changed.

Step 9: Click **Next** (figure 4.25, arrow B).

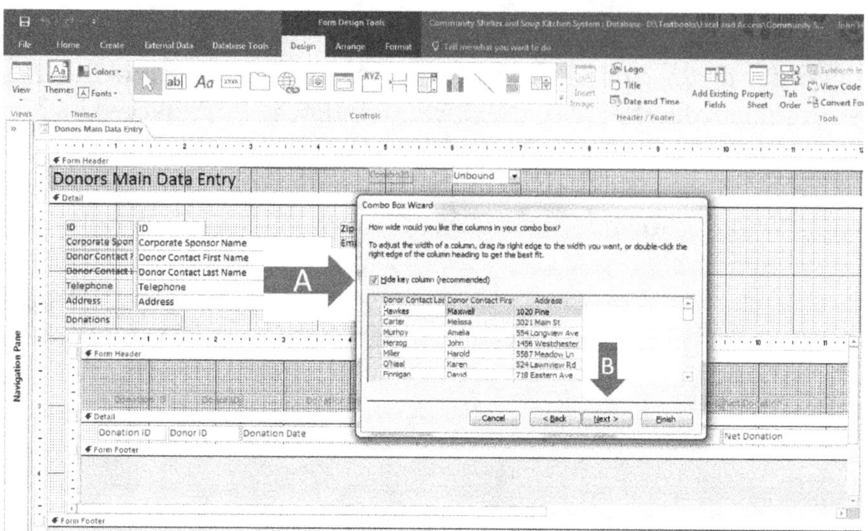

FIGURE 4.25   Creating a search combo box

Step 10: That will take you to the screen you see in figure 4.26. Make sure that the label is one that is appropriate. In this example, change the label to read *Find by Donor Last Name* (figure 4.26, arrow A).

Step 11: Click **Finish** (figure 4.26, arrow B).

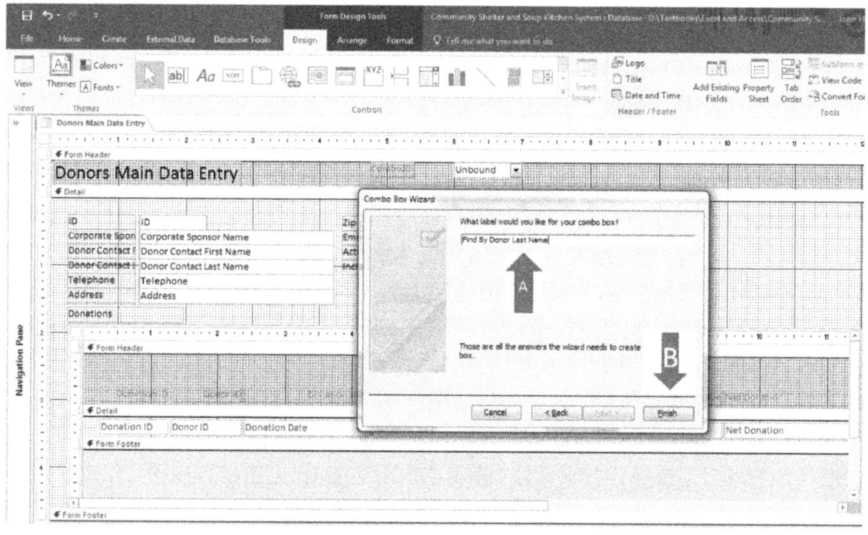

FIGURE 4.26   Creating a search combo box

Step 12: Your screen will now look like what you see in figure 4.27. Return to the **form view** by clicking the **Form View** button (figure 4.27, arrow).

NOTE: You must remember that **NO** combo box will work until you are in the **form view** of a form.

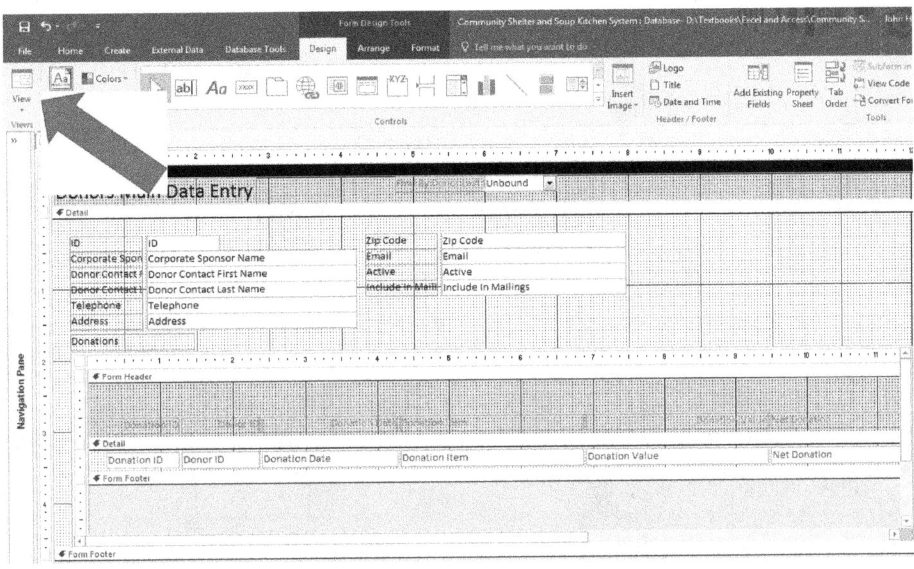

FIGURE 4.27 Creating a search combo box

Step 13: Tryout the new **combo box** by clicking on it and choosing any given *donor*. It should take you directly to that *donor's* record (figure 4.28a).

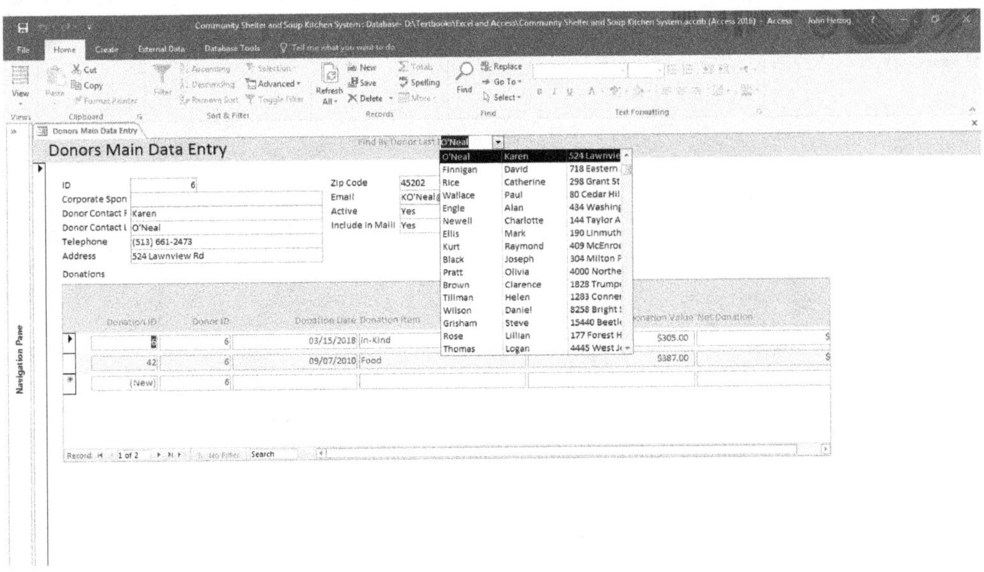

FIGURE 4.28a Creating a search combo box (completed)

SORTING A COMBO BOX

Please notice that in figure 4.28a, the names of the *donors* are not in alphabetical order; thus, the search box would be of little use. When you create one of the other **combo boxes**, there is a step in the **Combo Box Wizard** that will let you sort the list (which will be discussed later). However, the **combo box** we are now creating will not. In any case, there is a procedure used to sort lists that can be used on any **combo box**. To do so, take the following steps:

Step 1: Return to the form's **design view** (figure 4.28b).

Step 2: Select the search **combo box** (not the label) that you just created (figure 4.28b, arrow A).

Step 3: Click **Property Sheet** (figure 4.28b, arrow B) which will open the **property sheet**.

Step 4: Go to the **data tab** of the **property sheet** (figure 4.28b, arrow C).

Step 5: In the **data tab**, go to the **Row Source** text box. It will hold what is called the **Structured Query Language (SQL)** statement that begins with **SELECT** (figure 4.28b, arrow D).

Click the **Expression Builder** (the three-dotted box) in that same **Row Source** box (figure 4.28b, arrow E) and it will engage **SQL language** and thus take you to the **design view** of a query. Thus, it will take you to the screen you see in figure 4.28c.

NOTE: All queries in Access are written in a programming language that is called the **SQL language**. As you create queries in the way you have earlier, behind the scenes, Access is creating the query in the **SQL language** for you. You are seeing a glimpse of it in this text box. When you created this **combo box**, Access automatically creates and attaches a **query**, using the **SQL language** in the **combo box**. That **query** will show the list you want to use for your searches.

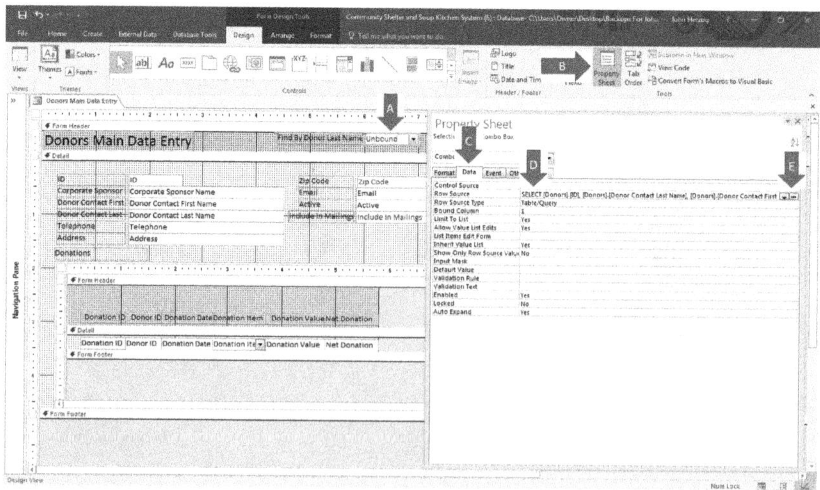

FIGURE 4.28b   Sorting a combo box

Step 6: This should look very familiar to you as it will take you to the **query design view** discussed in chapter 2. Simply change the sort order of the **query** to **ascending order** as you would any other **query** (figure 4.28c, arrow A).

Step 7: Click the **Close** button of the **query design** (figure 4.28c, arrow B). When asked if you want to **Save the changes made to the SQL statement and update the property**, click **Yes**.

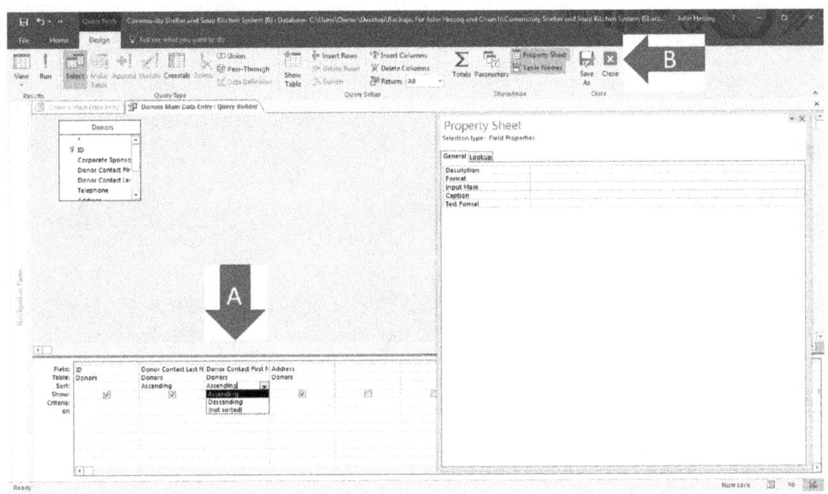

FIGURE 4.28c Sorting a combo box

Step 8: Return to the **form view** of the **form** as you have done before and open the **Search Box** now to see that the names are in alphabetical order (figure 4.28d).

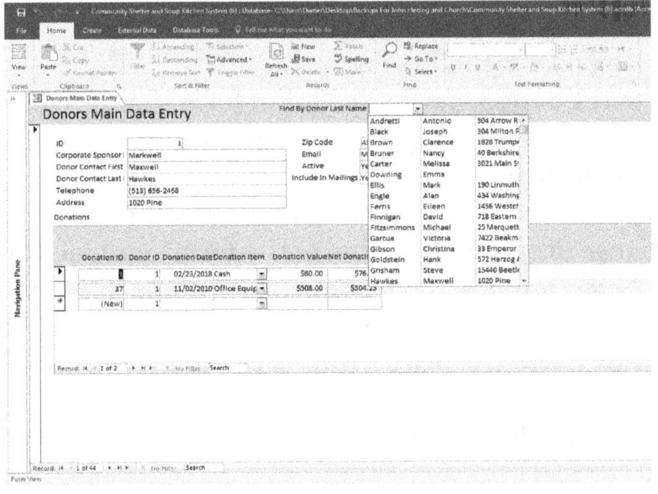

FIGURE 4.28d Sorting a combo box

CHANGING NAVIGATION BUTTON CAPTIONS

If you return to the **main form** in the **form view**, you will notice that there will always be a separate **navigation button** for the **main form** and one for the **sub-form** (figure 4.28e, double arrow). The problem is that it can be very confusing to the user as to which **navigation button** goes with what **form**.

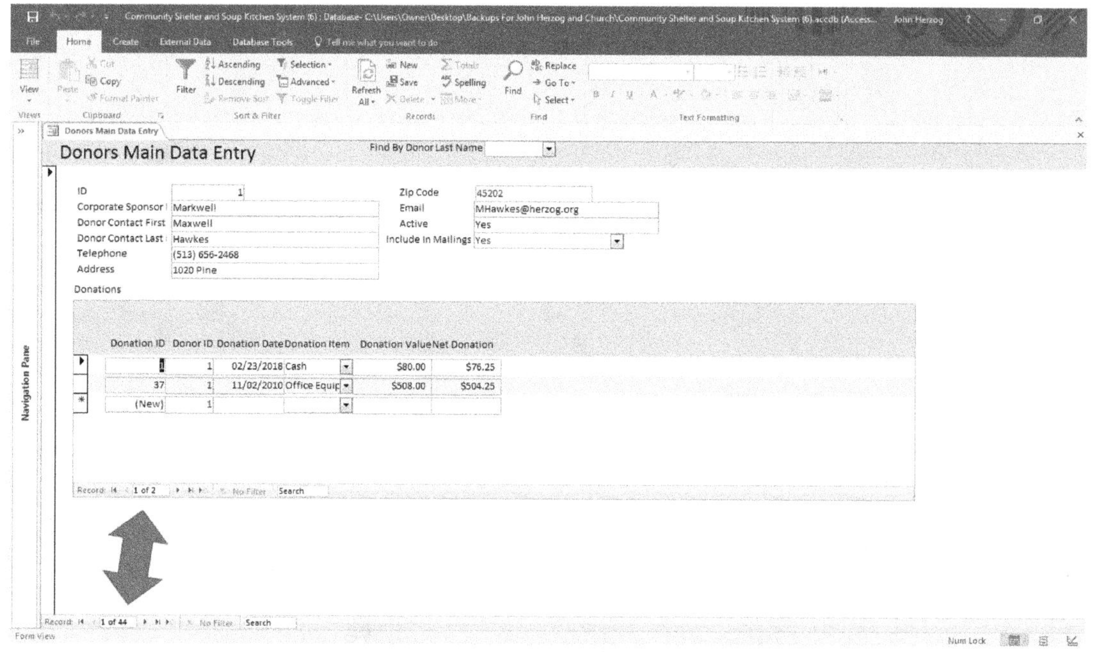

FIGURE 4.28e   Changing navigation captions

To fix the problem, take the following steps:

Step 1: Return to the **design view** of the **main form** (figure 4.28f).

Step 2: Double-click the **Form/Report Selection box** in the upper left of the **Design View Grid** (figure 4.28f, arrow A)

NOTE: Just as you can select a **control**, and then view or edit the **properties** of it, you can see and change the **properties** of an entire **form** (or **report**) by double-clicking the **Form/Report Selection box**. When you do, it will open the **property sheet** for the form and the **Property Sheet** icon in the **Form Design** ribbon will become active (figure 4.28f, arrow B). It will also place a tiny black box in the **Form/Report Selection box** (figure 4.28f, arrow A).

Step 3: Click the **All Tab** in the **property sheet** (figure 4.28f, arrow C) if it's not already active.

Step 4: In the **Navigation Caption box**, type the word *DONOR* (figure 4.28f, arrow D). It is probably best to type it in **ALL CAPS** in order to draw the users' attention to it while they are using the form.

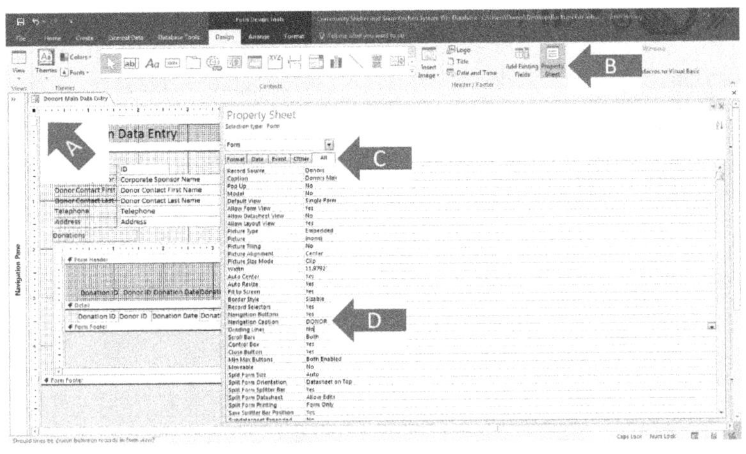

FIGURE 4.28f   Changing navigation captions

Step 5: Repeat this process for the **sub-form**. In the **design view**, double-click the **Form/Report Selection box** in the upper left of the **Design View Grid** of the **sub-form** (figure 4.28g, arrow A).

Step 6: In the **property sheet**, change the **navigation caption** to read *DONATION* (figure 4.28g, arrow B).

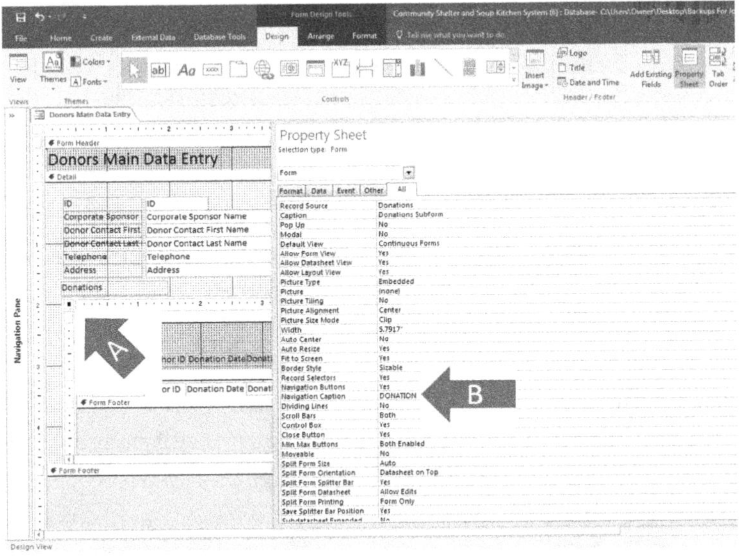

FIGURE 4.28g   Changing navigation captions

Step 7: Return to the **form view** and see that the lower **navigation button** is referring to the *DONOR* and the upper one is referring to the *DONATIONS* (figure 4.28h, double arrow). It will read *DONOR* 1 of 44 and *DONATION* 1 of 2. That means that you have selected the first of 44 *donors* (who in this example is *Maxwell Hawkes*) and that you are now in the first of his two *donations*.

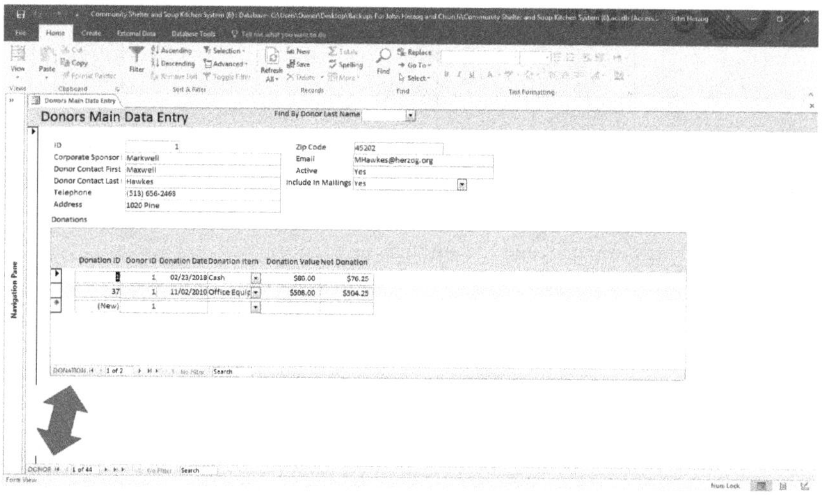

FIGURE 4.28h   Changing navigation captions

CREATING A COMBO BOX (DROP-DOWN LIST BOX) WITH CHOICES THAT DO NOT CHANGE REGULARLY (KNOWN AS A VALUE LIST)

Suppose you want the users of your database to enter only one of three choices into the *Include in Mailings* field. In this case, suppose those choices are the following:

- Yes
- No
- Newsletter only

If it is not likely that you would ever want these options to change in the life of the database, you can put a combo box in your **form** so that users will know that these are the only choices for what they can enter into that field.

This can be performed by using the following steps:

Step 1: Open the *Main Data Entry* **form** in the **design view** (figure 4.29).

Step 2: Delete the *Include in Mailings* field (figure 4.29, arrow) by clicking on/selecting it and then pressing the **delete key** on your keyboard.

NOTE: Any control and its label can be removed from a **form** or **report** in the **form/report design view** by clicking once on the control and then pressing the **Delete key**. If you click on its label and press the **Delete key**, only the label will be removed.

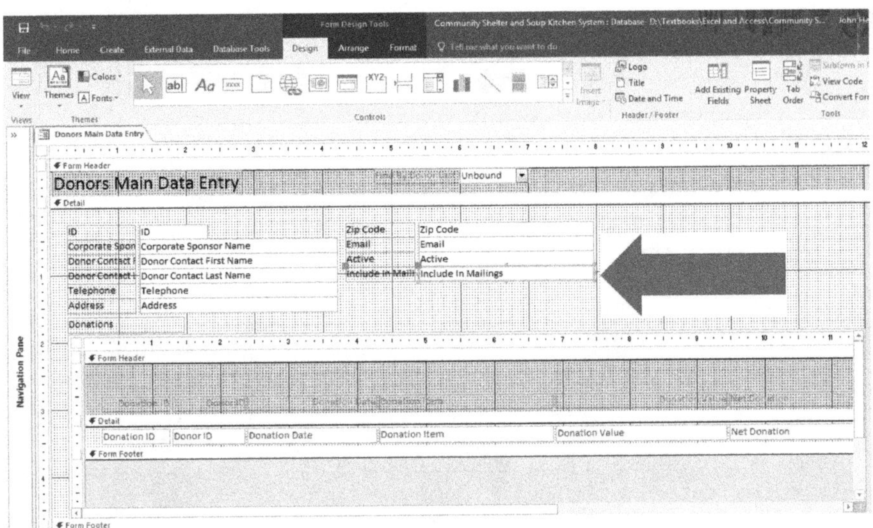

FIGURE 4.29   Creating a value list combo box

Step 3: Click on the scroll box in the **design controls** of the **form design view** (figure 4.30, arrow).

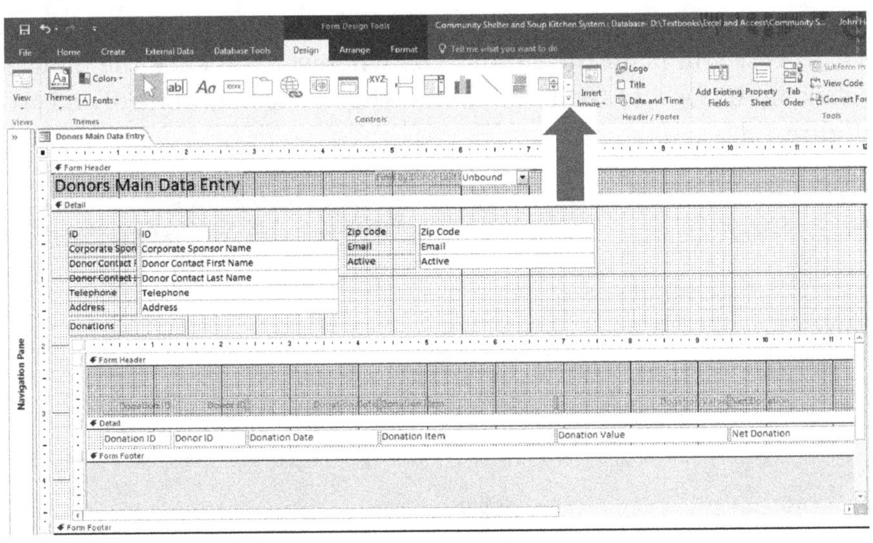

FIGURE 4.30   Creating a value list combo box

Step 4: When you do, it will open the menu. Make sure that the **Use Control Wizards** setting is **ON**. If it is, it will be recessed and shaded as you see in (figure 4.31, arrow A).

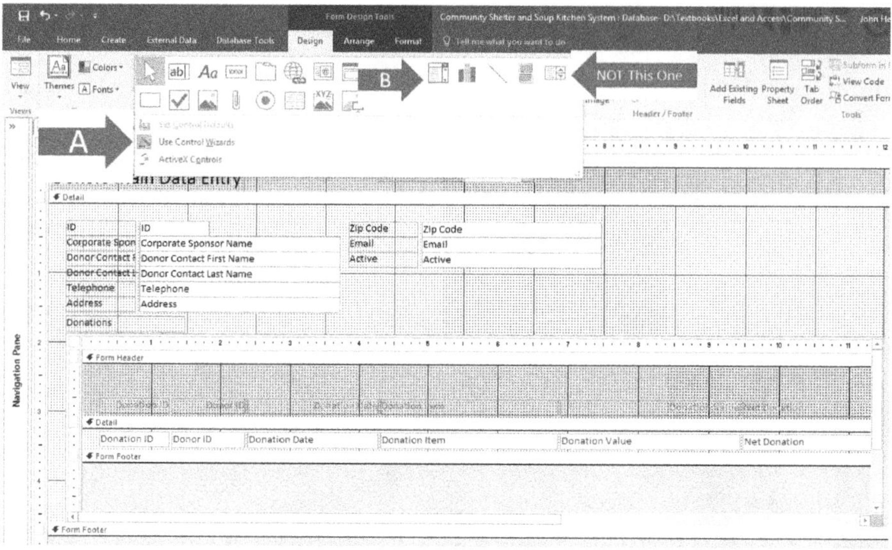

FIGURE 4.31 Creating a value list combo box

Step 5: Click on the **Combo box** icon in the **Control Toolbox** area (figure 4.31, arrow B).

Step 6: When you do, it will attach a tiny **combo box** to the mouse as you see in figure 4.32. **Access** will plant the **combo box** wherever you click the mouse. In this example, click on the **Design Grid** right below the *Active* field control and **NOT THE LABEL** (figure 4.32, arrow).

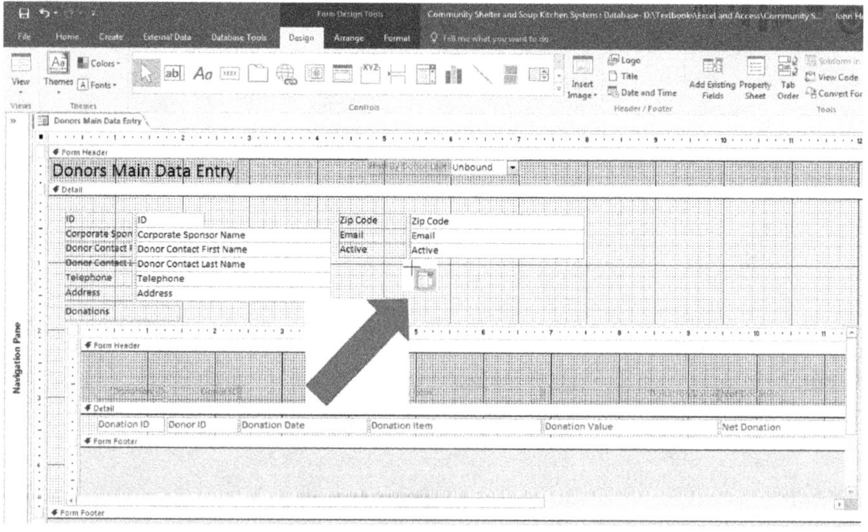

FIGURE 4.32 Creating a value list combo box

Step 7: When you do, the **Combo Box Wizard Menu pop-up menu** will appear (figure 4.33). This time, you will want to choose the **2nd option button** there that is labeled **I will type in the values that I want** (figure 33, arrow A). Why? Because the **combo box** you are creating will ask you to type in the choices (or values) that you want the list to display. As mentioned earlier, they will be *Yes, No,* and *Newsletter only.* Once you have chosen the **2nd Option button**, click **Next** (figure 4.33, arrow B).

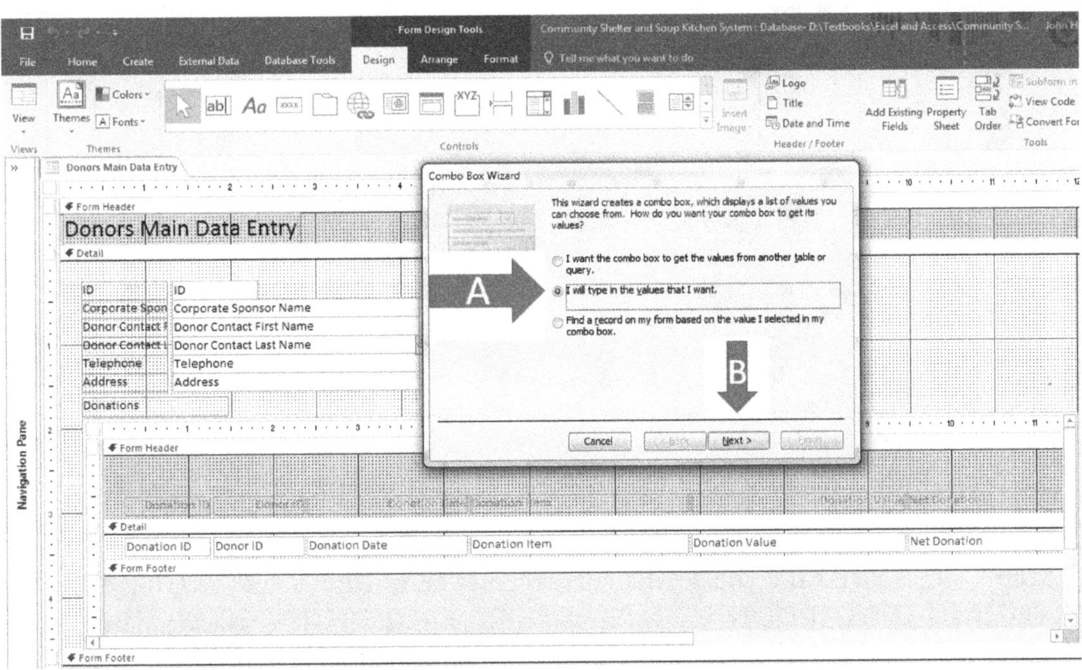

FIGURE 4.33 Creating a value list combo box

Step 8: That will take you to the screen you see in figure 4.34. In this menu, you are asked to enter the list of items that you want to put into the box. In this case, you will enter *Yes* in the box **BELOW** where it reads **Col1** (figure 4.34, arrow A). Once you have typed it, press **the Down Arrow Key (↓)** on your keyboard to go to the box below it. **DO NOT PRESS THE ENTER KEY AFTER YOU ENTER A LIST ITEM. IT WILL TAKE YOU TO THE NEXT SCREEN BEFORE YOU HAVE ENTERED ALL OF THE LIST ITEMS.** Once you have entered all of the list items, click **Next** (figure 4.34, arrow B).

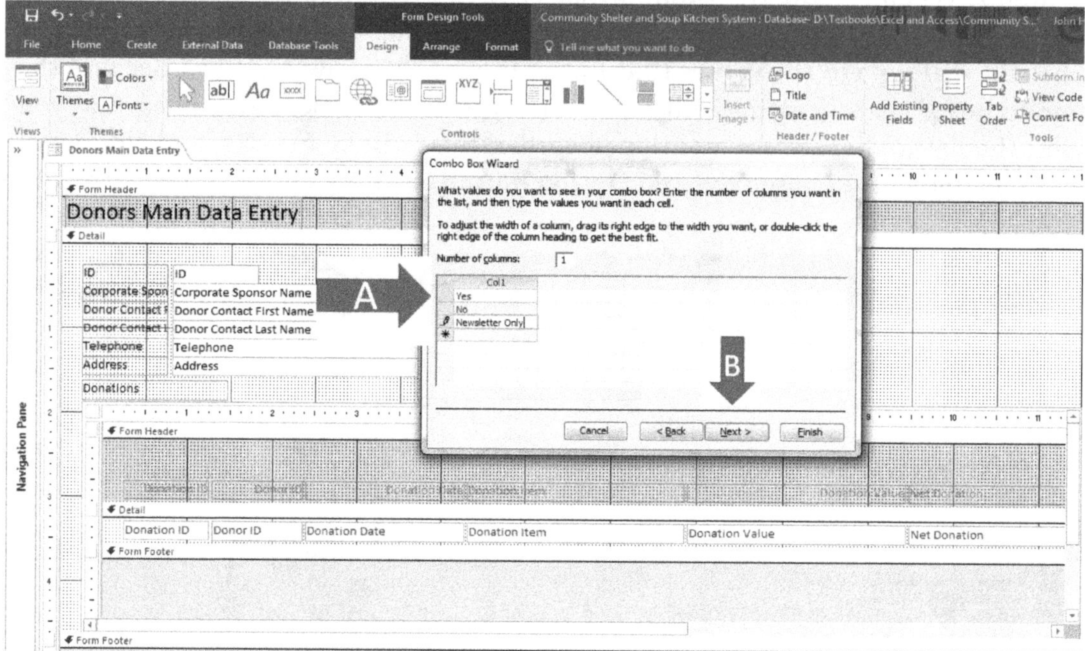

FIGURE 4.34   Creating a value list combo box

Step 9: That will take you to the screen you see in figure 4.35. THIS STEP IS CRITICAL!! IF YOU DON'T DO THIS STEP PROPERLY, THE COMBO BOX WILL NOT WORK PROPERLY, AND IT WILL BE OF NO USE! DO NOT USE THE "REMEMBER THE VALUE FOR LATER USE" OPTION EVEN THOUGH IT IS THE DEFAULT (figure 4.35, arrow labeled "DO NOT USE THIS ONE!!!"). YOU MUST CHOOSE THE FIELD TO WHICH THE LIST CHOICE WILL BE SENT AND PLACE THAT FIELD NAME IN THE BOX LABELED "STORE THAT VALUE IN THIS FIELD." In this case it is the *Include in Mailings* field in the drop-down list (figure 4.35, arrow B). MAKE SURE THAT OPTION BUTTON IS SELECTED (figure 4.35, arrow A). For example, if the user chooses the *Newsletter only* choice in the list, you must designate by this screen that the words *Newsletter only* must be sent to (or entered into) the *Include in Mailings* field.

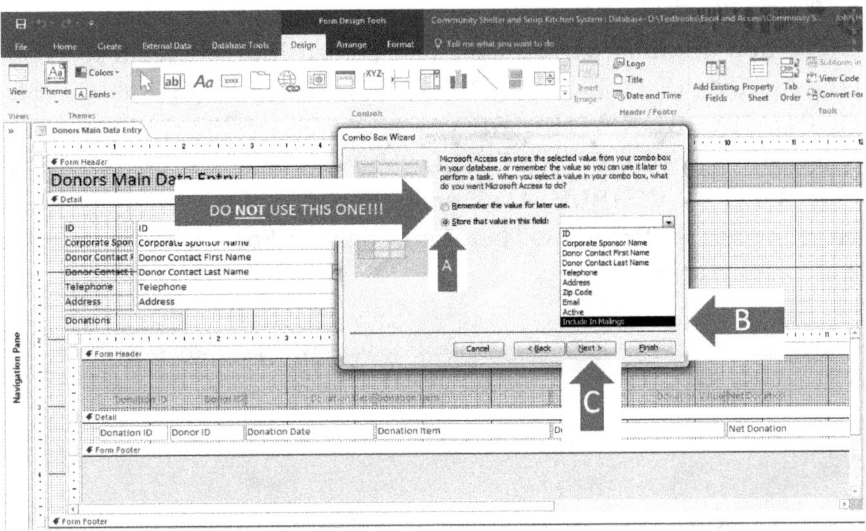

FIGURE 4.35   Creating a value list combo box

Step 10: Click **Next** (figure 4.35, arrow C). When you do, it will take you to the screen you see in figure 4.36. All that is needed here is for you to enter the caption of the label to accompany this **combo box**. You would likely use the field name as the label, which is *Include in Mailings* (figure 4.36, arrow A).

Step 11: Click **Finish** (figure 4.36, arrow B).

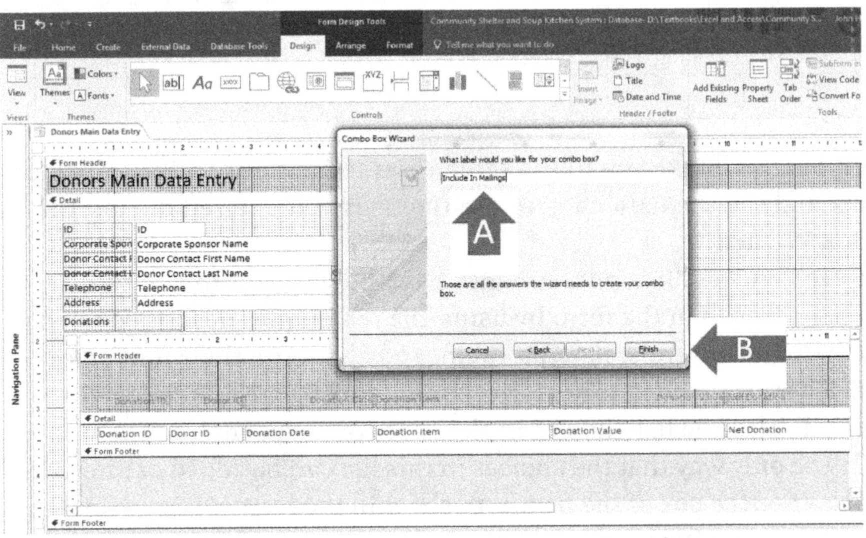

FIGURE 4.36   Creating a value list combo box

Step 12: That will take you to the screen you see in figure 4.37.

NOTE: The new combo box you have just placed into the form IS NOT TO BE UNBOUND. AGAIN, IF IT IS, THE DATA YOU ENTER WILL NOT BE SAVED AND THE COMBO BOX WILL BE OF NO USE! INSTEAD OF IT BEING UNBOUND, THE NAME OF THE FIELD TO WHICH THE ENTERED DATA IS STORED SHOULD BE VISIBLE. Also remember, that you cannot test a combo box without returning to the form view. Therefore, click the Form View icon in the Design ribbon (figure 4.37, arrow).

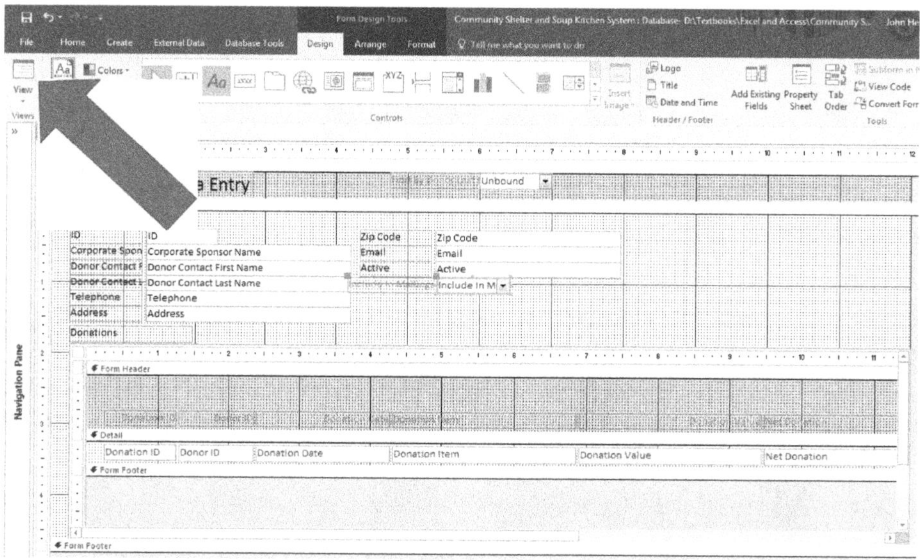

FIGURE 4.37   Creating a value list combo box

Step 13: That will take you to the screen you see in figure 4.38a. Click on the new **combo box** and see the data choices are the ones you entered during the use of the **Combo Box Wizard**.

NOTE: In figure 4.38a, adjustments have been made to the two combo boxes that have been placed in the form by using the techniques described to you earlier in resizing labels and controls, moving them, and changing their font colors.

NOTE: There is no need to sort this **combo box** because Access will assume that you want to display the list choices in the order that you entered them.

NOTE: The only way that the choices in this list can be added or altered is by doing so in the **Row Source** box of the **property sheet** in the **form design view** (figure 4.38b, arrow). Each new list item must be separated by a **semicolon** (;) (figure 4.38b, arrow). That is why it is not recommended to install a **value list** in a form if the choices will be changed regularly.

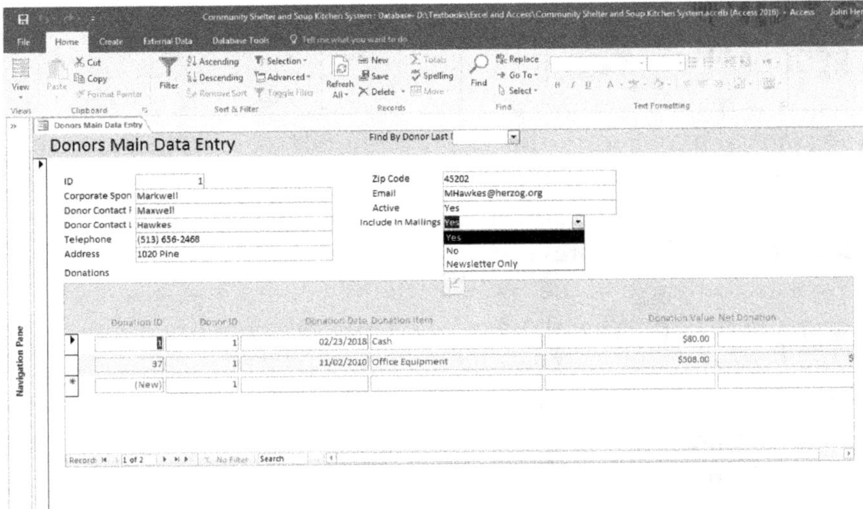

FIGURE 4.38a Creating a value list combo box

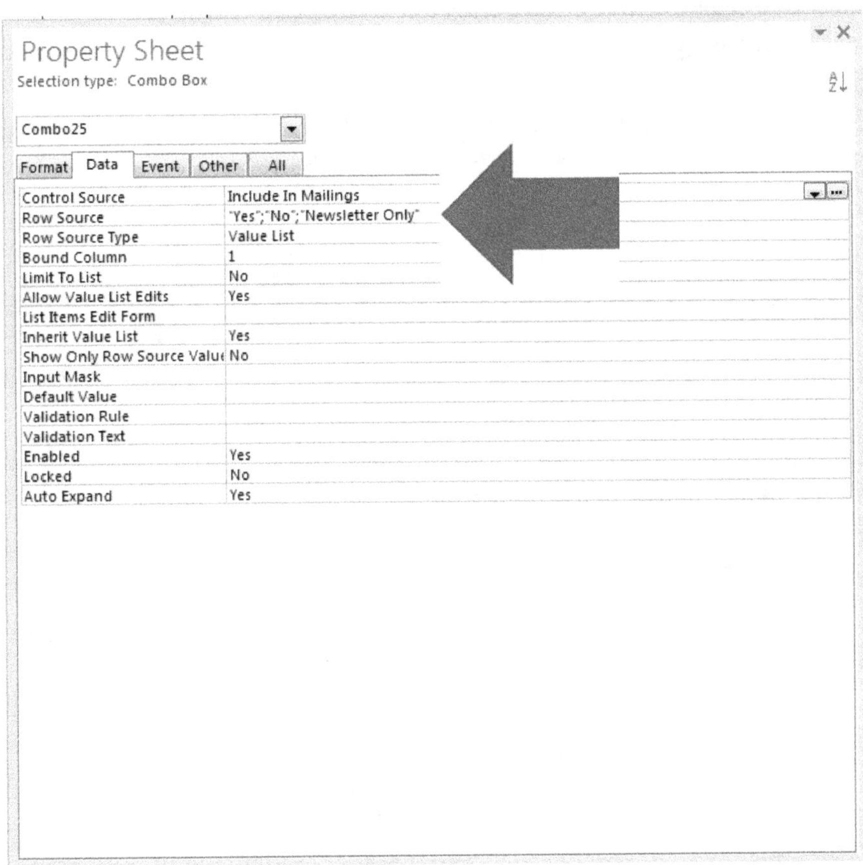

FIGURE 4.38b Creating a value list combo box

CREATING A COMBO BOX (DROP-DOWN LIST BOX) WITH CHOICES THAT DO CHANGE REGULARLY

Suppose you want to have a **drop-down list** of the *zip codes* from the *zip codes* table, so users can see what *zip codes* go with what cities. It is important to note here that using a **table** or **query** for such choices makes it possible for you to add *zip codes* for the **drop-down list** box (or **combo box**) **WITHOUT** having to change the design of the **form**. You need only to add the *zip codes* to the table as you would be doing with any data entry. On the other hand, when you create a **value list,** as we did with the *Include in Mailings* field, additions to that list **CANNOT** be made without changing the design of the form in the **design view.** Typically, end users of a form would never want to have to do that.

You can make a **combo box (drop-down list box)** that reads a **table** or **query** by using the following steps:

Step 1: Return to the **form design view** of the **main form** and click on (or select) the *zip code* field. Press the *Delete key* on your keyboard to remove it from the **form.**

Step 2: As before, click on the scroll box of the controls in the Design ribbon in the form design view (figure 4.39).

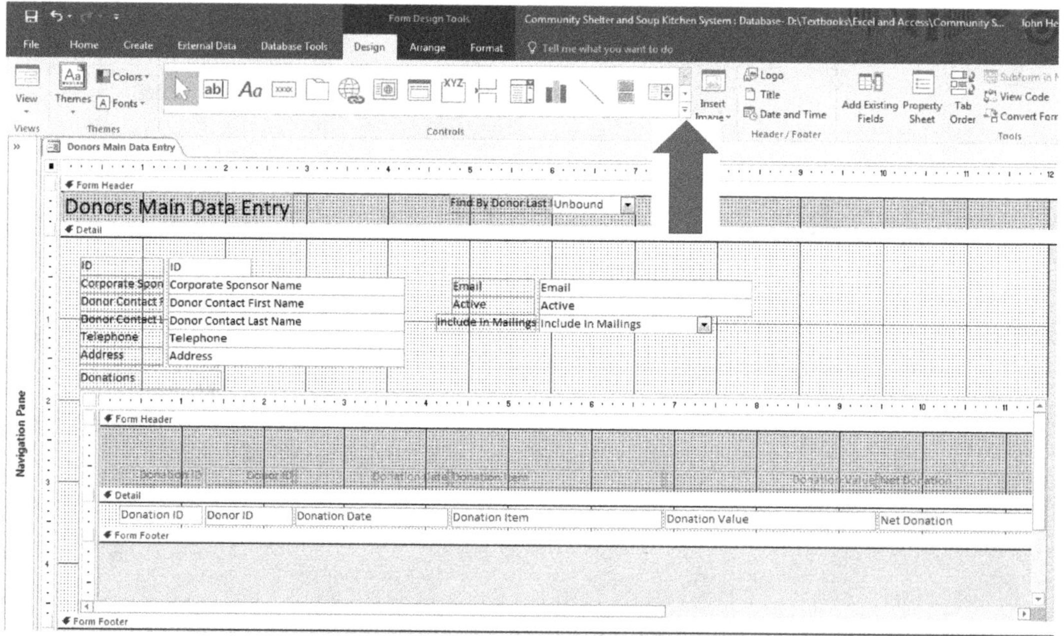

FIGURE 4.39 Creating a combo box that reads a table or query

Step 3: Make sure that the **Use Control Wizards** option is **ON** (figure 4.40, arrow A).
Step 4: Click on the **Combo Box** icon (figure 4.40, arrow B).

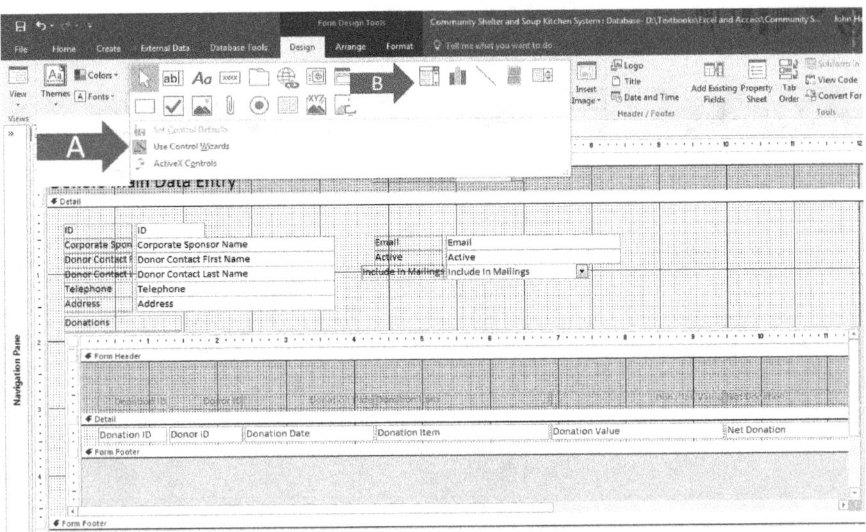

FIGURE 4.40   Creating a combo box that reads a table or query

Step 5: Once again, it will attach a tiny **combo box** to the arrow of your mouse, as you see in figure 4.41. Access will plant the **combo box** wherever you click the mouse. In this example, click directly above the *email* field (figure 4.41).

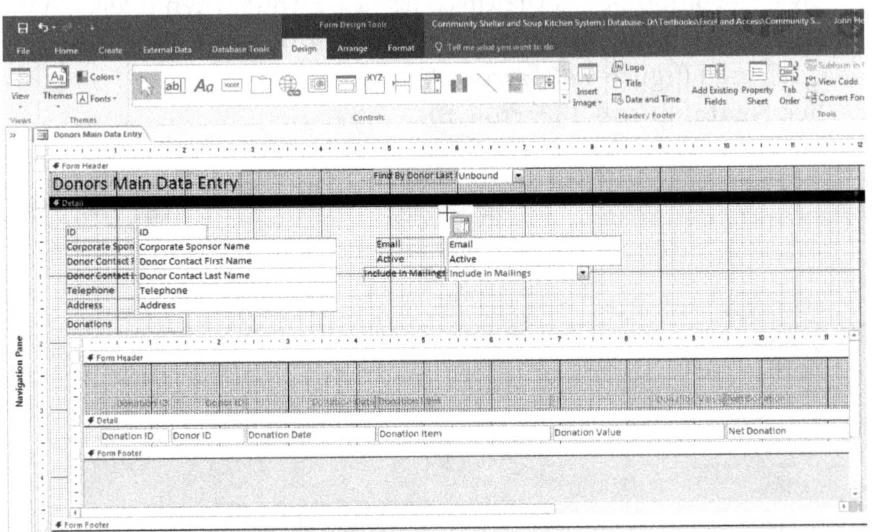

FIGURE 4.41   Creating a combo box that reads a table or query

Step 6: That will take you to the **Combo Box Wizard** (figure 4.42). Click the **1st option** that reads, **I want the Combo Box to get the values from another table or query** (figure 4.42, arrow A).

Step 7: Click **Next** (figure 4.42, arrow B).

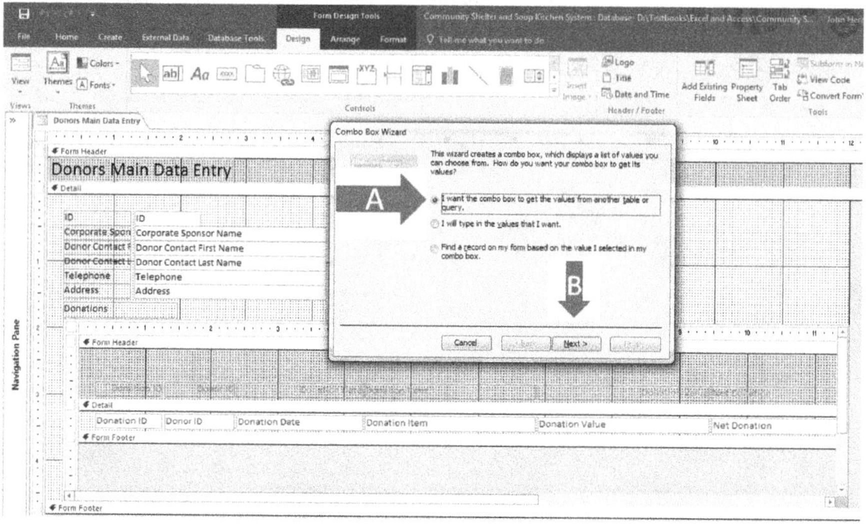

FIGURE 4.42 Creating a combo box that reads a table or query

Step 8: That will take you to the screen that you see in figure 4.43. The **table** that will be read by this **combo box** is the *zip codes* **table**, thus when it asks **Which table or query should be used for the values for your combo box?** make sure that the *zip codes* **table** is selected (figure 4.43, arrow A).

Step 9: Click **Next** (figure 4.43, arrow B).

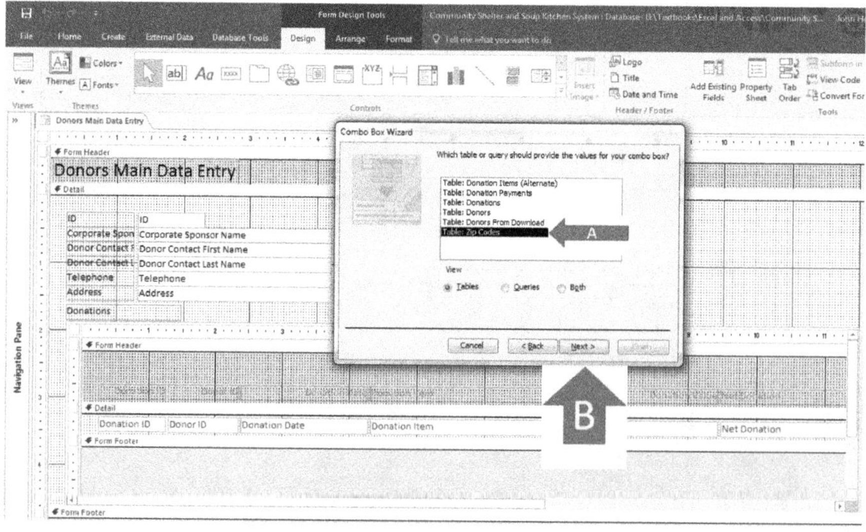

FIGURE 4.43 Creating a combo box that reads a table or query

Step 10: That will bring you to the menu that you see in figure 4.44. It is asking what fields you need to be seen in the **combo box**. In this example, **ALL** of them are needed, therefore, click the **double-arrow** (figure 4.44, arrow) to transfer them from the **Available Fields** window to the **Selected Fields** window.

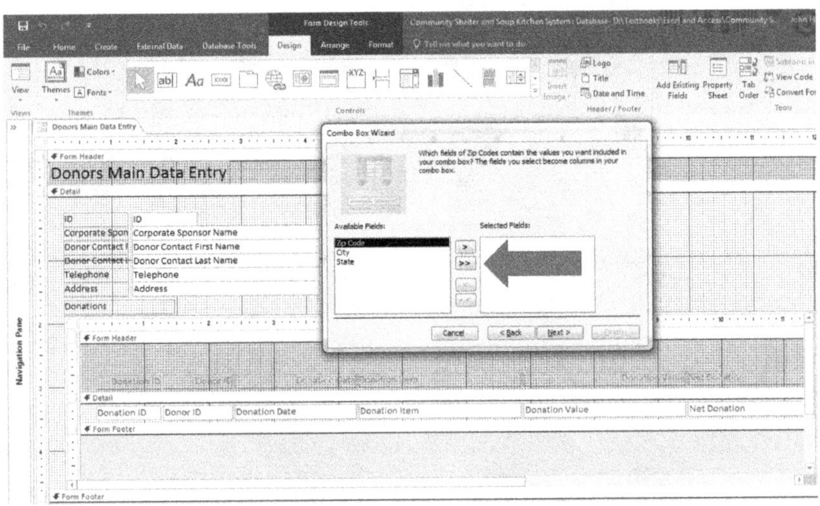

FIGURE 4.44 Creating a combo box that reads a table or query

When you click the **double arrow**, your screen will look like figure 4.45.

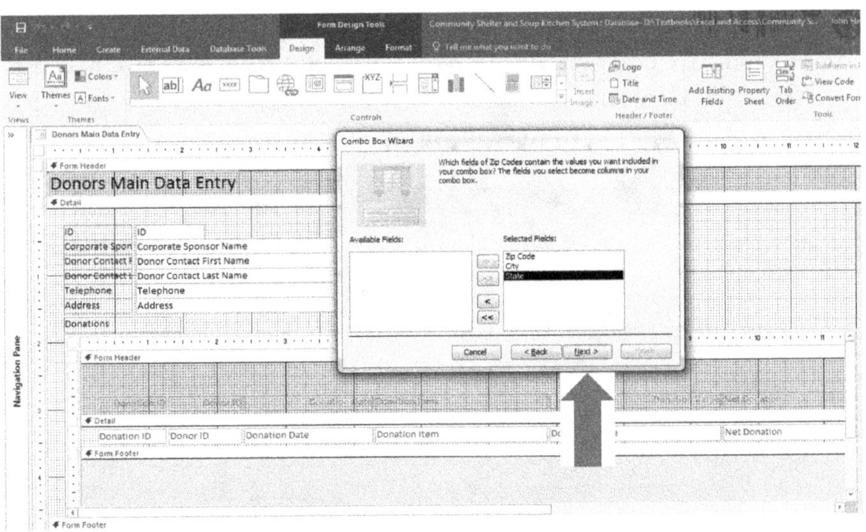

FIGURE 4.45 Creating a combo box that reads a table or query

Step 11: Click **Next** (figure 4.45, arrow).

Step 12: That will take you to the screen you see in figure 4.46. Make sure that you choose the *zip code* field as the one by which you will sort (figure 4.46, arrow A).

Step 13: Indicate the **sort order** that you want for the *zip code* field. In this example, use the default setting of **ascending order** (figure 4.46, arrow B).

Step 14. Click **Next** (figure 4.46, arrow C).

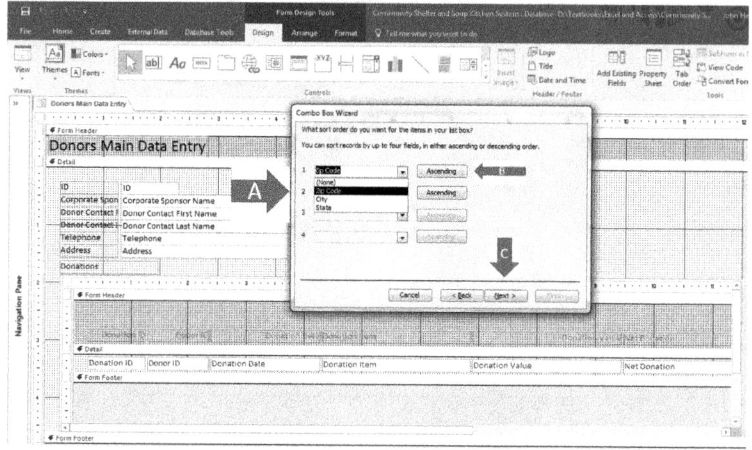

FIGURE 4.46　Creating a combo box that reads a table or query

Step 15: That would take you to the screen you see in figure 4.47. This would be a time that you do **NOT** want to leave the **Hide Key Column** box checked. The *zip code* field is the **key column**. If you hide it, you will not be able to see the *zip codes* at all when you click the box. Therefore, you will need to **UNCHECK** the **Hide Key Column** box (figure 4.47, arrow A).

Step 16: Resize the columns to look more like what you see in figure 4.47.

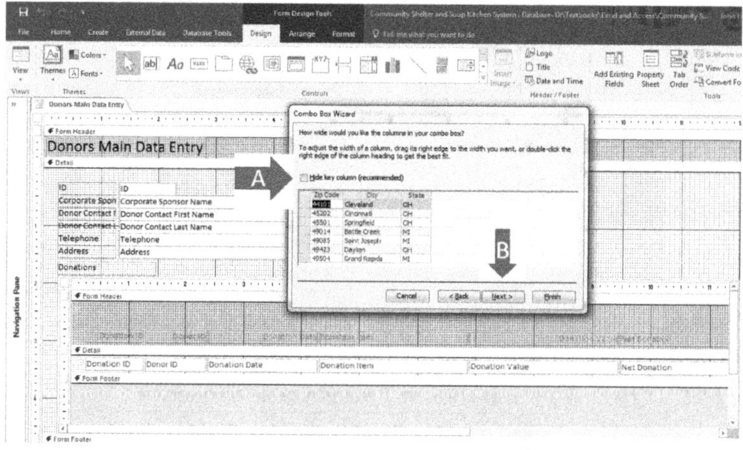

FIGURE 4.47　Creating a combo box that reads a table or query

Step 17: Click **Next** (figure 4.47, arrow B).

Step 18: That will take you to the screen you see in figure 4.48. It will ask you what field from the *zip code* **table** is to be sent to the *donors* **table**. Of course, the *zip code* field is what must be sent, and it is already selected, thus you need only to click **Next** (figure 4.48, arrow).

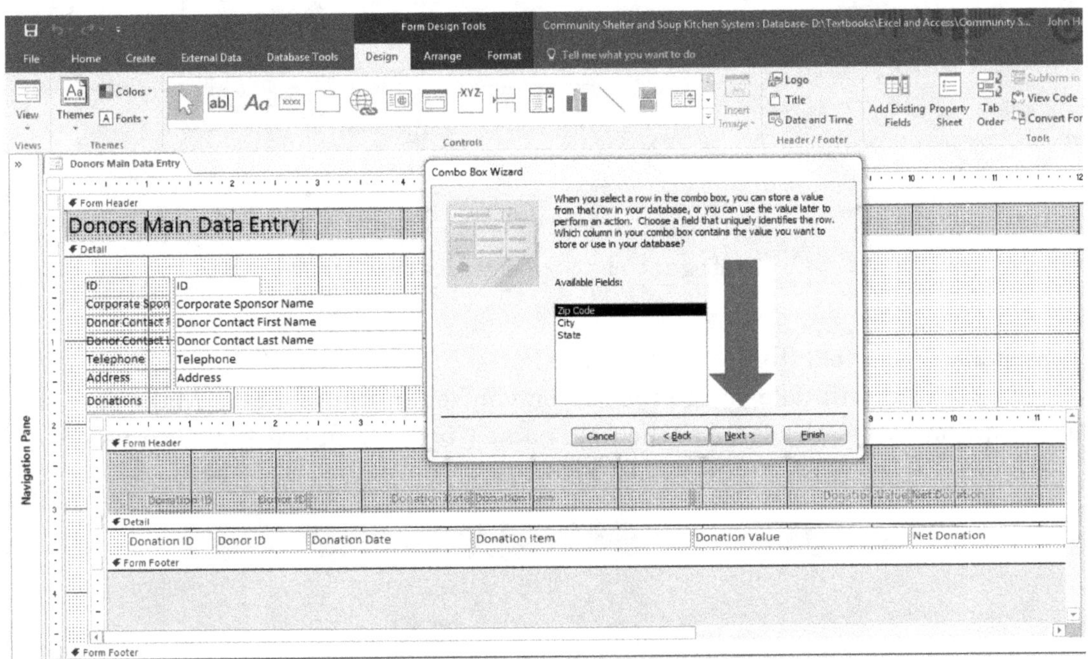

FIGURE 4.48   Creating a combo box that reads a table or query

Step 19: This will take you to the screen you see in figure 4.49. As before, THIS STEP IS CRITICAL!! IF YOU DON'T DO THIS STEP PROPERLY, THE COMBO BOX WILL NOT WORK PROPERLY AND WILL BE OF NO USE! THE DEFAULT IS SET WRONG! DO NOT USE THE "REMEMBER THE VALUE FOR LATER USE" OPTION, WHICH IS THE DEFAULT. YOU MUST CHOOSE THE FIELD TO WHICH THE CHOICE WILL BE SENT AND PLACE IT IN THE BOX LABELED "STORE THAT VALUE IN THIS FIELD" AND MAKE SURE THAT BOX IS CHECKED (figure 4.49, arrow A). For example, if the user chooses *zip code*, you must designate in this screen that the *zip code* you choose in the combo box must be entered into the *zip code* field in the *donors* table (figure 4.49, arrow B).

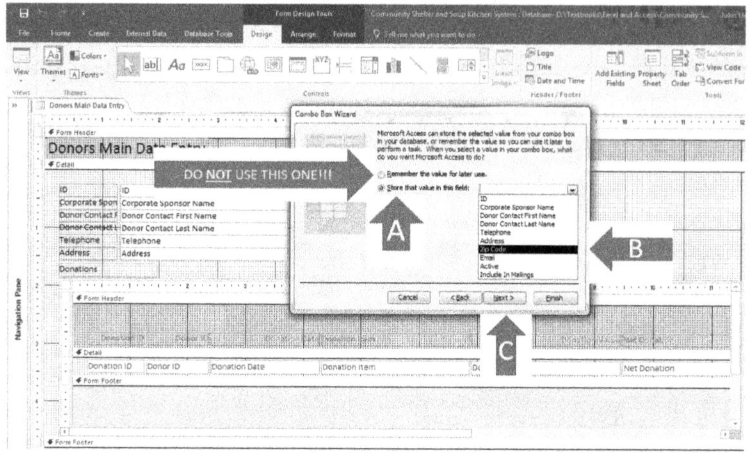

FIGURE 4.49   Creating a combo box that reads a table or query

Step 20: Click **Next** (figure 4.49, arrow C).

Step 21: That will take you to the screen you see in figure 4.50a. Make sure the label caption will be the same as the field name (*zip code*) (figure 4.50a, arrow A).

Step 22: Click **Finish** (figure 4.50a, arrow B).

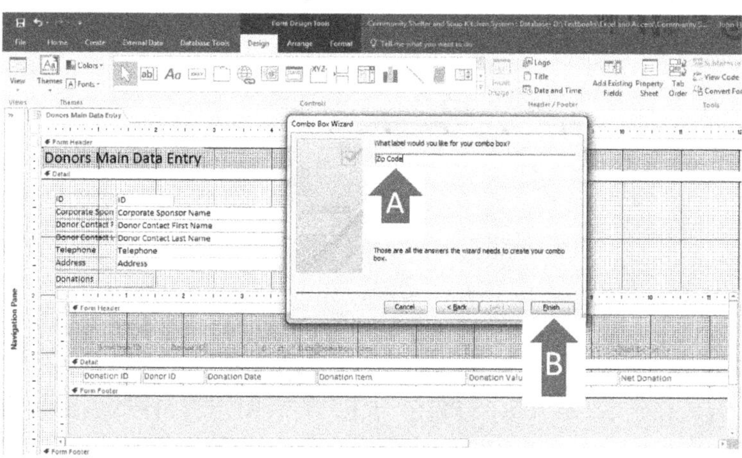

FIGURE 4.50a   Creating a combo box that reads a table or query

NOTE: The new combo box you have just placed into the form IS NOT TO BE UNBOUND. AGAIN, IF IT IS, THE DATA YOU ENTER WILL NOT BE SAVED AND THE COMBO BOX WILL BE OF NO USE! INSTEAD OF IT BEING UNBOUND, THE NAME OF THE FIELD TO WHICH THE ENTERED DATA IS STORED SHOULD BE VISIBLE IN THE COMBO BOX. In this case, that would be *zip code* (figure 4.50b, arrow).

FIGURE 4.50b   Creating a combo box that reads a table or query

Step 23: Return to the **form view** and test the **combo box**. It should look like what you see in figure 4.51.

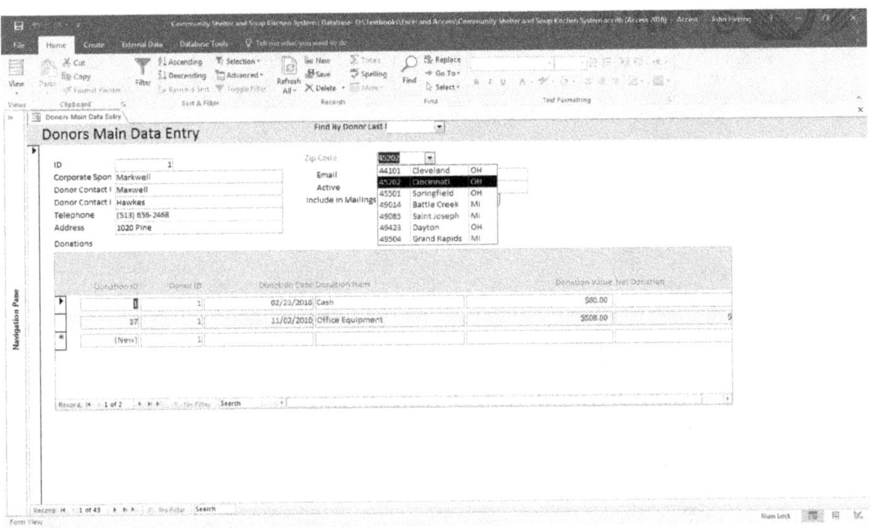

FIGURE 4.51   Creating a combo box that reads a table or query

MODIFYING AND ADDING A SUM CALCULATION TO THE SUB-FORM

While you are in the design view of the **main form,** many of the types of changes we have made can also be made in the **sub-form.** However, it is often cumbersome to do so while in the **main form.** When making such changes, it is often best to close the **main form** and then open the **sub-form** in the **design view** by itself.

NOTE: You cannot open the **sub-form BY ITSELF** while the **main form** is still open.

To open the **sub-form** (or any form) in the **design view**:

Step 1: Go to the **navigation pane**. If it's not already open, press the **F11 key** (figure 4.52)

Step 2: Right click on the *Donations sub-form* (figure 4.52).
Step 3: That will give you a **short-cut menu** (figure 4.52, arrow).

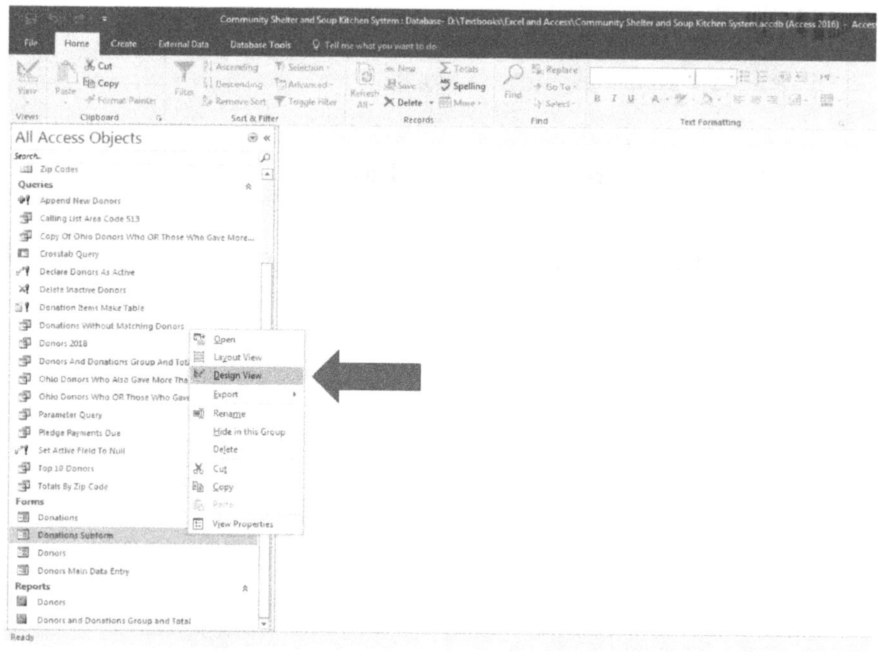

FIGURE 4.52   Changing form design and adding a sum

Step 4: Click **Design View** (figure 4.52, arrow). That will take you to the **form design view** as you see in figure 4.53.

SHRINKING CONTROLS AND THEIR LABELS

At this point, we see that many of the fields and their labels are much longer than they have to be, causing a user to have to scroll to the right or left to see all of the fields. You can shrink them by performing the following steps:

Step 1: Place your mouse in the **ruler bar** (figure 4.53) directly above the **control** and **label** you want to select until you see a tiny arrow that points downward (figure 4.53). It will select anything to which the arrow points.

NOTE: You can also click and drag the tiny arrow right or left across the **ruler bar** at the top of the form. That would select the area of the **grid** in the **form design** to which the arrow is pointing. This is an alternate means of selecting numerous controls in the **form design**. You can also click on a control and then press the **Shift key** as you click on as many other controls that you wish, which will select all of those you click. You can also place your mouse on the vertical **ruler bar** on the left of the form and click the tiny arrow to select anything in the **form design** to which the arrow is pointing. If you do so, and click and drag the arrow up and down, it will also select everything to which the arrow pointing. If you press, **Ctrl A** on your keyboard, that will select everything in the form.

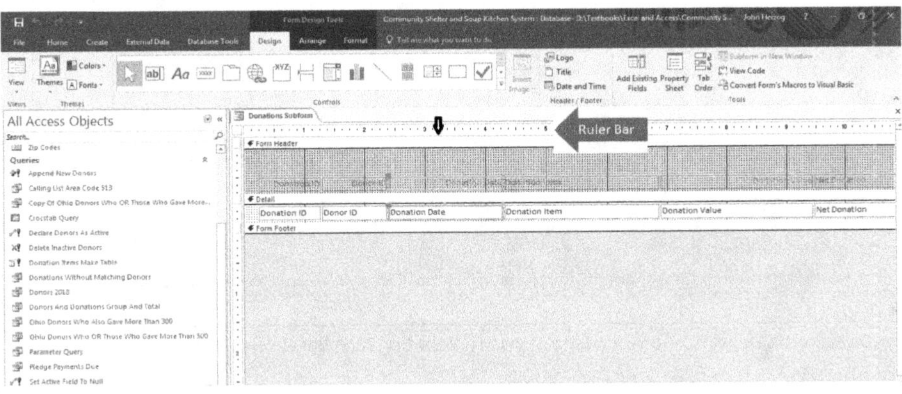

FIGURE 4.53   Changing form design and adding a sum

Step 2: Press and hold the **Shift key** and press the **left cursor control arrow key on your keyboard** ← until the **controls** and **labels** shrink the size you want. Shrink them as much as you can without covering their captions. Repeat this process for all of the other **controls** and their **labels** (figure 4.54).

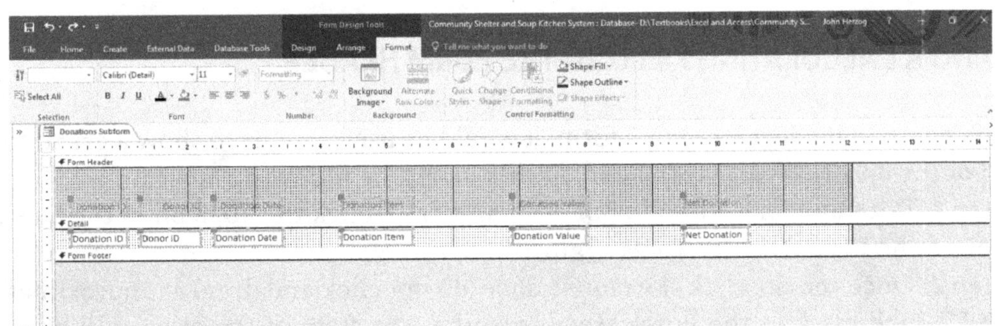

FIGURE 4.54   Changing form design and adding a sum

At this point, using the methods already discussed in this chapter, let us now do the following:

1. Make the fonts of those **controls** to color of **black**.
2. Align the fonts to the left, by using the **Arrange**, **Left** menu.
3. Shore up the **detail** area of the form from the bottom and from the left.

After accomplishing these things, your **sub-form** should look like figure 4.55.

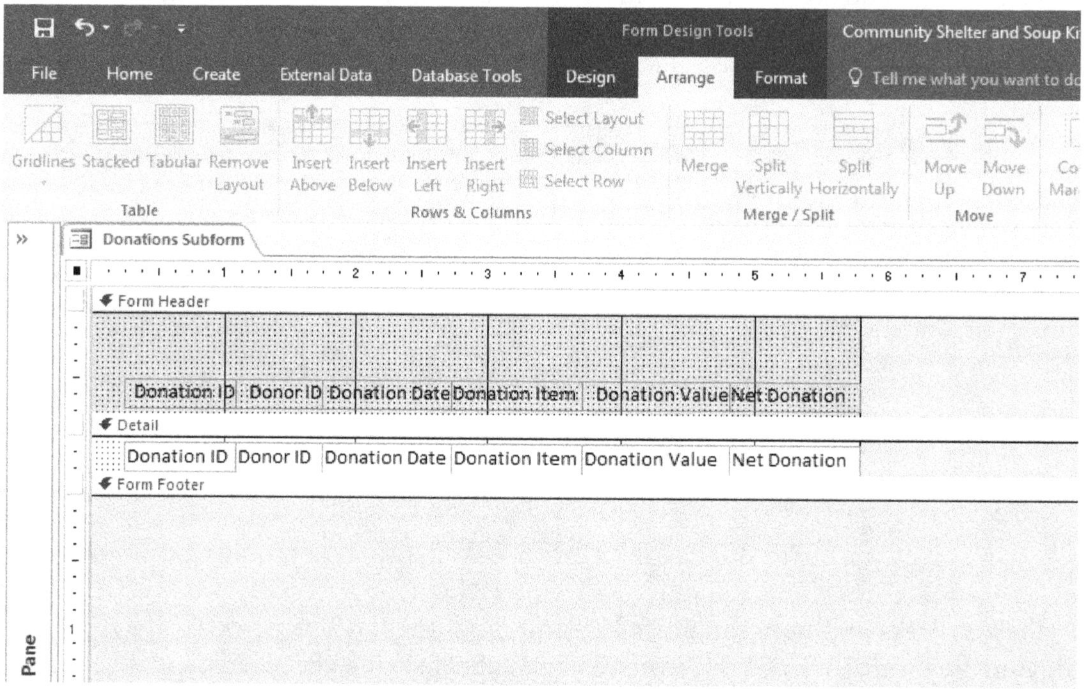

FIGURE 4.55   Changing form design and adding a sum

ADDING A CALCULATING FIELD TO THE FORM HEADER

At this point, in the **header** of this form, it would be useful for us to show a grand total of all of the *donation values*. You can do this using the following steps:

Step 1: With only one click (do not click and drag) click the **ab|** icon in the **Design** ribbon in the **Form Design Tools** (figure 4.56, arrow).

Step 2: Once you do, click the mouse once (do not click and drag) at approximately the **3.75-inch** mark in the **Form Header** so that the stem of the cross will be even with the edge of the **Donation Value** field **label** (figure 4.56).

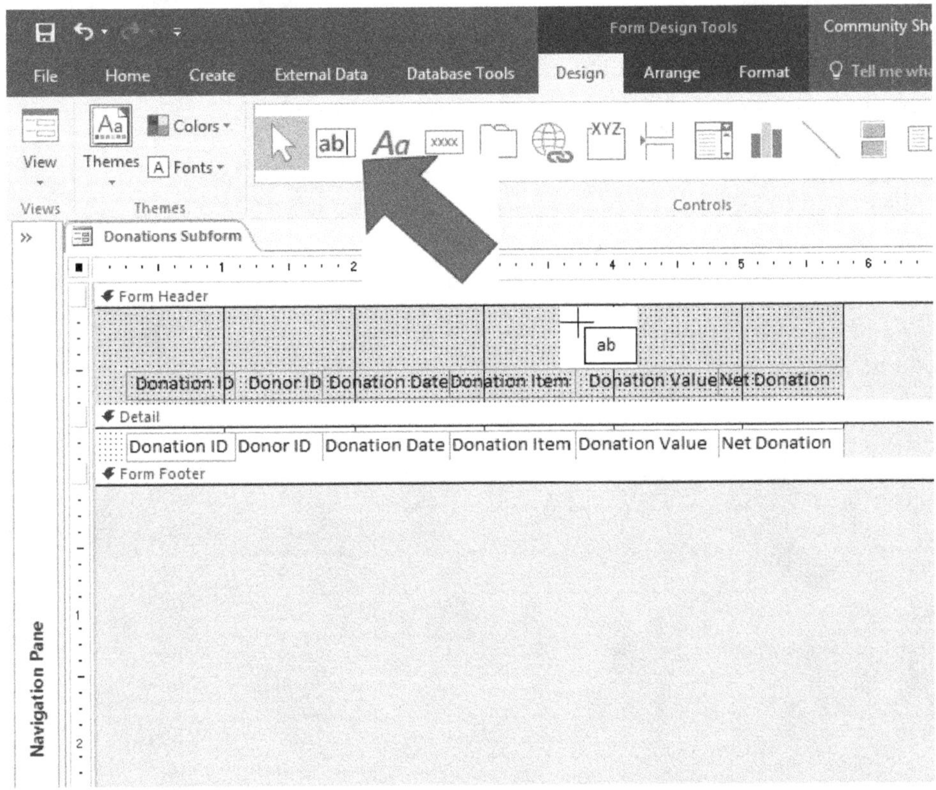

FIGURE 4.56   Changing form design and adding a sum

Step 3: When you do, it will give you an **unbound field** (or **unbound text box**) in the **Form Header** where you clicked (figure 4.57).

Step 4: Click inside the **unbound field** once or twice until it will allow you to enter text into it. Type the following expression inside the **text box**:

=SUM([Donation Value])

NOTE: Access uses many functions such as this that are also used in **Microsoft Excel** and in many cases, they work the same. The only difference is that Access will ask you to use a field name, (surrounded by square brackets) as an argument (rather than a cell range as Microsoft Excel would do).

Step 5: As shown in chapter 3 (Figures 3.10 and 3.11), you can also click on the **unbound field** and open the **properties sheet**. There you can put in a formula such as the one here, and the principals for doing so are the same. The option of changing the **format** to **currency** using the same methods is also available. You can also press and hold the **Shift key** and tap the **Right Cursor Control Arrow key** → on your keyboard in order to widen the field to your liking. You can apply the **black** font color to it as well.

Step 6: Click inside the label to the **unbound field** and change the caption to read *Total Donations*. Apply the **black** font color to it also. Your screen should look a lot like figure 4.57.

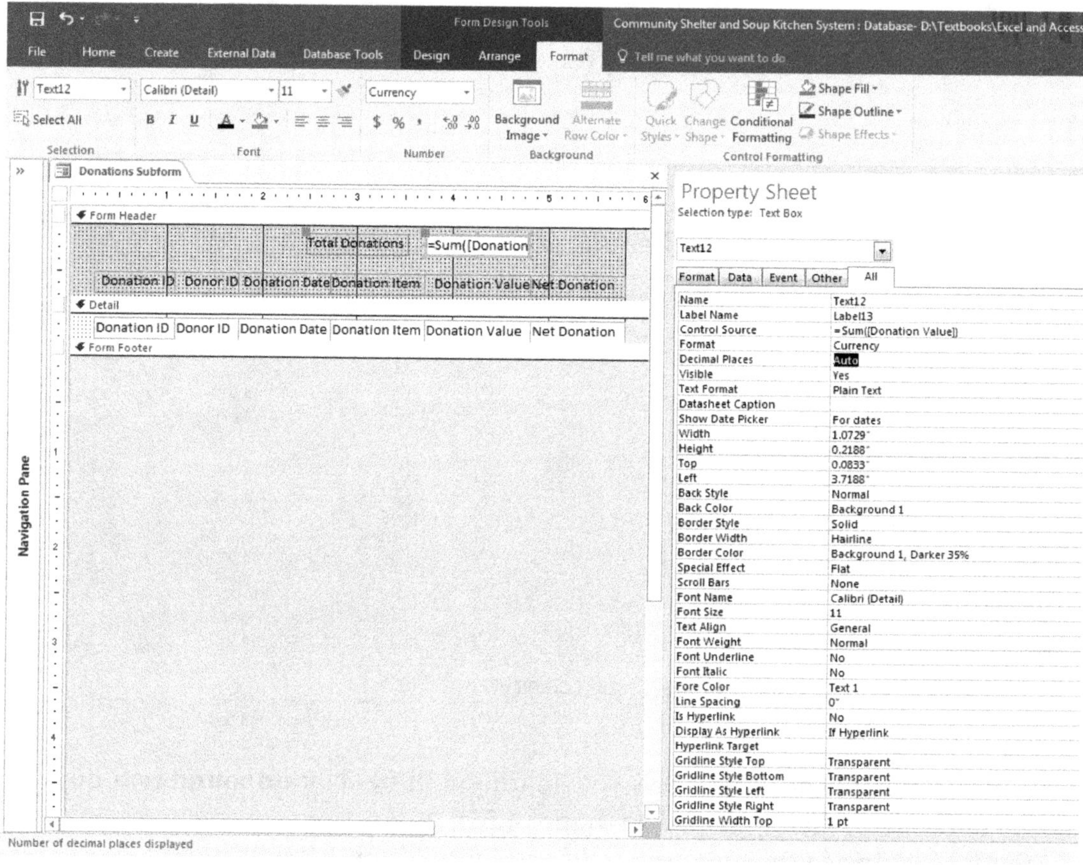

FIGURE 4.57   Changing form design and adding a sum

Step 7: Return to the **form view** and see the results (figure 4.58).

NOTE: You cannot see the results of a calculation without viewing the **form** in the **form view** or the form **layout view**. In the same way, you cannot see the results of a calculation in a **report** without viewing the **report** in the **report layout view**, the **report view**, or the **print preview**.

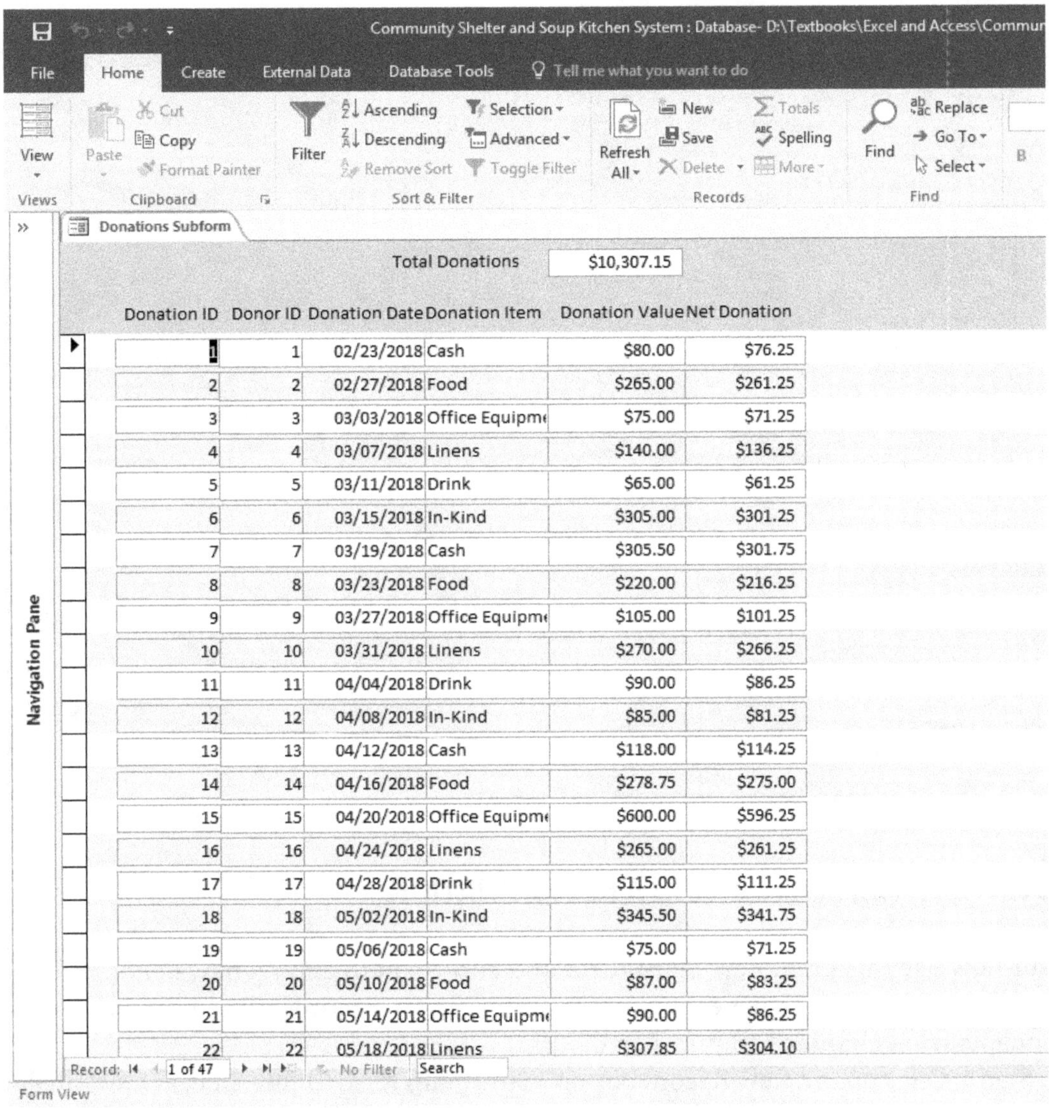

FIGURE 4.58   Changing form design and adding a sum

Please notice that the *total donations* calculated field you just created tallies the *donation value* of **ALL** donations. Later, when you view the **main form** again, that will change.

Step 8: Close the form and **save** the form when prompted.

Step 9: Open and view the **main form** named *Donors Main Data Entry* (figure 4.59).

The *donation totals* are now reflecting the *donations for the donor* **OF THE CUR-RENT RECORD**. Notice also how much cleaner the **sub-form** has become as a result of the resizing and realigning we did earlier.

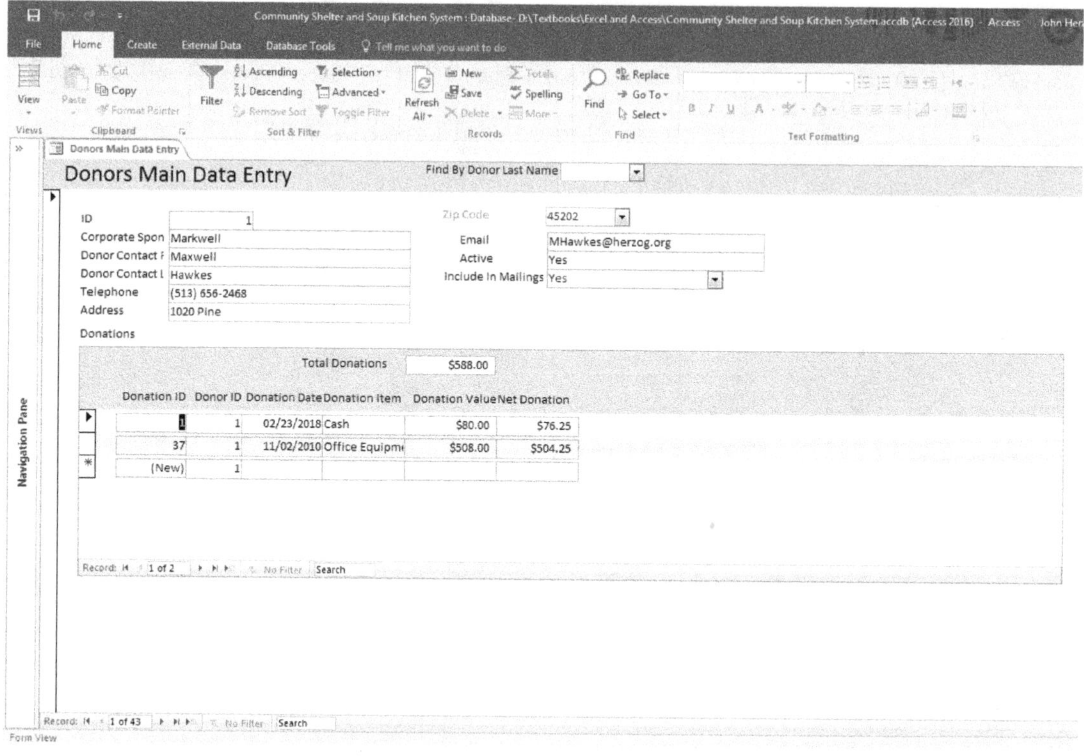

FIGURE 4.59 Seeing a subform sum beneath each record

TO ENTER A NEW DONOR AND A NEW DONATION

Let's now use this great new **main form** and **sub-form** to enter a new *donor* and two new *donations* from that new *donor*. To enter a new **record**, please remember the concepts from chapter 1 (figure 1.16a).

Step 1: Click on the New Record button in the main form (figure 4.60, arrow Labeled *Click Here To Add New Donor*). YOU MUST START A NEW DONOR FIRST. IF YOU DON'T, THE DONATION WILL NOT BE ATTRIBUTED TO ANY DONOR.

NOTE: The **navigation buttons** in a **form** work the same as they do in a **table**.

Step 2: Enter the *donor contact last name* as *Downing* and enter the *donor contact first name* as *Emma*. Notice as you do that that the **sub-form** has automatically set the default of the *donor ID* for her first donation with her *ID* number (figure 4.61, arrow).

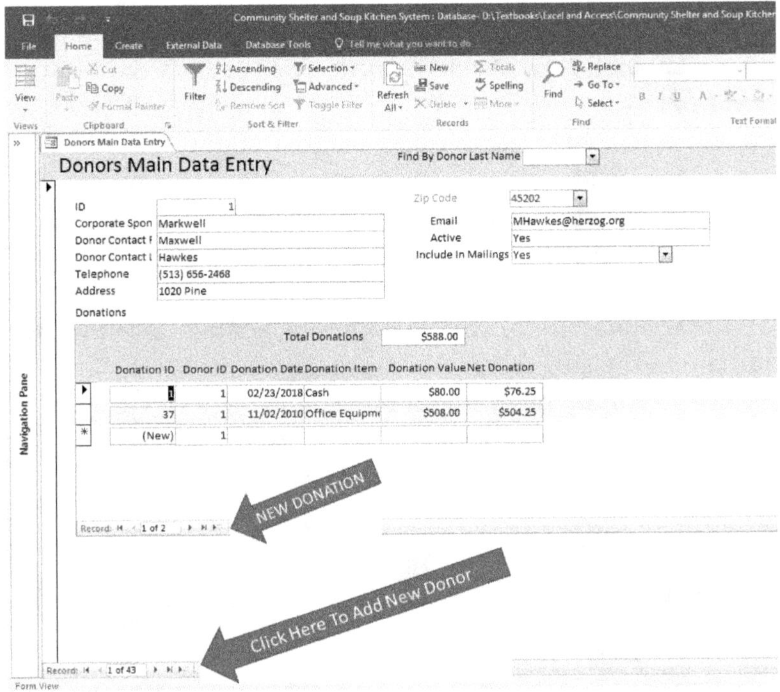

FIGURE 4.60   Adding records in a main and subform

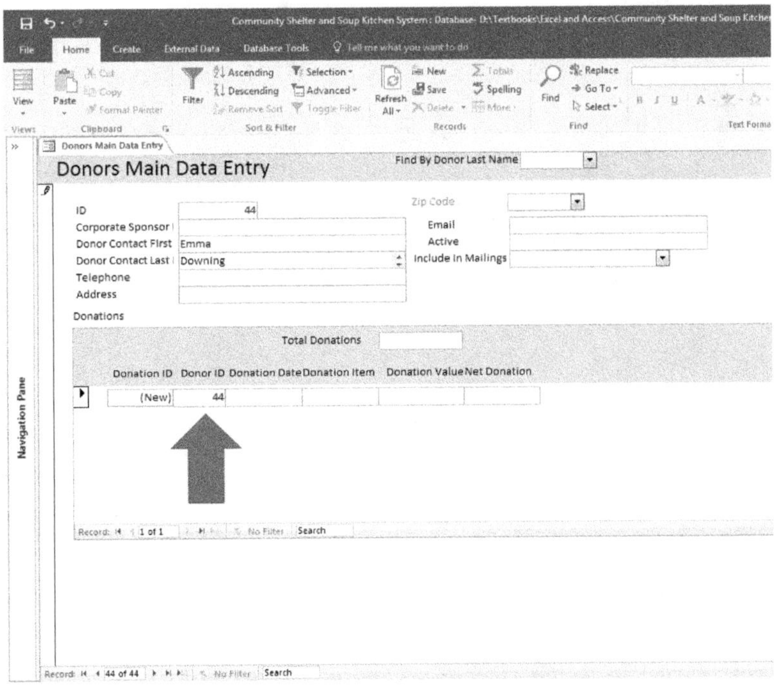

FIGURE 4.61   Adding records in a main and subform

Step 3: Start a new *donation* by clicking the **New Record** icon inside the **sub-form** (figure 4.60, arrow Labeled *New Donation*) and enter today's date into the *donation date* field (figure 4.62).

NOTE: If you press and hold the **Ctrl** key and then tap the **Colon** (:) key while you are in the *donation date* field, Access will automatically enter today's date.

Step 4: Click in the *Donation Item* field and enter *Bible* (figure 4.62).

Step 5: Click in the *Donation Value* field and enter *$20*. Notice that the *net donation* will always be calculated (figure 4.62).

Step 6: Start a new *donation* for this same donor, *Emma Downing*, by clicking the **New Record button** in the **sub-form**, which is also marked in figure 4.60.

Step 7: Click in the *Donation Date* field and enter today's date again (figure 4.62).

Step 8: Click in the *Donation Item* field and enter *Cash* (figure 4.62).

Step 9: Click in the *Donation Value* field and enter *$100* (figure 4.62).

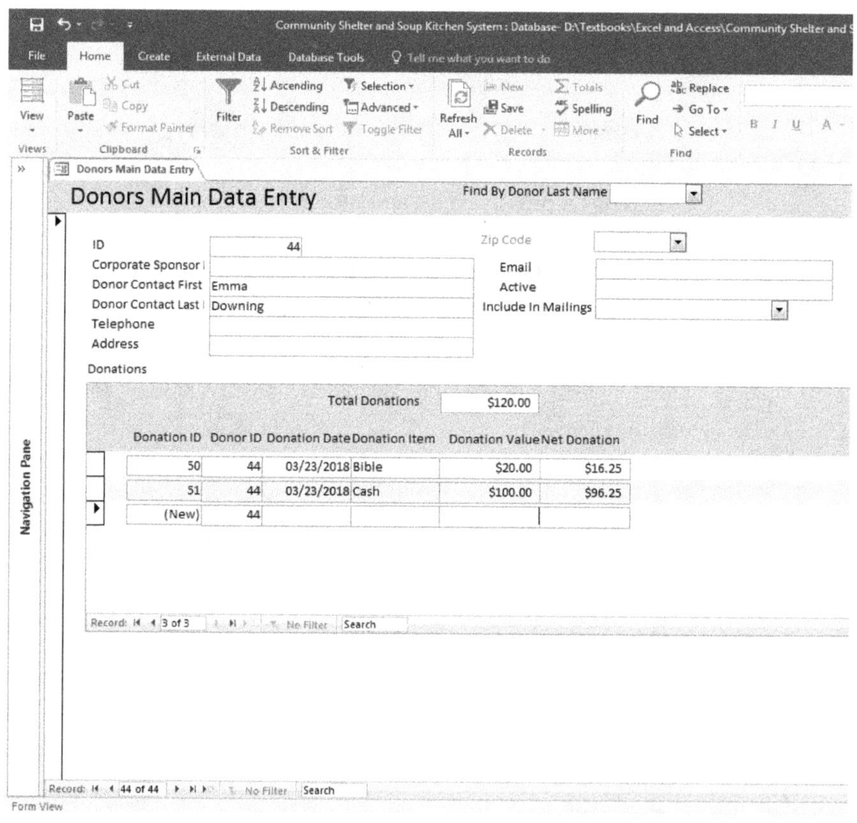

FIGURE 4.62 Adding records in a main and subform

Step 10: Close the donation main data entry form.

AN ALTERNATE WAY TO ADD A COMBO BOX (DROP-DOWN LIST BOX) THAT LOOKS UP DATA FROM ANOTHER TABLE

Suppose we now want to use the *Item* table in a **combo box (drop-down list box)** in the *Donations* **sub-form**. It can be done in a different way without using the **Combo Box Wizard**. You can do so with the following steps:

Step 1: Open the *donations sub-form* in the **design view**.

Step 2: Right-click on the *Donation Item* field and choose the **Change To** option in the **shortcut menu** that is then displayed (figure 4.63, arrow A).

Step 3: Choose **Combo Box** in the submenu that is then displayed (figure 4.63, arrow B).

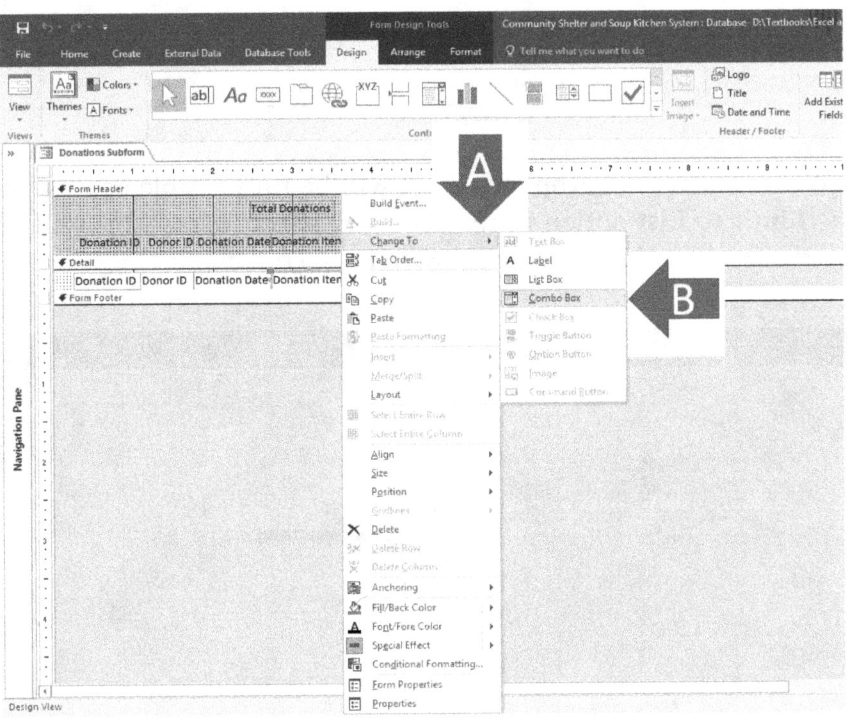

FIGURE 4.63   Alternate way of making a combo box

Step 4: When you do, it will then have been converted to a **combo box** (figure 4.64, arrow A).

Step 5: While the field is still selected click **Property Sheet** (figure 4.64, arrow B).

Step 6: Make sure that the **property sheet tab** labeled **Data** is selected and click **the drop-down list box** that will give you a list of all of the queries and tables in this database (figure 4.64, arrow C). Select the **Table** named *Donation Items* (figure 4.64, arrow D).

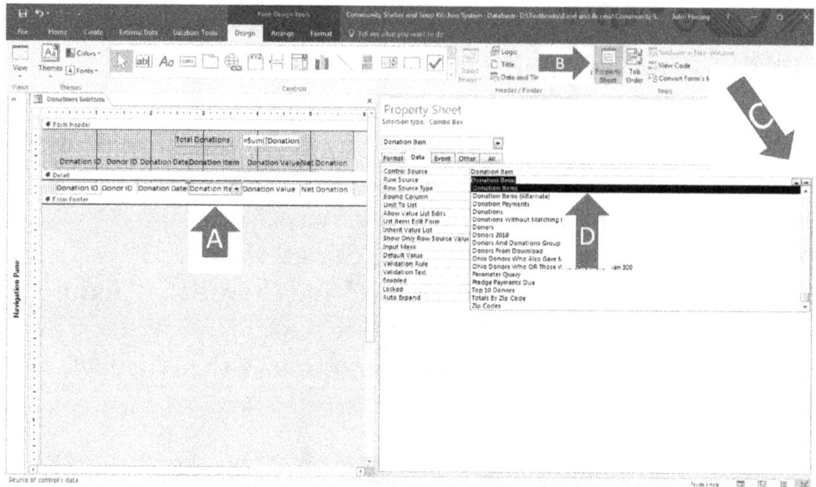

FIGURE 4.64   Alternate way of making a combo box

Step 7: When you do, the **property sheet** will be fully visible again (figure 4.65). Change the **Limit to List** option in the **property sheet** to **Yes** (this will make sure that **ONLY** the items in that list can be entered into this field).

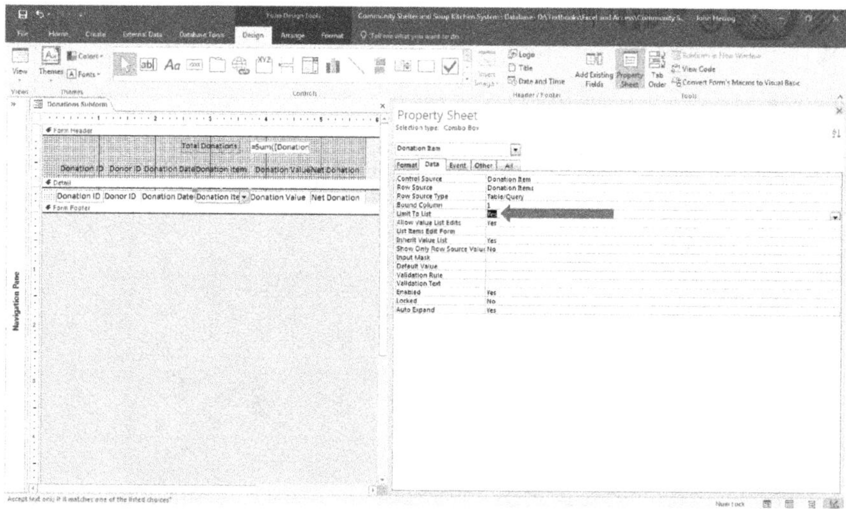

FIGURE 4.65   Alternate way of making a combo box

Step 8: Close and save the *donations sub-form* and return to the **form view** of the *donation main data entry* in order to see the list and how it now displays a list of all of the possible items that can be donated to the *community shelter and soup kitchen* (figure 4.66).

NOTE: Whenever you convert a field control to a **combo box** using this method, it will also change its **tab index** and it will have be adjusted. That means that as you press **Enter** or **Tab** while entering data in the **form**, the cursor will not go to the fields in the order you want. The way to adjust it, along with changing all **tab indexes** for a form's **tab order** is discussed next.

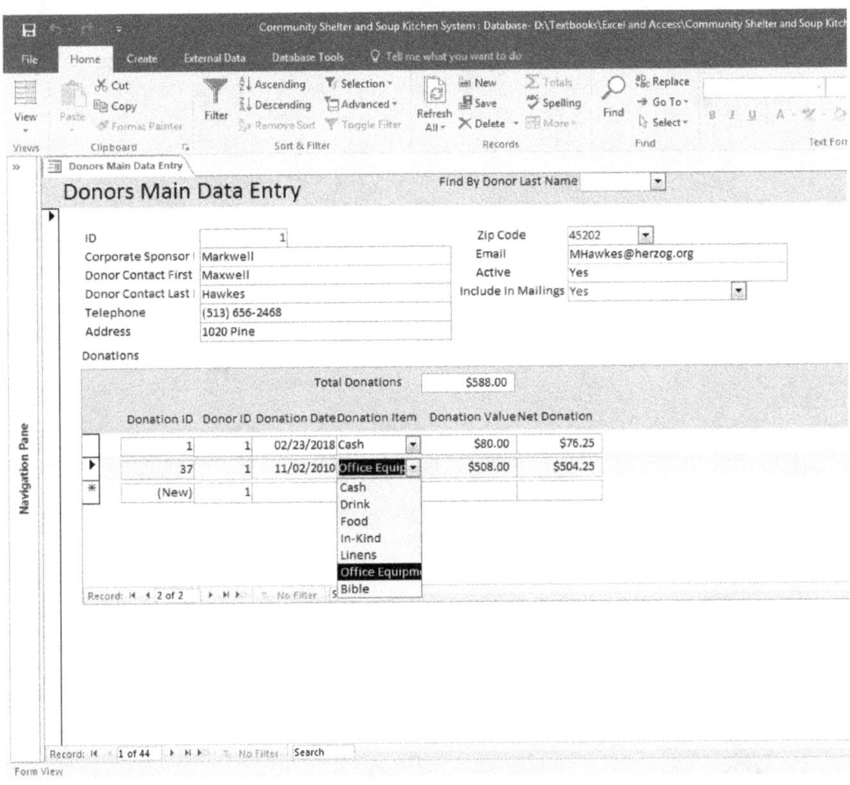

FIGURE 4.66 Alternate way of making a combo box

CHANGING THE TAB ORDER OF A FORM (METHOD 1)

As you enter data into a **form**, the **tab order** in a **form** determines where your cursor will go after you press **Tab**, **Enter**, or any one of the **cursor control arrow** →←↓↑ keys. It is very difficult to enter data if it does not move in an order that is easy to follow. By default, after creating the **form**, the **tab order** will often be in the order that the fields are in when they are sent to the **Selected Fields** window as discussed way back in figures 4.2 and 4.3.

After removing fields and replacing them with **combo boxes,** as we have done in this form, the last controls placed in the form will be the last fields that Access will tab to when users press the **Tab**, **Enter**, or **cursor control arrow** keys. Hence, the **tab order** in our form is not currently correct.

NOTE: When a form is first created, Access will often put the **tab order** in the complete opposite order that was described earlier. If they do, each field will have to have the **tab index** changed, because users will likely want the cursor to move to fields from left to right.

To change the **tab order** of fields:

Step 1: In the **form design view**, click on the first field and open the **property sheet** (figure 4.67).

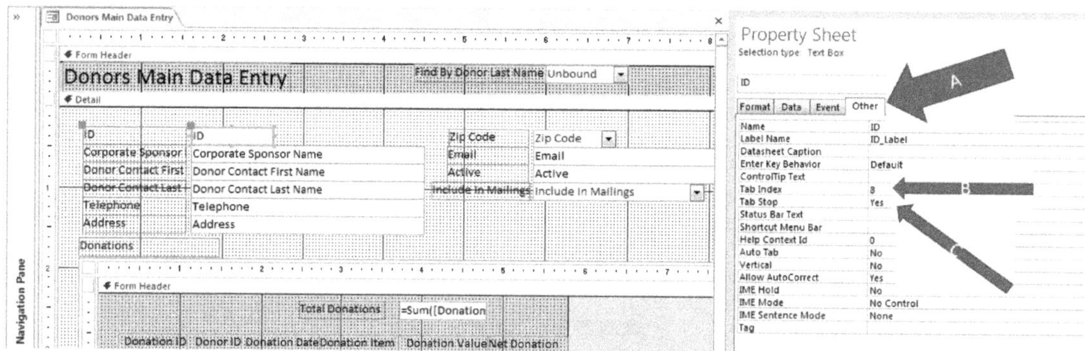

FIGURE 4.67 Changing tab order method 1

Step 2: In the **property sheet**, click the **Other** tab (figure 4.67, arrow A).

Step 3: If the field you are in is the first one in which you want the cursor to be when you start entering data in the form, the **tab index** needs to be set to *zero (0)* (figure 4.67, arrow B).

Step 4: If the field you are in is the second one in which you want the cursor to be when you are entering data in the form, the **tab index** needs to be set to *one (1)*. This process must then continue until all fields are in the correct **tab order**.

NOTE: You would not likely want the cursor to stop at some fields or controls, such as **calculated fields** or **auto-number** fields, because data cannot be entered or changed in such fields. Therefore, it is best to tell Access that you don't want to allow such fields to have a **tab stop**. To do this, change the **Tab Stop** option in the **Other Tab** from **Yes** to **No** (figure 4.67, arrow C).

CHANGING THE TAB ORDER OF A FORM (METHOD 2)

Using Method 1 to change a **form's tab order** would take a very long time if your **form** consists of dozens of different fields. Therefore, another way to change the **tab order** of a **form** is to do the following:

Step 1: Open the **form** in the **design view**.

Step 2: Right click anywhere on a blank area of the **design grid** in the **detail area** of the **form** (figure 4.68).

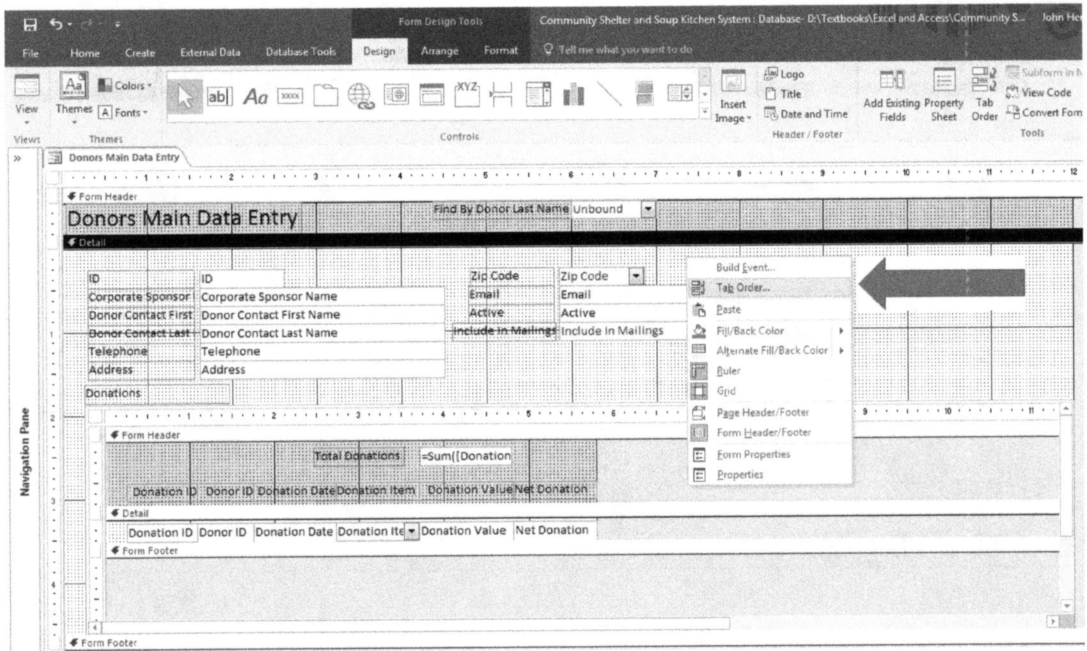

FIGURE 4.68   Changing tab order method 2

Step 3: In the **shortcut menu** that will appear, click **Tab Order ...** (figure 4.68, arrow)

The **Tab Order pop-up menu** will then appear (figure 4.69a). Click on the tiny gray box to the left of the field name whose **tab order** you want to change, and then click and drag the field above the field name of what it should precede, or below the field name you want it to follow (figure 4.69a, arrow).

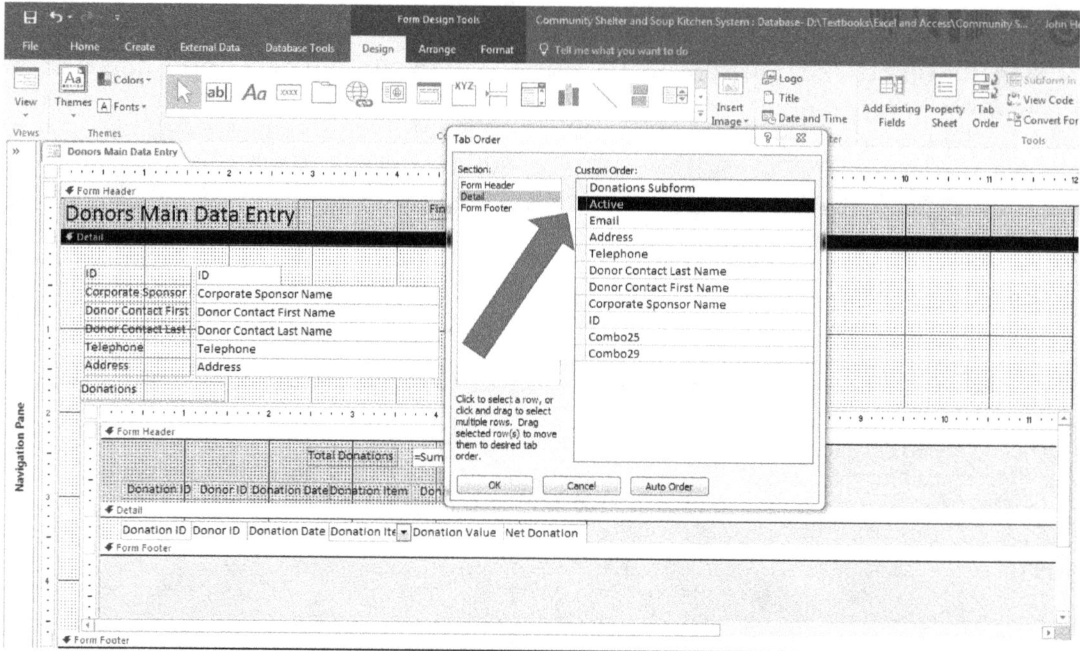

FIGURE 4.69a   Changing tab order method 2

NOTE: The two **combo boxes** that were added earlier in this chapter are not named after their field name. In this example, Access automatically named them for you as *Combo25* and *Combo29*. If you want to rename them as their field names are, they can be changed in the **property sheet** by you in the **Name Property box** (figure 4.69b, arrow). For example, figure 4.69b shows a **text box** named *Text12*. To change it, you can click in the **Name** property box and name the control for the field whatever you like, **AS LONG AS THERE ARE NO OTHER CONTROLS USING THE SAME NAME**. It is not necessary to rename the label.

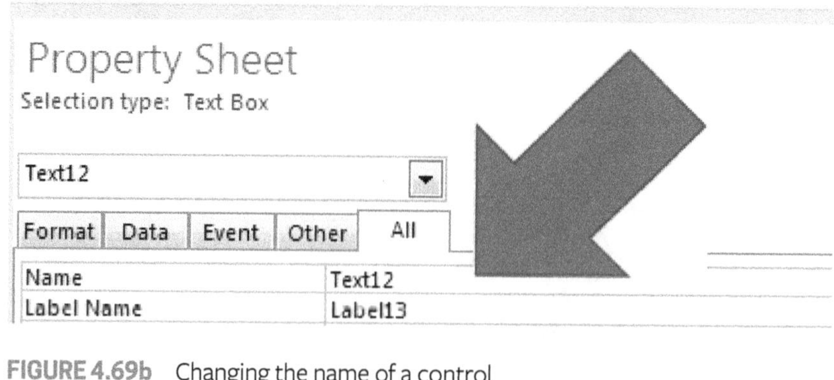

FIGURE 4.69b   Changing the name of a control

NOTE: The **tab order** for **form footers** and **form headers** must be done separately as they will each have their own **tab order**. Ordinarily, **page headers** and **page footers** for forms do not need a **tab order**, because you cannot enter data in them. If your headers or **footers** have **command buttons**, users often want them to have a specific tab order. That will be discussed in a later chapter. **Page headers** and **page footers** for forms will only appear in a **printout** of a form and do not need a **tab order**.

TO APPLY AN INPUT MASK

To eliminate the need for entering parenthesis around area codes for telephone numbers, dashes for Social Security numbers, slashes for dates, and other types of typical data entry redundancies, you can apply what is called an **input mask** to a field. You can apply it to a field in the **properties** area of a table's design or a form's design. If you apply it in the **table's design**, Access will automatically apply it to that field in all subsequent **forms** and **reports** reading the table with that field. If you add the **input mask** to a field in the **table** after the **form** is created, however, it would not show in the **forms** that are already reading that table unless you apply the **input mask** separately in each of those **forms**. In this example we will use the *telephone* field for the *donor main data entry* form, and we will do it using the following steps:

Step 1: Open the *Donors Main Data Entry* form in the **design view**.

Step 2: Click on the *Telephone* field control (but do not select its accompanying label) (figure 4.70, arrow A).

Step 3: Open the **property sheet** in the **design view** tab of the **form design tools** (figure 4.70, arrow B).

Step 4: Click the **Data** tab in the **property sheet** (figure 4.70, arrow C).

NOTE: Other properties can be changed in the **property sheet** of a **form** that can also be set in the **table properties**, such as **default values** and **validations rules** as you can see in figure 4.70.

Step 5: Click the **three-dotted box** (often called the **Wizard/Expression Launcher**) that is inside the box that is labeled **Input Mask** (figure 4.70, arrow D).

Step 6: When you do, the **Input Mask Wizard** will appear. Make sure the **Telephone** option is already chosen in the **Input Mask/Data Look** window (figure 4.70, arrow E).

Step 7: Click **Finish** (figure 4.70, arrow F).

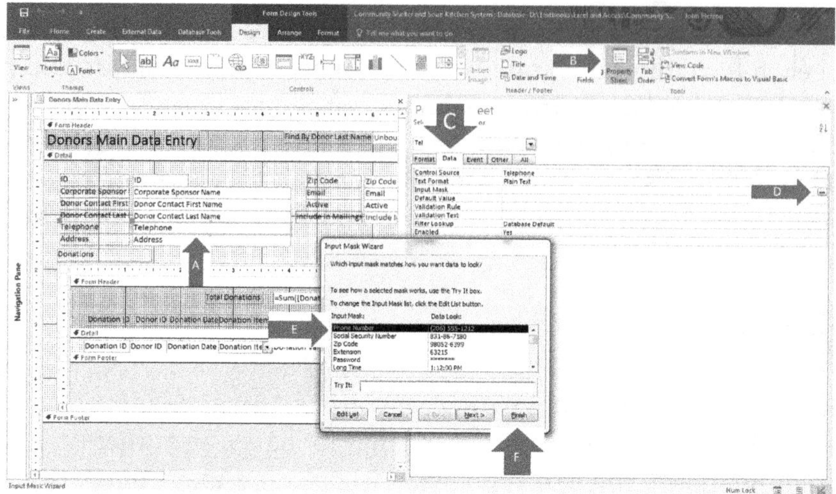

FIGURE 4.70   Input masks

Step 8: Notice that there is coding entered into the **Input Mask** box that will put the parentheses, spaces, and dashes where you would want them in a given telephone number (figure 4.71, arrow).

NOTE: If parentheses, spaces, or dashes are already entered in the phone numbers of the field, the **input mask** you create will **ONLY** apply them automatically when you type a **NEW** telephone number in the field or if you **RETYPE** the one that is already there. It will not remove the parentheses, spaces, or dashes that are already entered.

NOTE: If you get a flavor for what the coding is for a telephone **input mask**, you can type it in the **input mask** box freehand without using the **Input Mask Wizard**.

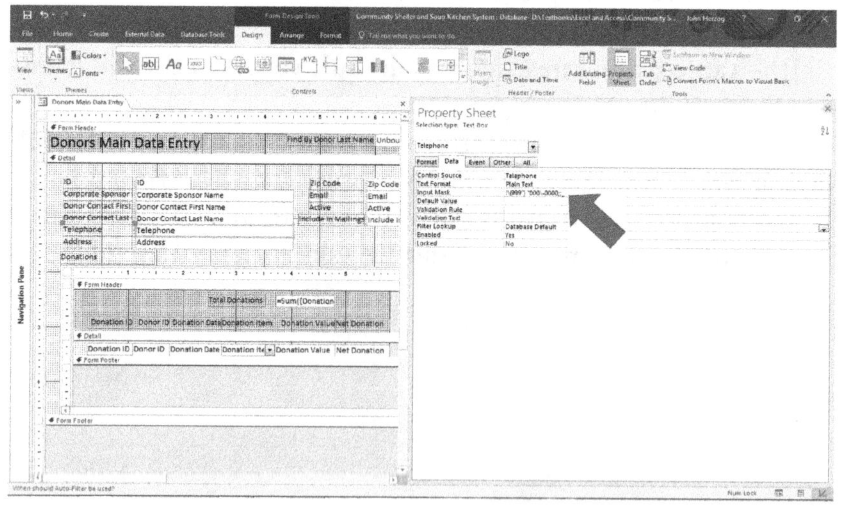

FIGURE 4.71   Input masks

Step 9: Return to the **form view** and enter a new *telephone* number in the *telephone* field of *Emma Downing* (figure 4.72).

NOTE: Access will begin to put the parentheses, spaces, and dashes in the field as soon as you click in it.

NOTE: When using an **input mask** you cannot leave any of the spaces blank.

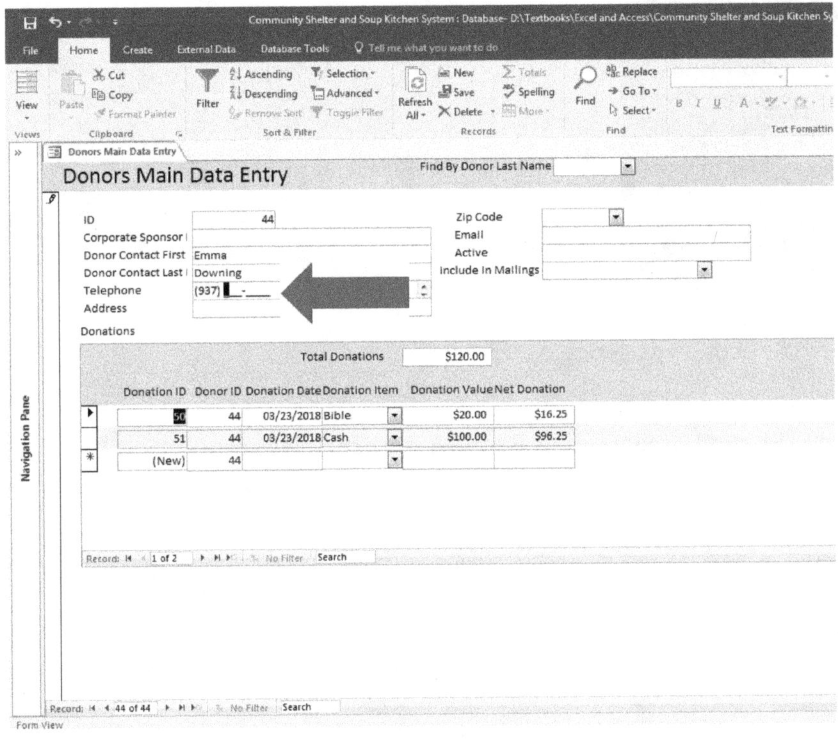

FIGURE 4.72   Input masks

ADDING A PASSWORD TO AN ACCESS DATABASE

A database is often full of information that you want to keep private, such as Social Security numbers, bank account numbers, and much more. A *donation* database is certainly going to have information that must also be kept secure. You may also want to keep people from editing your system who are not authorized to do so. If you want to secure the database with a **password**, do the following steps.

Step 1: Close the database **AND** Microsoft Access.

Step 2: Reopen Access.

Step 3: At the first screen, click **Open Other Files** (figure 4.73, arrow).

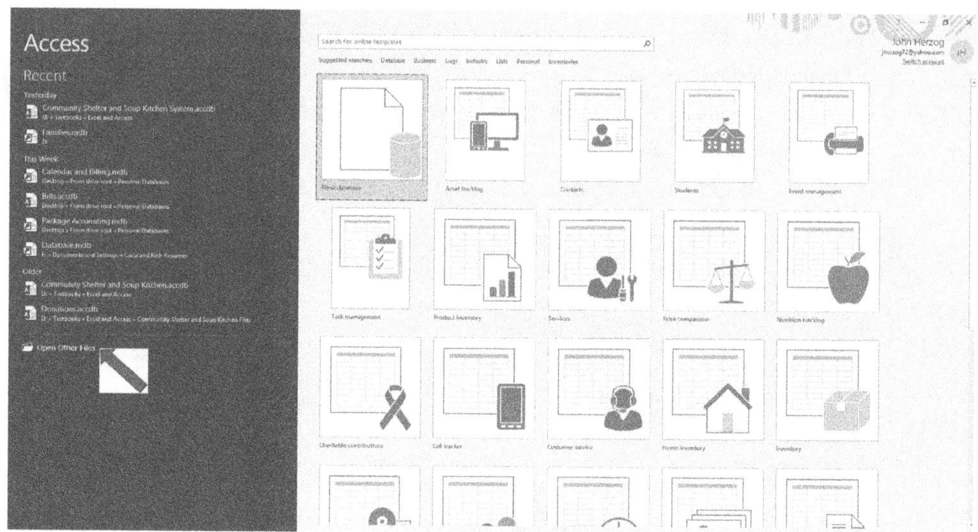

FIGURE 4.73   Adding a password to a database

Step 4: At the next screen click **Browse** (figure 4.74, arrow).

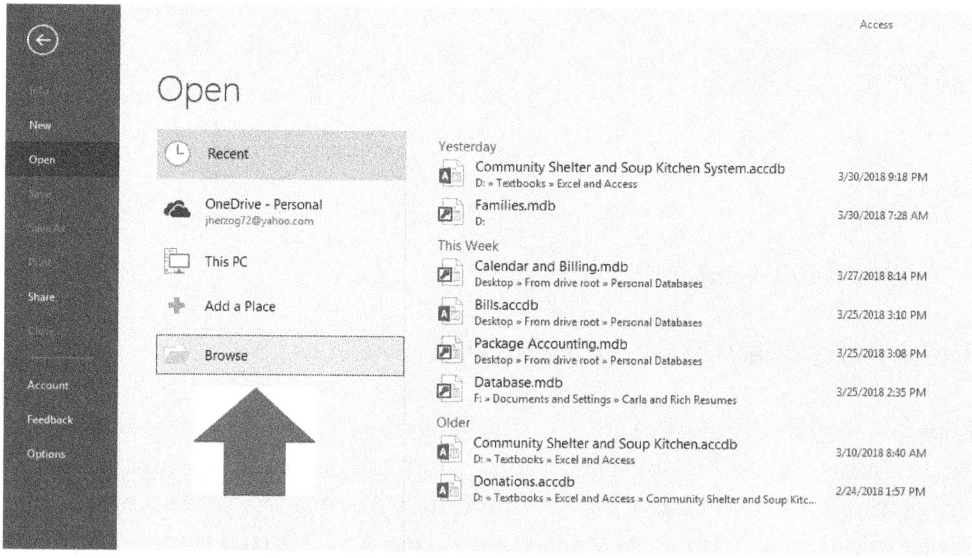

FIGURE 4.74   Adding a password to a database

Step 5: Locate the folder where your database is stored (figure 4.75). When you do, click **ONCE** (**DO NOT DOUBLE-CLICK**) on the name of the database (figure 4.75, arrow A).

Step 6: Click the tiny arrow to the right of the **Open Command** button (figure 4.75, arrow B). A **drop-down list** will appear (figure 4.75, arrow C).

Step 7: Click **Open Exclusive** (figure 4.75, arrow C).

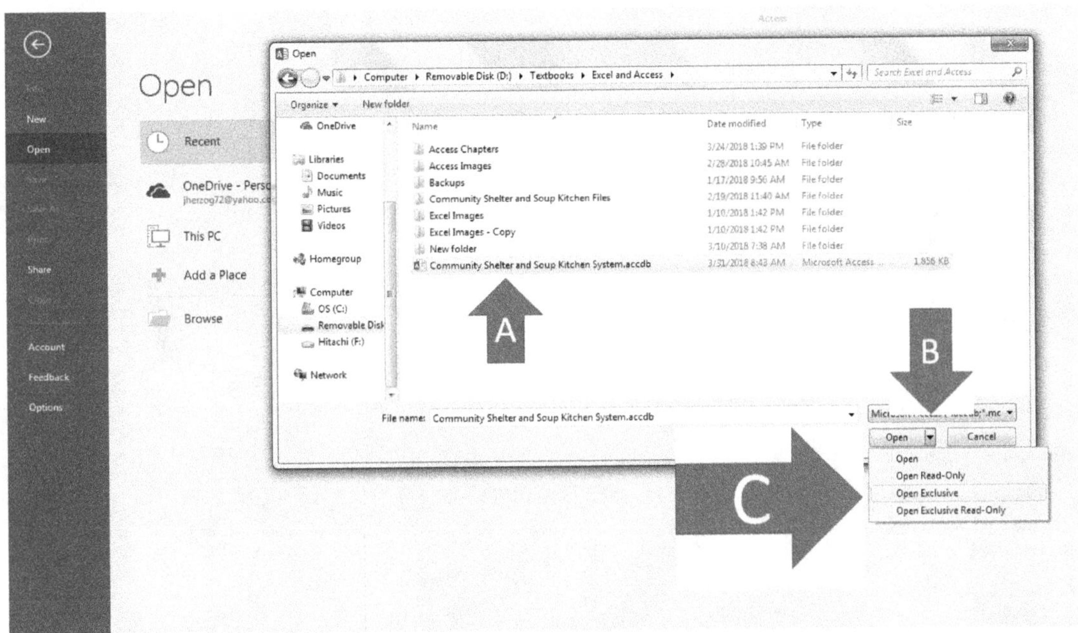

FIGURE 4.75   Adding a password to a database

Step 8: When you do, the database will appear to open as usual (figure 4.76); however, you now have the ability to set a password in the database that you did have before opening it in the **Open Exclusive** mode. Click the **File** ribbon (figure 4.76, arrow).

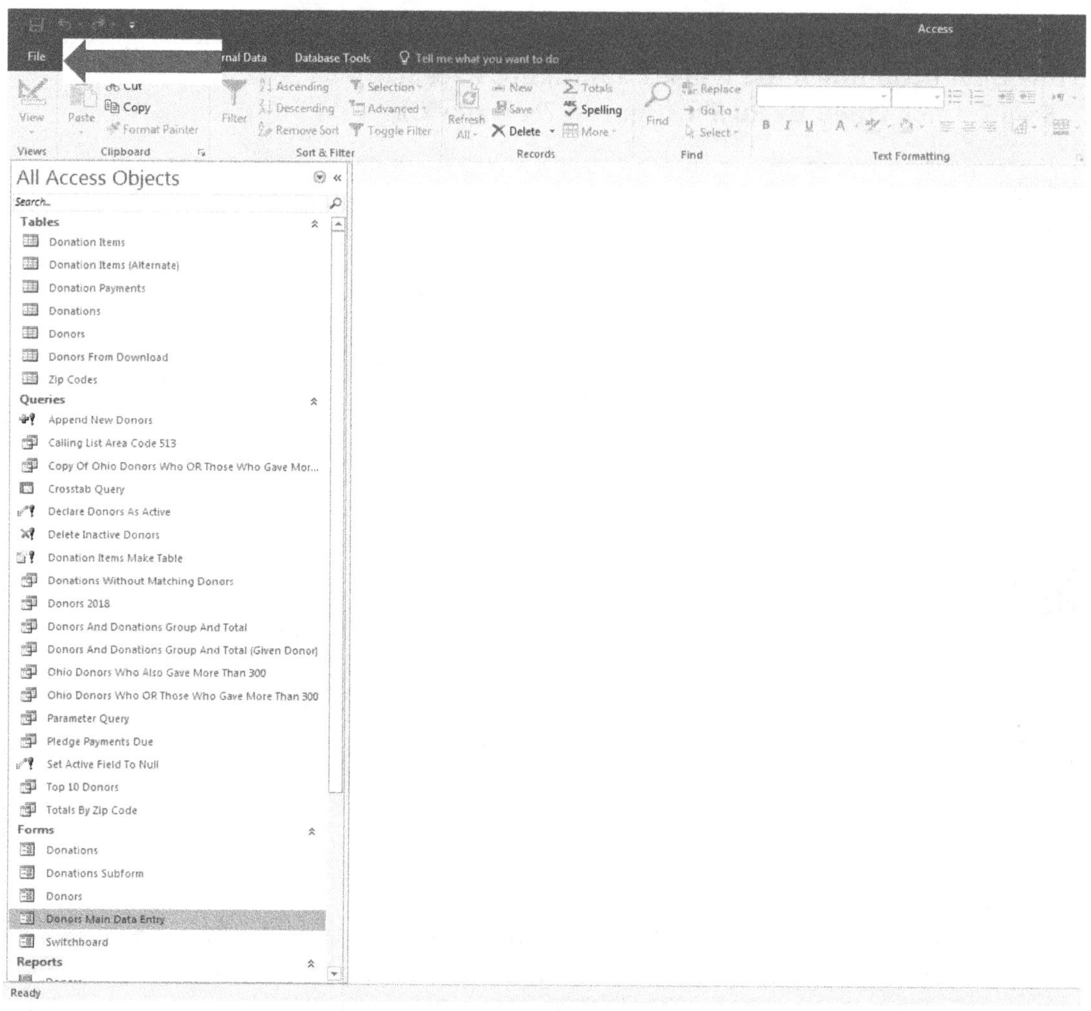

FIGURE 4.76 Adding a password to a database

Step 9: At the next screen click the **Encrypt with Password** button (figure 4.77, arrow).

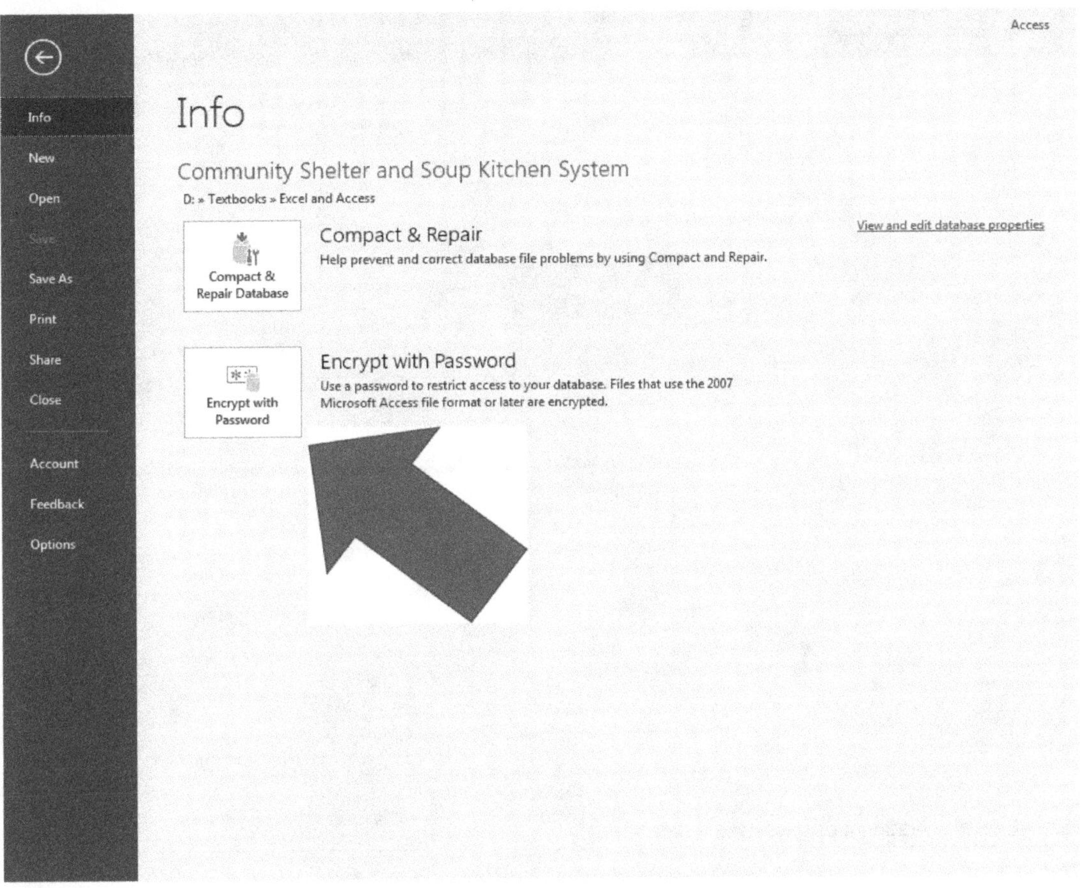

FIGURE 4.77 Adding a password to a database

Step 10: At the next screen (figure 4.78), type the password you want to use for your database in the box labeled **Password** (figure 4.78, arrow A) and then repeat it (for confirmation purposes) in the box labeled **Verify** (figure 4.78, arrow A). In this example, use the password of *herzog* (using all lower-cased letters).

Step 11: Click **OK**.

NOTE: When you click OK, you may get a message that reads "Encrypting with a block cipher is incompatible with row level locking." Row-level locking with be ignored (figure 4.78). Just click OK (figure 4.78, arrow B).

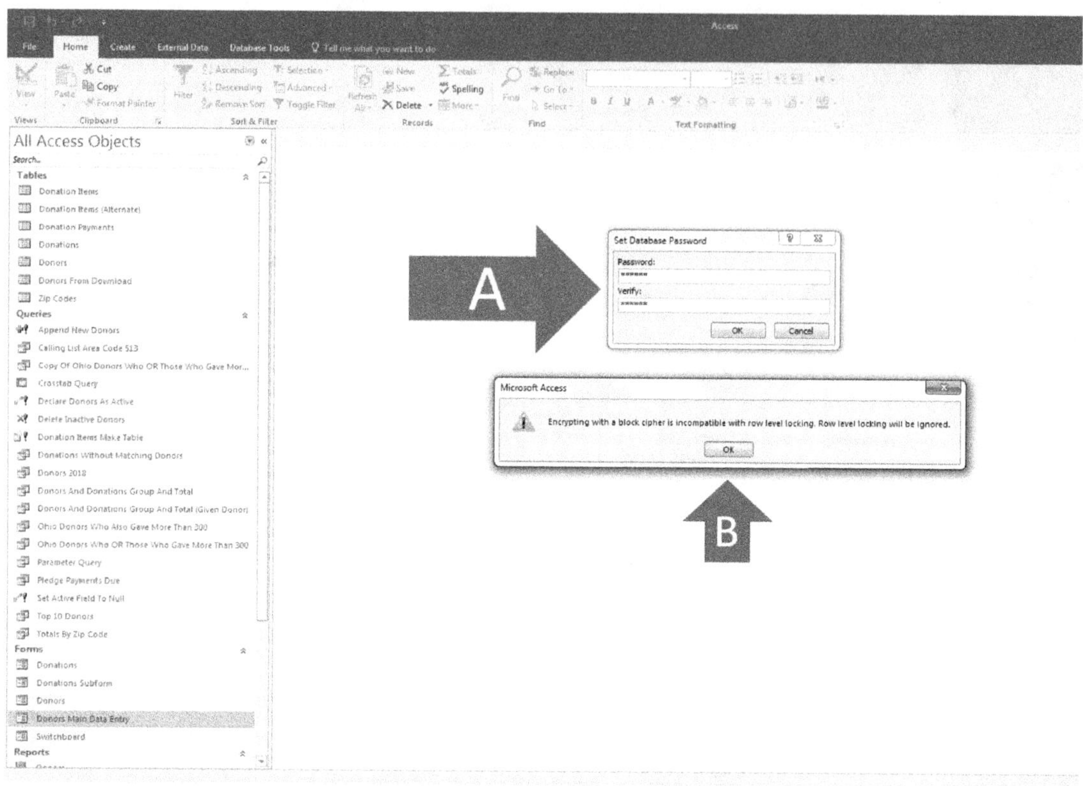

FIGURE 4.78 Adding a password to a database

Chapter 4: Assignment (Summary of Tasks)

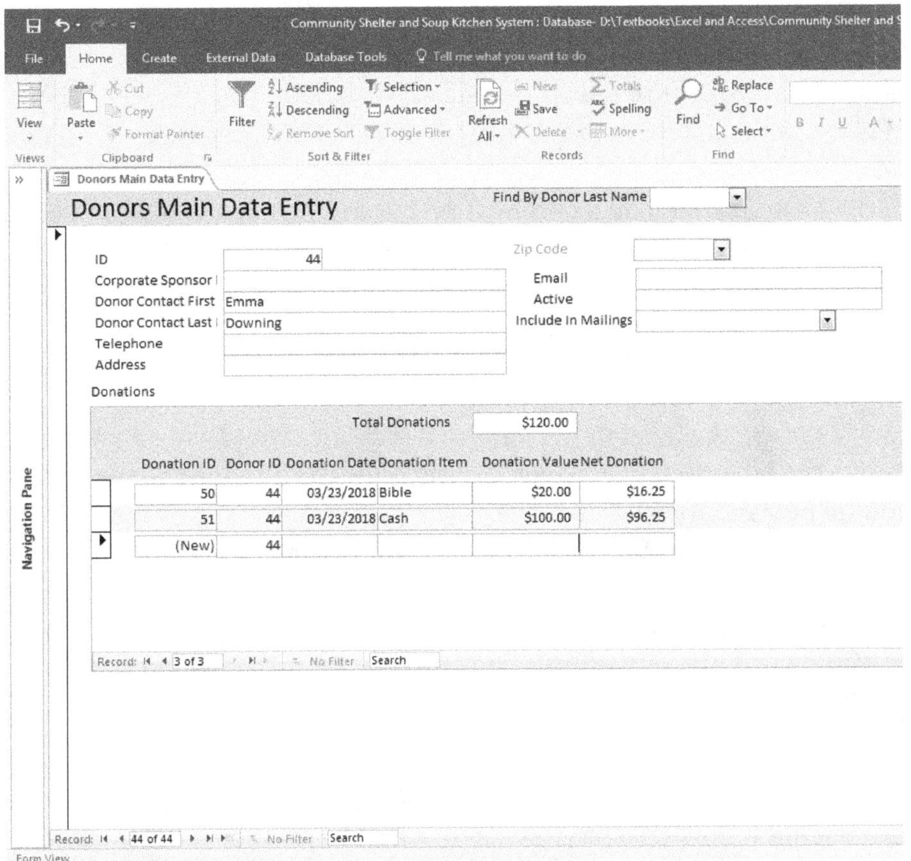

FIGURE 4.79　Assignment main subform sample

1. In the *Community Shelter and Soup Kitchen System* database, create a **main** and **sub-form** with all of the fields from the *donors* table in the **main form** and all of the fields from the *donations* table in the **sub-form**.

2. Name the **main form** *Donors Main Data Entry* and name the **sub-form** *Donations Sub-Form*.

3. Make all of the fields in the **main** form and in the **sub-form** the same height as the shortest field.

4. In the **main form**, align all of the fields to the top to be even with the first fields at the top of the **detail area** of the form.

5. Resize **labels** and **controls** in both the **main** and **sub-form** so that there is no wasted space in them. Shore up wasted space in the **header, footer,** and **detail** area of the **main form** and the subform in **form design view.**
6. Change the **font color** of all labels and controls in both the **main** and **sub-form** to black.
7. Move the **sub-form** so that it is beneath its label.
8. Make sure the label of the sub-form is as close as it can be to the fields of the **main form.**
9. In the **form header**, add a **combo box** that can be used as a search box to find specific records.
   a. When using the **Combo Box Wizard**, and when asked what fields are to be displayed in the **combo box**, use the *donor contact last name, donor contact first name,* and *donor address.*
   b. Hide the **key** field.
   c. Make sure the label to the **combo box** reads, *Find Donor By Last Name.*
10. Resize the label so it can be completely seen and change the **font color** of the **combo box** and its label to **black.**
11. Delete the *Include in Mailings* field in the **main form.** Create a **value list combo box** that will give the user the ability to enter **ONLY** the following options into the *Include in Mailings* field:
    a. Yes
    b. No
    c. Newsletter only

NOTE: Make sure that the **Limit to List option** is chosen in the **property sheet** of this new **combo** box.
12. Make sure that the label of *Include in Mailings* field also reads *Include in Mailings.* **MAKE SURE THIS COMBO BOX IS NOT UNBOUND!**
13. Delete the *zip code* field/control of the **main form** and make it a **combo box** reading the *zip code* table and make sure its label reads as *zip code.*
    a. Make sure the combo box uses all of the fields in the *zip code* table.
    b. Make sure the *zip code* **key** field is **NOT** hidden.
    c. Make sure that the **combo box** is sorted by the *zip code* field.
    d. MAKE SURE THIS COMBO BOX IS NOT UNBOUND!

NOTE: Make sure that the **Limit to List option** is chosen in the **property sheet** of this new **combo box.**
14. Change the *donation item* field in the **sub-form** to a **combo box** that reads the *Donation Item* table.

15. In the subform, add a **text box** above the *donation value* field that will give a **SUM** of all *donation values*. Give it a caption of *Total Donations* and make sure it uses the **currency format**. Your **main** and **sub-form** should look like figure 4.79.

16. Using the **main form** (**NOT THE TABLE**), enter *Emma Downing* as a new *donor* (you do not need to enter any more information about her than that) and in the **sub-form**, enter two *donations* for *Emma* as follows:
    a. Enter the *donation date* of the first *donation* as today's date.
    b. Enter the *donation item* of the first *donation* as a *Bible*.
    c. Enter the *donation value* of the first *donation* as *$20*.
    d. Enter the *donation date* of the second *donation* as today's date.
    e. Enter the *donation item* of the second *donation* as *Cash*.
    f. Enter the *donation value* of the second *donation* as *$100*.

17. Make sure the **tab order** of the **main form** and **sub-form** are in the order as they appear in those forms, from top to bottom and from left to right.

18. Make sure there is no **tab stop** on the *net donation* field in the sub-form.

19. Apply an **input mask** to the *telephone* field in the **main form**.

20. Give the database a password of *herzog* (using all lower-case letters).

Chapter 4: Assignment (Alternate) (Summary of Tasks)

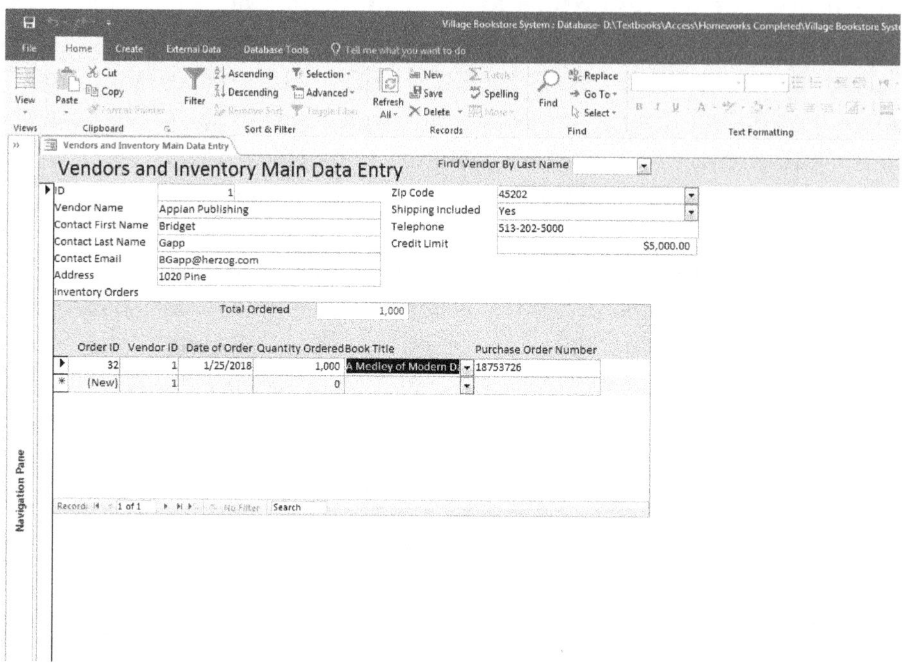

FIGURE 4.80 Alternate assignment main subform sample

1. In the *Village Bookstore Database System*, create a **main** and **sub-form** with all of the fields from the *vendors* table in the **main form** and all of the fields from the *inventory orders* table in the **sub-form**.
2. Name the **main form** Vendors and Inventory Main Data Entry and name the **sub-form** *Inventory Orders Sub-Form*.
3. Make all of the fields in the **main** form and in the **sub-form** the same height as the shortest field.
4. In the **main form**, align all of the fields to the top, to be even with the first fields at the top of the **detail area** of the form.
5. Resize **labels** and **controls** in both the **main** and **sub-form** so that there is no wasted space in them. Shore up wasted space in the **header**, **footer**, and **detail** area of the **main form** and the subform in **form design view**. Remove the word *Vendor* from the labels of the *vendor contact first name*, *vendor contact last name*, *vendor contact email*, *vendor address*, *vendor telephone*, and *vendor credit limit* fields.
6. Change the **font color** of all labels and controls in both the **main** and **sub-form** to **black**.

7. Move the **sub-form** so that it is beneath its label.
8. Make sure the label of the **sub-form** is as close as it can be to the fields of the **main form**.
9. In the **form header**, add a **combo box** that can be used as a search box to find specific records.
   a. When using the **Combo Box Wizard**, and when asked what fields are to be displayed in the **combo box**, use the *vendor contact last name, vendor contact first name*, and *vendor address*.
   b. Hide the **Key** field.
   c. Make sure the label to the **Combo Box** reads, *Find Vendor by Last Name*.
10. Resize the label so it can be completely seen and change the **font color** of the **combo box** and its label to b**lack**.
11. Delete the *Shipping Included* field in the **main form**. Create a **value list combo box** that will give the user the ability to enter only the following options into the *Shipping Included* field:
    a. Yes
    b. No
    c. On special orders only

NOTE: Make sure that the **Limit to List option** is chosen in the **property sheet** of this new **combo box**.

12. Make sure that the *Shipping Included* field label also reads as *Shipping Included*. **MAKE SURE THIS COMBO BOX IS NOT UNBOUND!**
13. Delete the *zip code* field/control of the **main form** and make it a **combo box** reading the *zip code* table and make sure its label reads as *Zip Code*.
    a. Make sure the combo box uses all of the fields in the *zip code* table.
    b. Make sure that the *zip code* **key** field is **NOT** hidden.
    c. Make sure that the **combo box** is sorted by the *zip code* field.
    d. MAKE SURE THIS COMBO BOX IS NOT UNBOUND!

NOTE: Make sure that the **Limit to List option** is chosen in the **property sheet** of this new **combo box**.

14. Change the *book title* field in the **sub-form** to a **combo** box that reads the *Book Title* table.
15. In the **sub-form**, add a **text box** above the *quantity ordered* field that will give a **SUM** of all *quantities ordered*. Give it a caption of *Total Ordered* and make sure it uses the **standard format**. Your **main** and **sub-form** should look like figure 4.80.
16. Using the **main form** (**NOT THE TABLE**), enter *Herzog Publishing* as a new vendor (you do not need to enter any more information about *Herzog Publishing*

than that) and in the **sub-form**, enter two *inventory orders* for *Herzog Publishing* as follows:

a. Enter the *date of order* of the first *order* as today's date.
b. Enter the *quantity ordered* of the first *order* as *400*.
c. Enter the book title of the first order as A Medley of Modern Day Miracles I.
d. Enter the *date of order* of the second *order* as today's date.
e. Enter the *quantity ordered* of the second *order* as *600*.
f. Enter the book title of the second order as A Medley of Modern Day Miracles II.

17. Make sure the **tab order** of the **main** and **sub-form** are in the order as they appear, from top to bottom and left to right.
18. Make sure there is no **tab stop** on the *stocking fee* field in the subform.
19. Apply an **input mask** to the *vendor telephone* field in the **main form**.
20. Give the database a password of *herzog* (all lower-case letters).

Making a Main Menu for Transaction Data Entry and Reports

IN THIS CHAPTER you will learn how to do the following:

1. Create a switchboard (or main menu) for a database
2. Create **macros**
3. Make a database to become more like an application

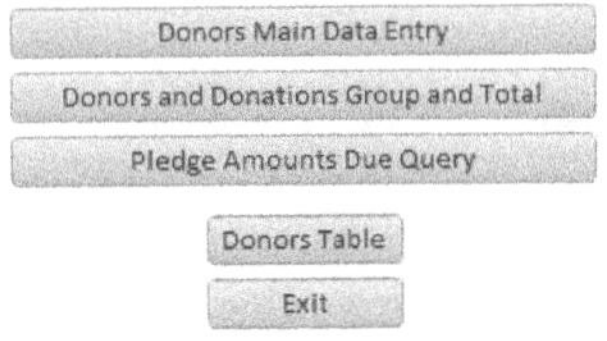

FIGURE 5.1   Switchboard sample

CHAPTER 5 TASK: CREATING A SWITCHBOARD AND MACROS

At this point, if any user ever opened the *community shelter and soup kitchen system* database you have created, as it was by the end of chapter 4, most of them would have had no idea as to what to do when they opened it. Anyone who would have wanted to enter data or view reports would have had no clue as to how to do so. Therefore, is it important for you to create a **main menu** (often called a **switchboard**) that would be self-evident, that will tell the users how to open and use the forms and reports you have created for their use (as shown in figure 5.1).

There are many ways to do this. Earlier versions of Access have had quick access to creating what has been called a **switchboard**. It is also possible to create such **switchboards** or **main menus** from scratch, which is the method we will use today. There are often times you will need to make such forms within a database that are used strictly for directing users to other Access database objects, and that is why we will use that method, here.

TO MAKE THE INITIAL MAIN MENU (OR SWITCHBOARD) FROM SCRATCH

A **main menu** (or **switchboard**) is nothing more than a blank form with **command buttons** attached. In order to start to create a **Switchboard**, do the following steps:

Step 1: Click the **Create** ribbon (figure 5.2, arrow A).

Step 2: Click **Blank Form** (figure 5.2, arrow B).

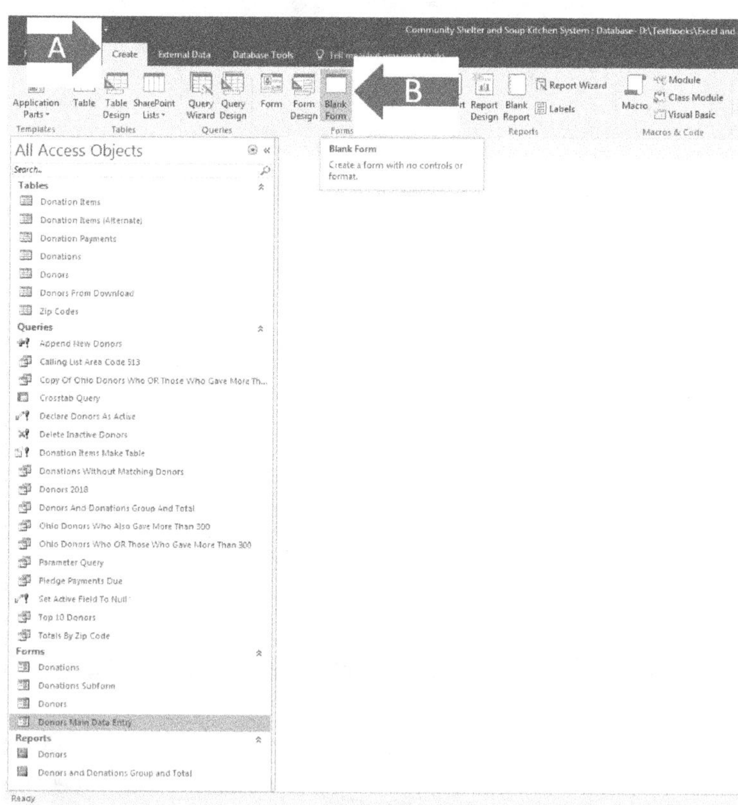

FIGURE 5.2   Creating a switchboard

Step 3: That will take you to the screen shown in figure 5.3.

Please notice that Access will instantly create a new form by default, and it will name it *Form1*. It will also add a **record selector** and a **navigation button**, neither of which is needed by a **switchboard** because it will only have **command buttons** on it and will usually not be reading any records. Therefore, the **Record Selectors** and the **Navigation button** will only confuse the user **AND CONFUSING THE USER IS WHAT WE ARE ALWAYS TRYING TO COMPLETELY AVOID IN ANY WAY THAT WE CAN WHEN WE CREATE A DATABASE.**

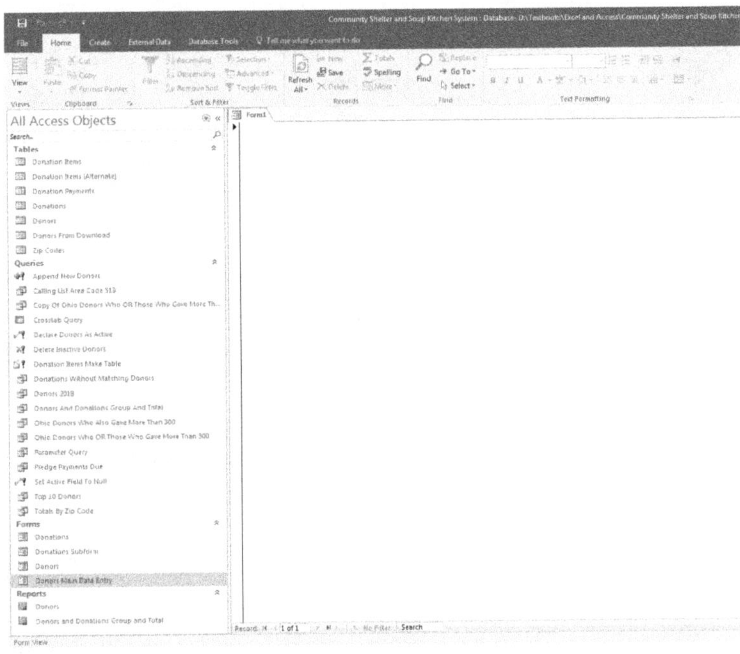

FIGURE 5.3   Creating a switchboard

TO REMOVE THE RECORD SELECTOR AND NAVIGATION BUTTON FROM A FORM

To remove the **Record Selector** and **Navigation button** from a form, do the following steps:

Step 1: Go to the **design view** of the form.

Step 2: Double-click the **Form Selection box** in the upper left of the **Design View grid** (figure 5.4, Arrow A). When you do, it will open the **property sheet** for the form.

NOTE: As mentioned in an earlier chapter, just as you can select a **control** and view or edit the **properties** of it, you can see the **properties** of an entire **form** (or **report**) by double-clicking the **Form/Report Selection box**. When you do, it will open the **property sheet** for the form and it will also place a tiny black box in the **Form Selection box** (figure 5.4, arrow A).

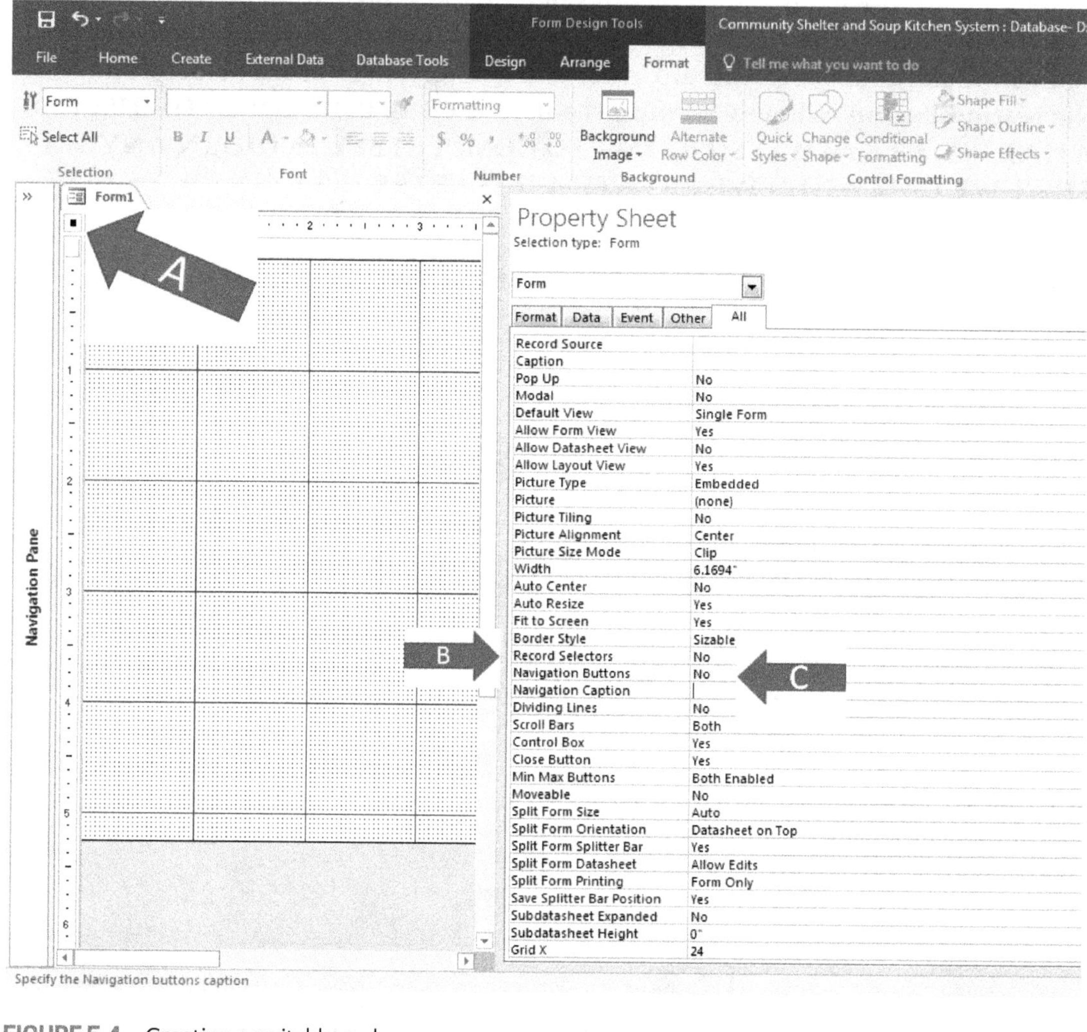

FIGURE 5.4 Creating a switchboard

Step 3: In the **All** tab of the **property sheet**, make sure that the **Record Selectors** and **Navigation button** settings are changed to **No** (figure 5.4, arrows A and B).

Step 4: Close the property sheet.

Step 5: Save the **form** and name it *Switchboard*.

TO ADD COMMAND BUTTONS

Command buttons are what you will use to add to the **switchboard** that users can click to open any object, such as a **table**, **form**, **query**, or **report**. **Command buttons** **CANNOT** be put into **tables** or **queries**. Access will permit you to put them in **reports**, but they simply won't work. **THE BUTTONS THEMSELVES DO NOT DO ANY-THING.** Once you place them in the **switchboard** you will then need to create what is called a **macro**. **Macros** are objects within a database that will carry out a set of commands that you will give in a certain order or sequence, such as opening a **form** and then going to a **new record** in that **form**. Once you create a **macro**, you will then have to attach it to a **Command button** that will run it once the **Command button** is clicked or selected.

The concept is similar to what happens when electricians put a switch on a wall. Switches are of no use if there are no wires attached. **Command buttons** are like electric switches with no wires attached. **Command buttons** are only useful if they have **macros** attached.

Clicking, double-clicking, or selecting a **Command button** is called an **event**. That is important to remember as we begin this process.

At this point we will add separate command buttons that will do the following:

a. Open the *Donor Main Data Entry* form
b. Open the *Donors and Donations Group and Total* report
c. Open the *Pledge Payments Due* query
d. Open the *donors* table
e. Close the database

You can do so by taking the following steps:

Step 1: While still in the design view of the **switchboard**, click once (do not click and drag) on the **Button** icon in the **Controls toolbox**, in the **Design** ribbon within the **form design tools** (figure 5.5, arrow A). When you do, the mouse will have a cross and a small button attached to it (figure 5.5, arrow B).

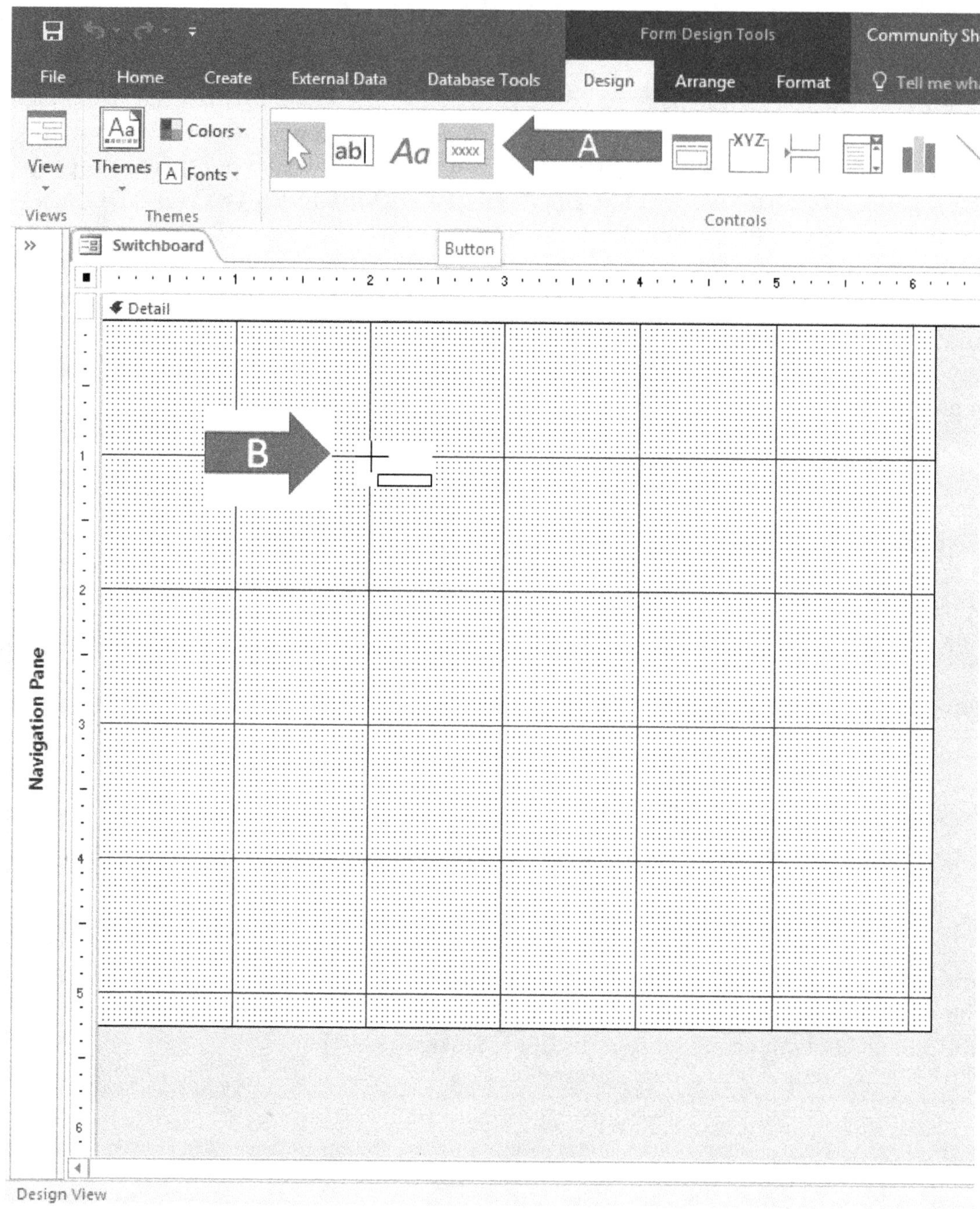

FIGURE 5.5   Adding command buttons to a switchboard

Step 2: Click your mouse once (do not click and drag) on the **Form grid** where the **2-inch** and **1-inch** lines intersect (figure 5.5, arrow B).

Step 3: The **Command Button Wizard** will then appear as you can see in figure 5.6. At this point you will not need to use the **Command Button Wizard**; therefore, click **Cancel** (figure 5.6, arrow).

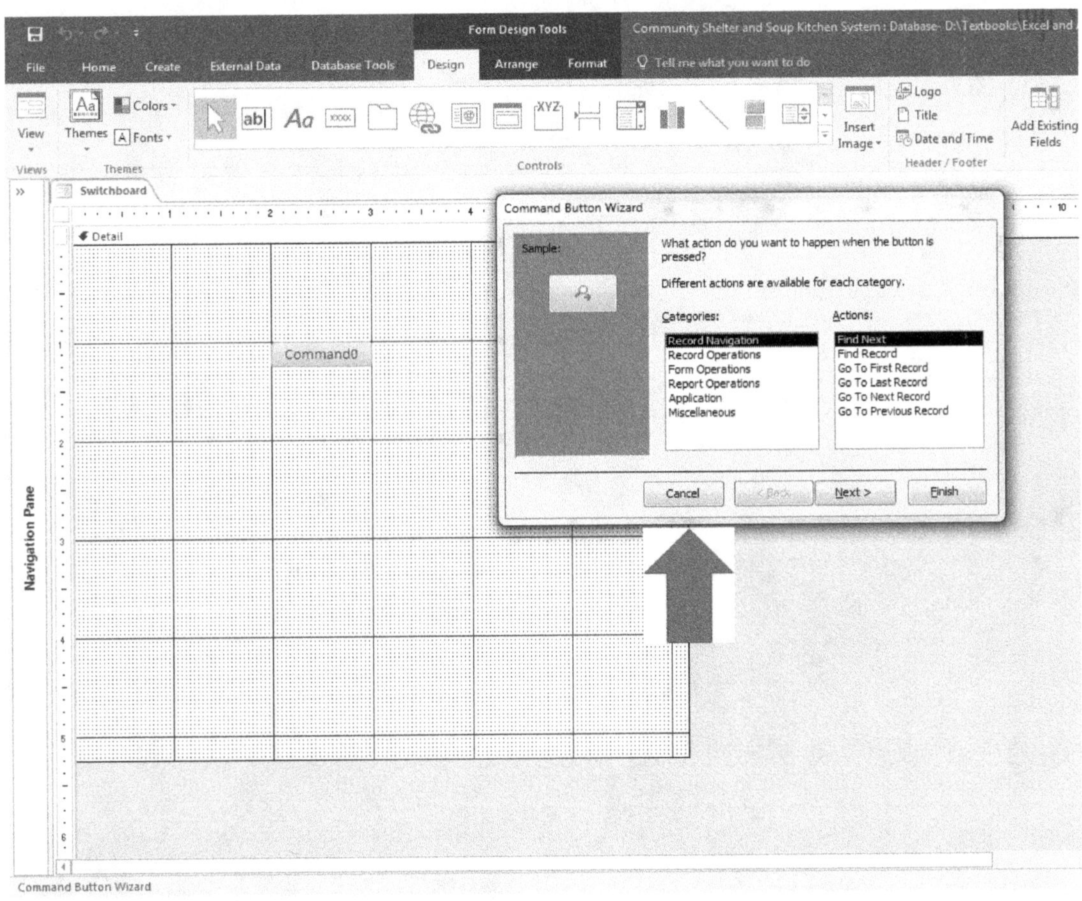

FIGURE 5.6 Adding command buttons to a switchboard

Step 4: When you do, notice that a **Command button** is now placed in your **switchboard**. Click on the **Command button** and then click **Copy** in the **Home** ribbon (figure 5.7, arrow).

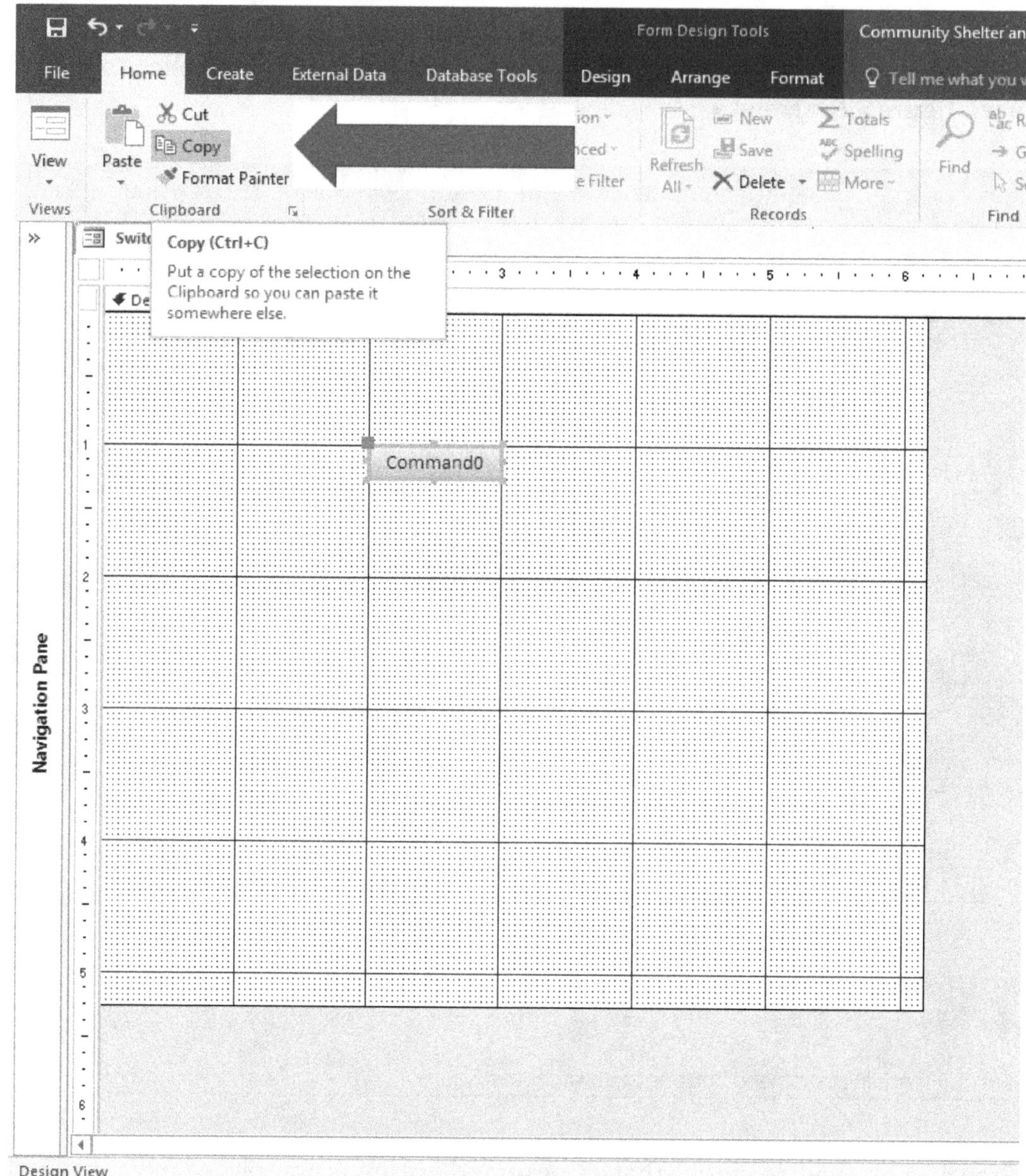

FIGURE 5.7 Adding command buttons to a switchboard

Step 5: Click the **Paste** command two (2) times (figure 5.8, arrow). There will now be three **Command buttons** in the **Switchboard** as shown in figure 5.8.

NOTE: The captions on your **Command buttons** that Access puts on them by default may differ from what you see here (*Command0*), but they are not relevant because you will be changing the captions in a few moments.

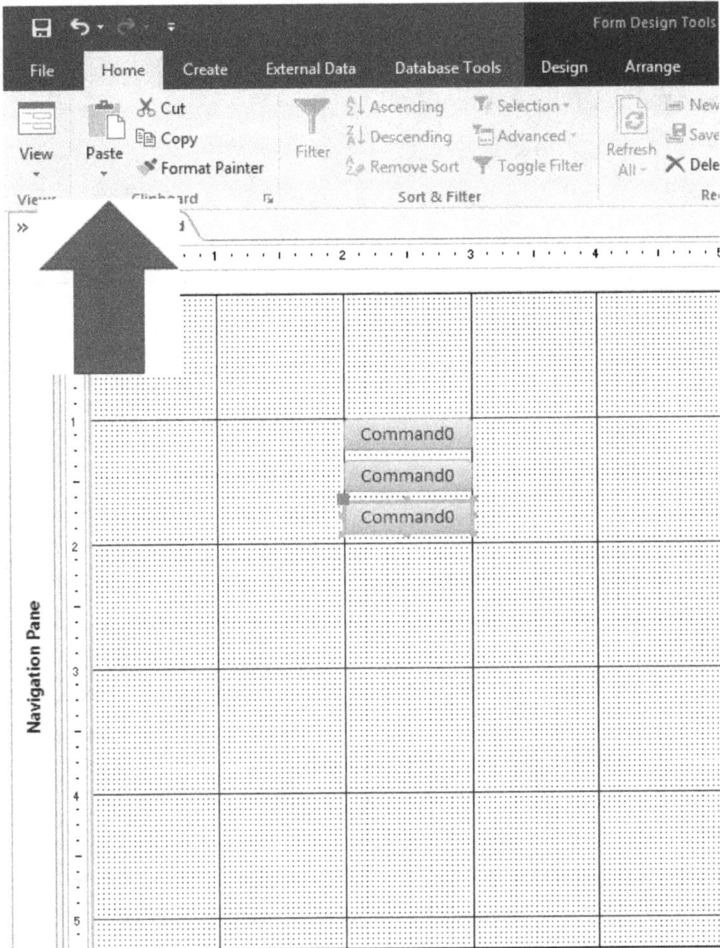

FIGURE 5.8   Adding command buttons to a switchboard

TO ADJUST THE WIDTH OF COMMAND BUTTONS

You can use the same methods described earlier for changing widths of form controls in order to adjust the width of **Command buttons** by doing the following:

Step 1: Select the buttons (by either drawing a box around them or by clicking on each one of them while holding down the **Shift key** of your keyboard).

Step 2: Resize the **Command buttons** by pressing and holding the **Shift key** while also pressing and holding down the **Right Arrow Cursor Control key** → on your keyboard. Continue to do so until the **Command buttons** are approximately three (3) inches wide.

Your screen will look like figure 5.9.

Step 3: Click outside or away from the buttons to remove the highlighting.

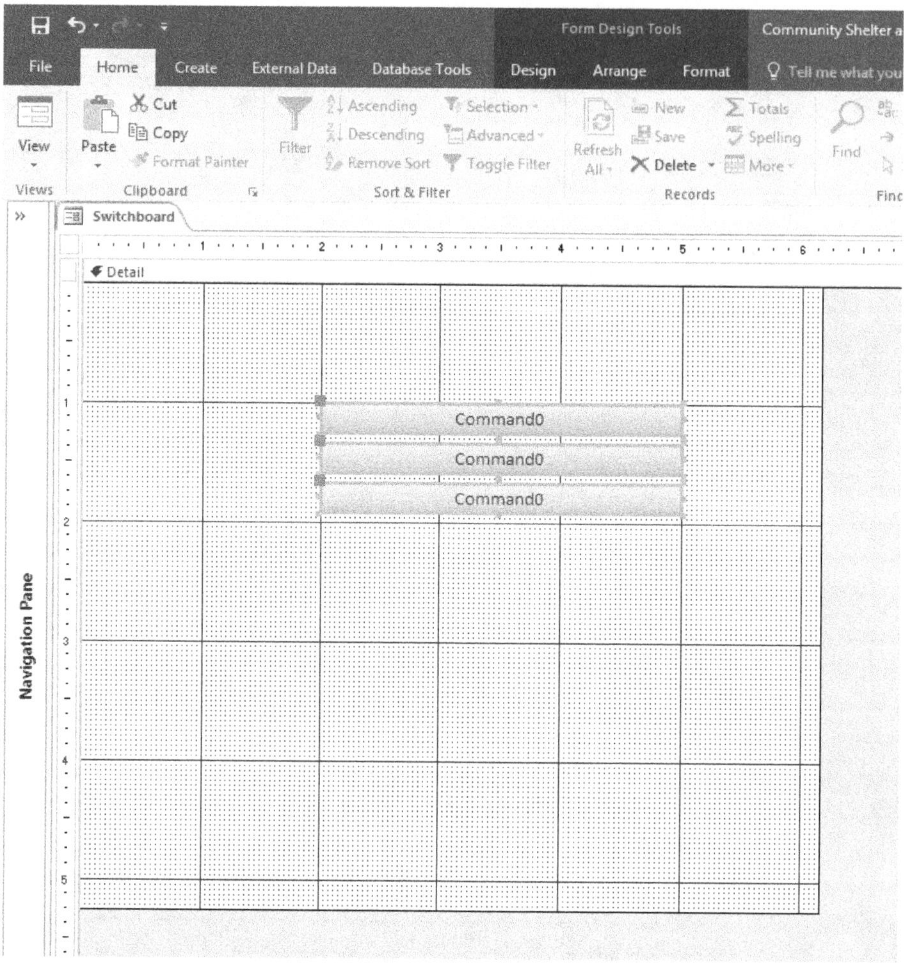

FIGURE 5.9   Adding command buttons to a switchboard

TO CHANGE THE CAPTION OF A COMMAND BUTTON

Now let's make the menu as user-friendly as possible. Change the caption of each button as follows:

Step 1: Click two (2) times on the top button and then press the **Backspace** and/or the **Delete** keys on your keyboard until the existing caption is gone.

Step 2: Type Donors Main Data Entry inside the box.

Step 3: Press **Enter** on your keyboard.

Step 4: Repeat this process with the remaining **Command buttons**, giving the caption of *Donors and Donations Group and Total Report* and *Pledge Amounts Due Query* to them in that order.

Your screen should look very much like figure 5.10.

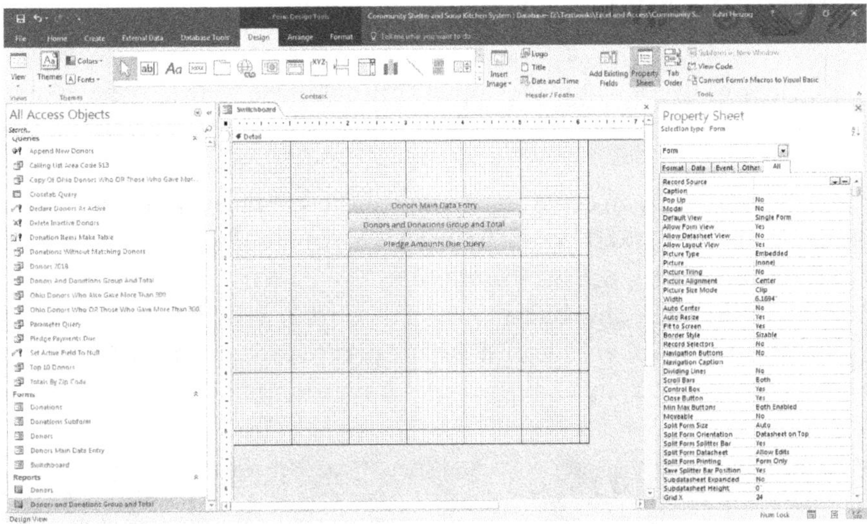

FIGURE 5.10   Changing captions on command buttons

Step 5: Now repeat this entire process to put two **SMALLER Command buttons** beneath the ones you have just placed on the **switchboard**, but **DO NOT** resize them. Change the caption of the first, new, smaller **Command button** to read *Donors* table and the other **Command button** should read *Exit*. Your screen should look a lot like figure 5.11a.

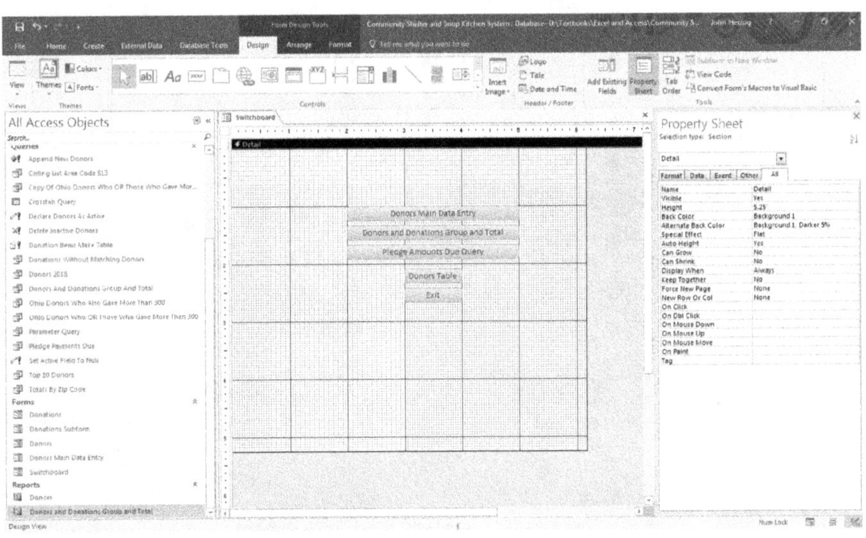

FIGURE 5.11a   Changing captions on command buttons

Step 6: Save the **switchboard**.

ADDING A TITLE TO A SWITCHBOARD

It needs to be clear to users when they open this database that they have opened the right one. A **title** will make that happen. To add a **title** to the **switchboard** you can **take the following** steps:

Step 1: While remaining in the **design view** of the **switchboard**, click (do not click and drag) the **Aa** (or the **Label**) button in the **Controls toolbox** of the **Design** ribbon in the **form design tools** (figure 5.11b, arrow A).

Step 2: When you click the **Label** icon, the cursor will have a cross and an A attached to it (figure 5.11b, arrow B). Place it just below the **detail bar** at approximately the **1.5 inches** mark, by clicking there once (do not click and drag).

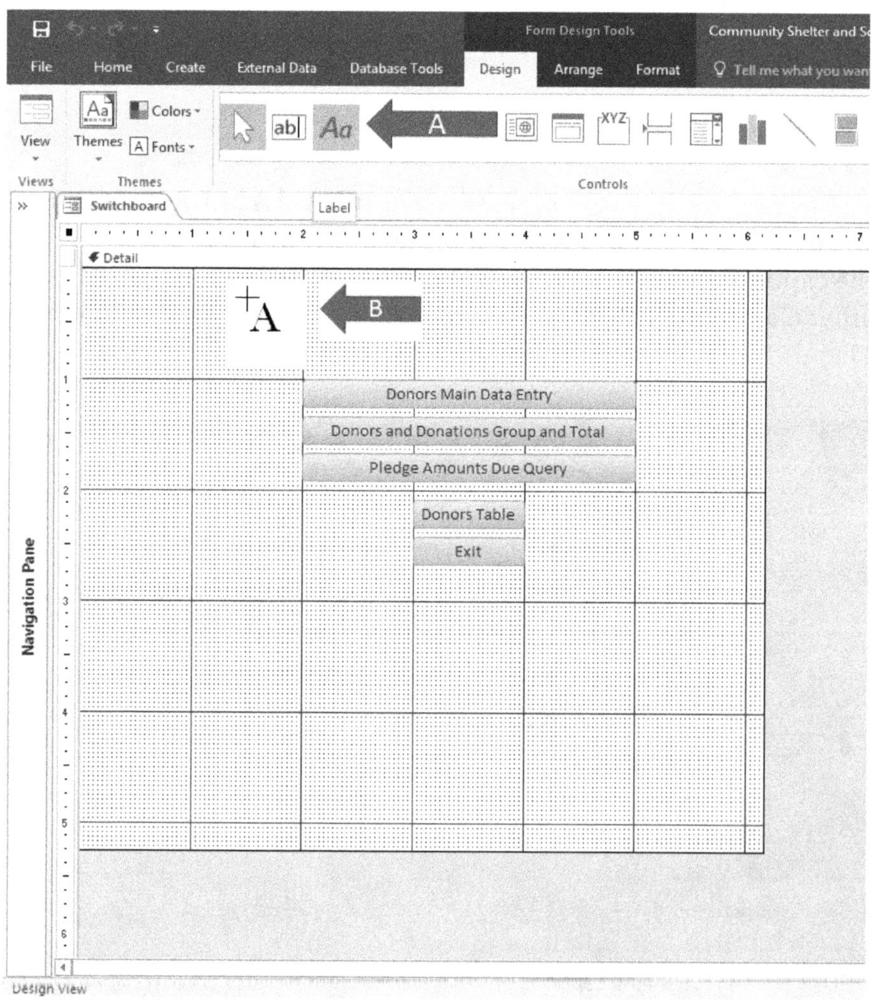

FIGURE 5.11b   Adding a title to a switchboard

Step 3: When you do, a tiny box with a cursor blinking in it will appear at the place where you clicked and spotted the label (figure 5.11c, arrow).

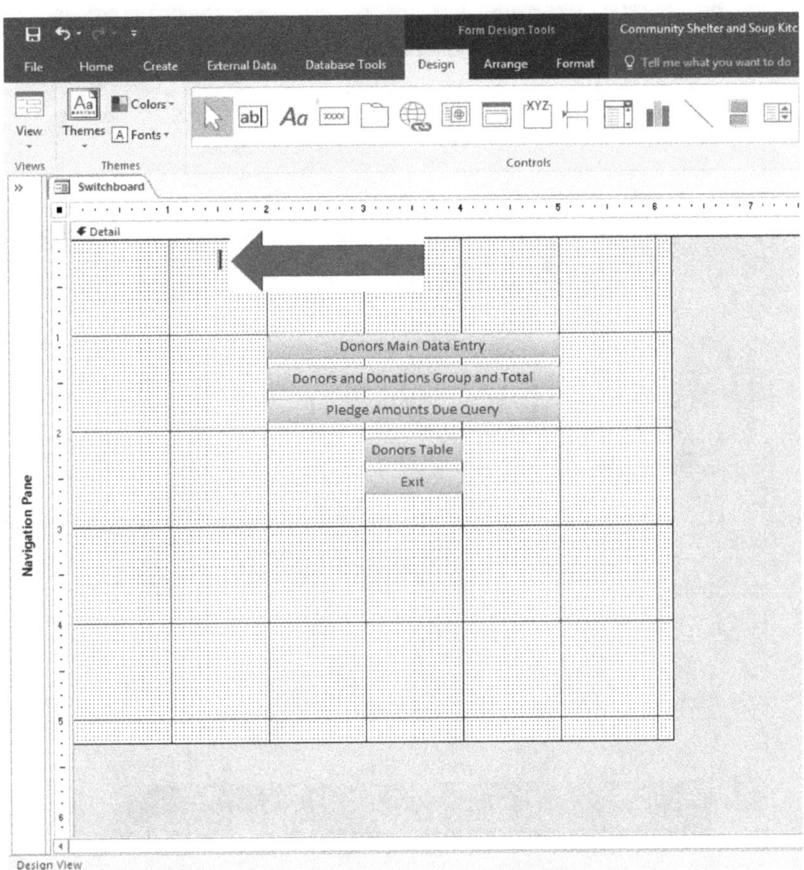

FIGURE 5.11c   Adding a title to a switchboard

Step 4: Inside the **Label** type the words *Community Shelter and Soup Kitchen Donation Tracking System*.

Step 5: Press **Enter**.

Step 6: At this point your screen will look a lot like figure 5.11d. Resize the label and change the font as you have done in previous discussions. In this example, do the following:

a. Change the font to **black**.
b. Change the font size to a size **20**.
c. Resize the label to make sure that all of the words are completely visible in two rows.

Step 7: When you do, your screen should look like figure 5.11e. Close and save the **switchboard**.

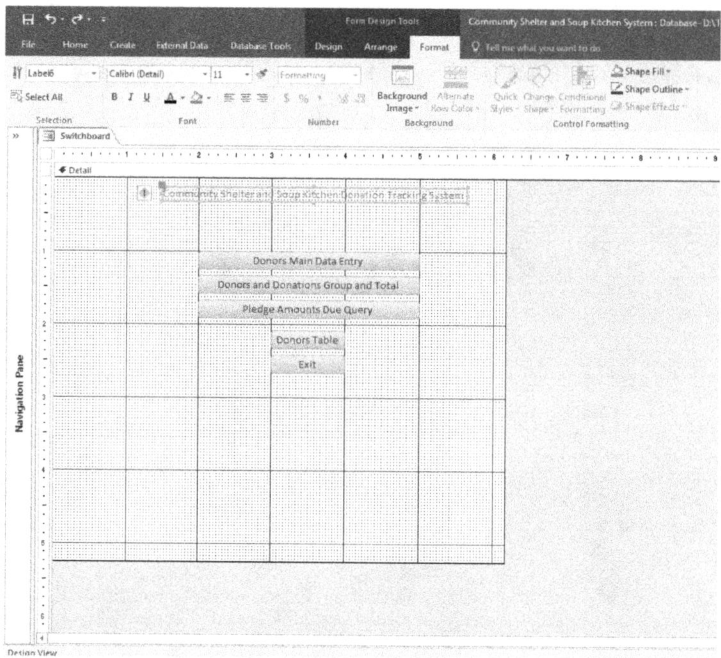

FIGURE 5.11d   Adding a title to a switchboard

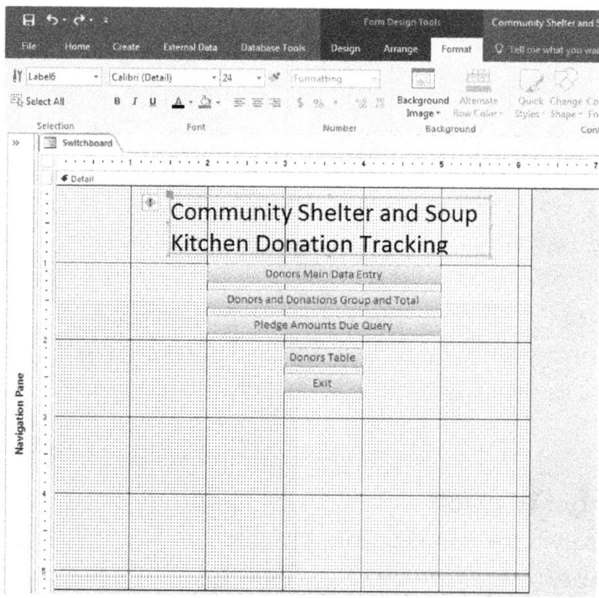

FIGURE 5.11e   Adding a title to a switchboard

NOTE: This procedure for creating a **title** or **caption** can be used to put text wherever you want it, whether it be in a **form** or **report**.

CREATING A MACRO TO OPEN FORMS

Now it's time to create the **macros** that will open these objects we have been creating all of this time in the database that we have just discussed. We will also create a **macro** that will close the database.

NOTE: Macros can be created in one of two ways: By making each of them as an **access objec**t or by creating an **embedded macro**. It is not recommended that you create **embedded macros** and they will be discussed again later.

Let's first create the **macro** that will open the *Donors Main Data Entry* form. You can do so by taking the following steps:

Step 1: Click the **Create** ribbon (figure 5.12, arrow A).

Step 2: Click **Macro** (figure 5.12, arrow B).

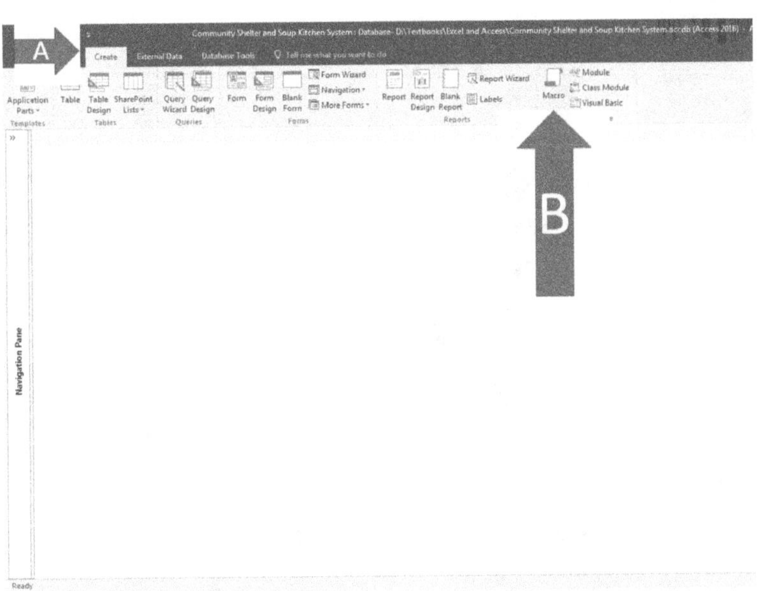

FIGURE 5.12   Creating a macro to open forms

Step 3: That will take you to the **design view** of a **macro**. It will look like the menu you see in figure 5.13. Click on the **drop-down list box** in the screen (figure 5.13, arrow A) and see all of the many, many commands you can do.

NOTE: It would be a good thing for you to scroll through the list to see the dozens of commands that a **macro** can do. For now, our focus will be simply opening objects and closing the database, because they are the commonly used.

In the **drop-down list box** type **OpenForm** (with no space) (figure 5.13, arrow B). After you do, please notice that it will take you to the **OpenForm** command in the list (figure 5.13, arrow C). Also notice that when we create **macros** that will open **queries**, **reports**, and **tables**, we can choose options for doing so with the same process we just used for opening a form (figure 5.13, arrow D).

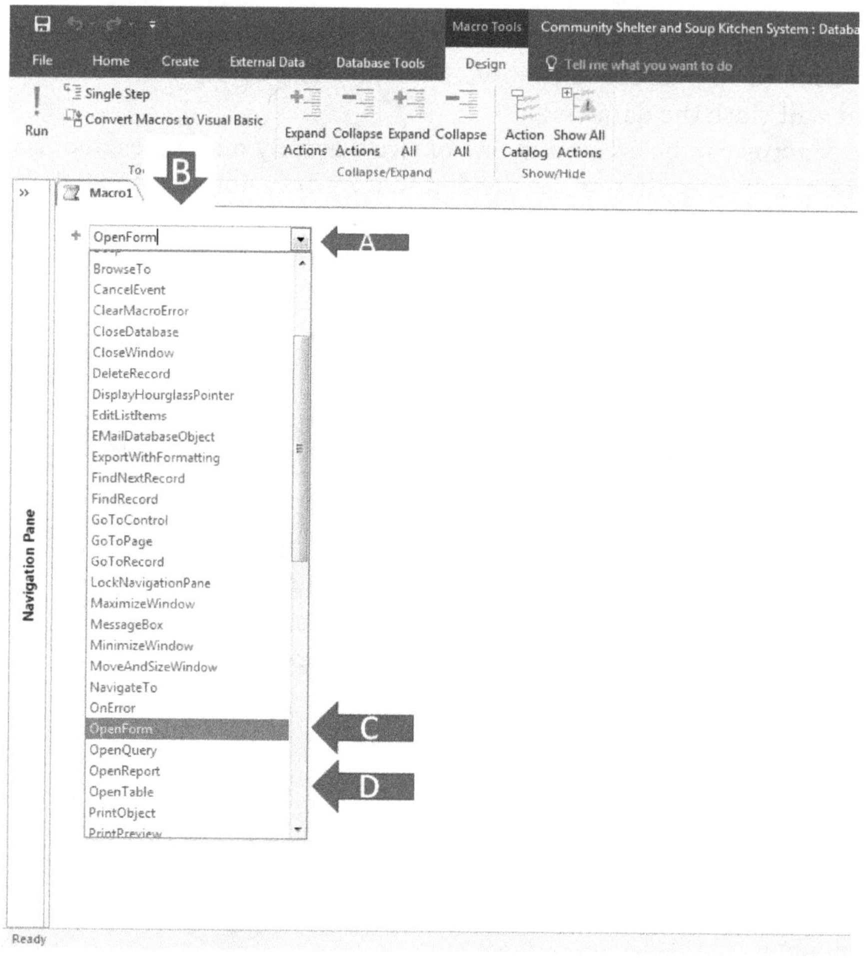

FIGURE 5.13   Creating a macro to open forms

Step 4: Press **Enter** on your keyboard.

Step 5: That will take you to a screen that looks a lot like figure 5.14.

Step 6: Using the **drop-down list box** (figure 5.14, arrow), select the **form** you want to have open when this **macro** is run (and it will run when we attach the **macro** to a **Command button** later). In this example, choose the *donors main data entry* **form**. **NEVER** enter a form name in the **Form Name box** that is not shown in the **drop-down list box**!

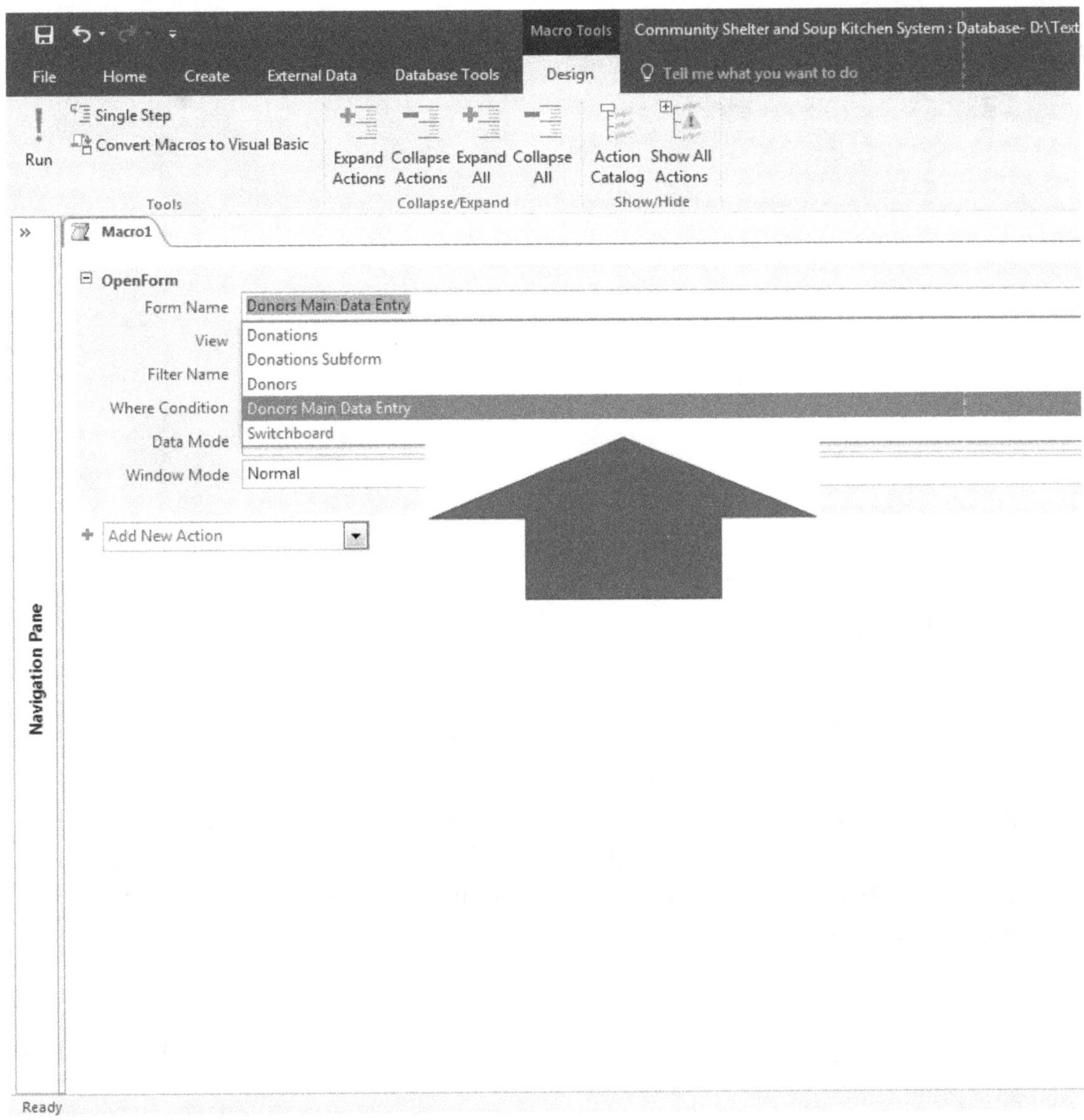

FIGURE 5.14   Creating a macro to open forms

Step 7: After choosing the *donors main data entry* form, you will see the screen seen in figure 5.15. Please notice the box that reads **Add New Action** below the **OpenForm command** that we just entered (figure 5.15, arrow). To do the next action, click the arrow to the right of that **Add New Action** box (figure 5.16, arrow B).

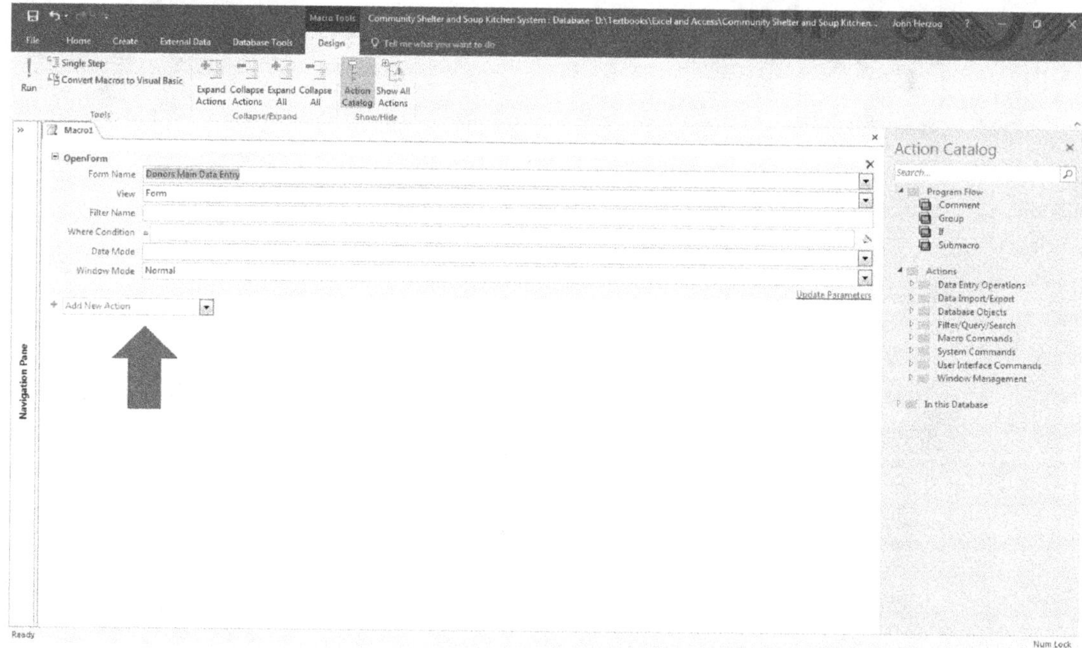

FIGURE 5.15   Creating a macro to open forms

Step 8: When you do, you will see a **drop-down list box** (figure 5.16). In the **drop-down list box**, click the **GoToRecord** command (figure 5.16, arrow A). After doing so, as happens when any **action** is chosen, it will place it in the **Add New Action** box (figure 5.16, arrow B).

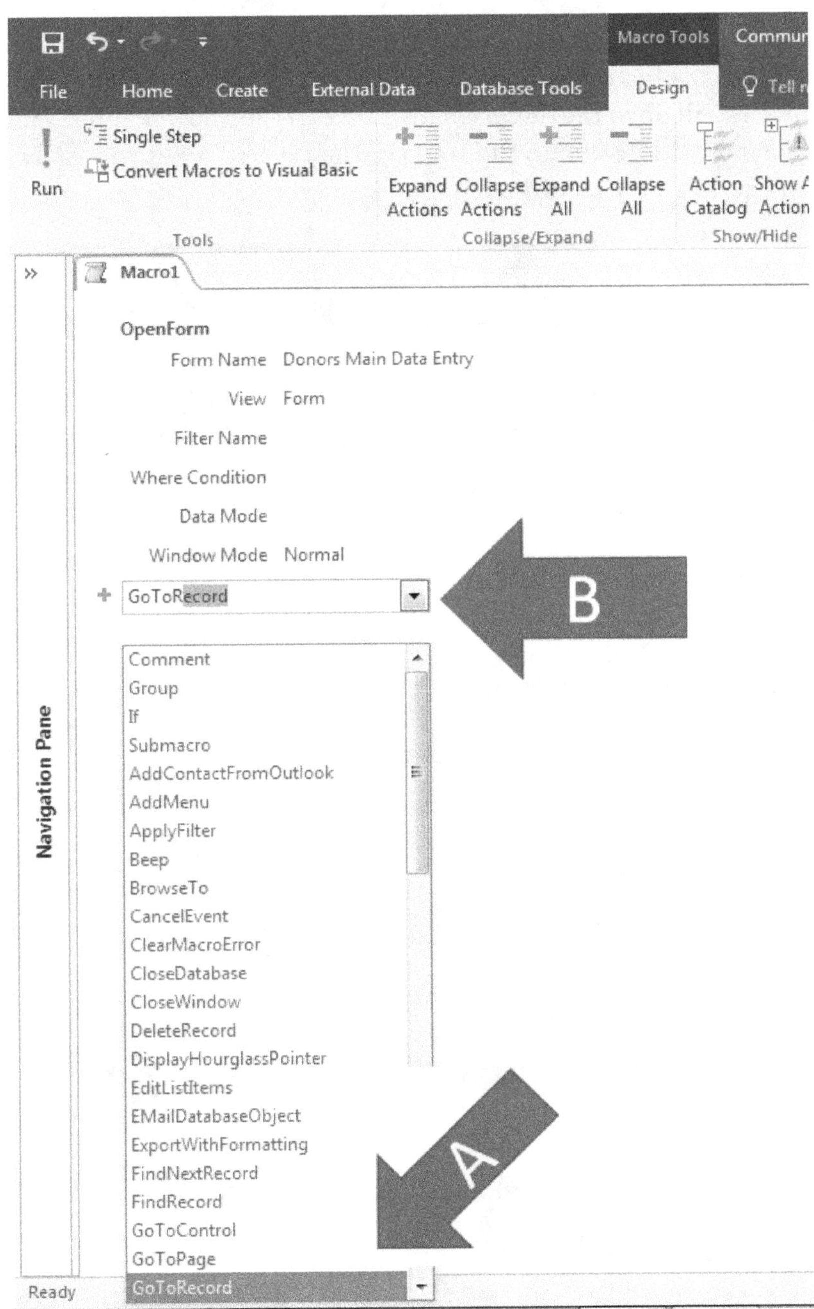

FIGURE 5.16   Creating a macro to open forms

Step 9: That will take you to what you see in figure 5.17. Enter the **object type** as a **form**, enter the **object name** as *Donors Main Data Entry*, and in the **drop-down list box** labeled **Record**, select **New**.

NOTE: Anytime you make a **macro**, it will do the actions you enter in the order you entered them from the top of the **Macro Design** menu to the bottom of it. Therefore, in this example, it will open the *donors main data entry* form first and it will then take the user to a new record. If you do the steps in the reverse order you will get an error as Access can't go to a new record until the form is open.

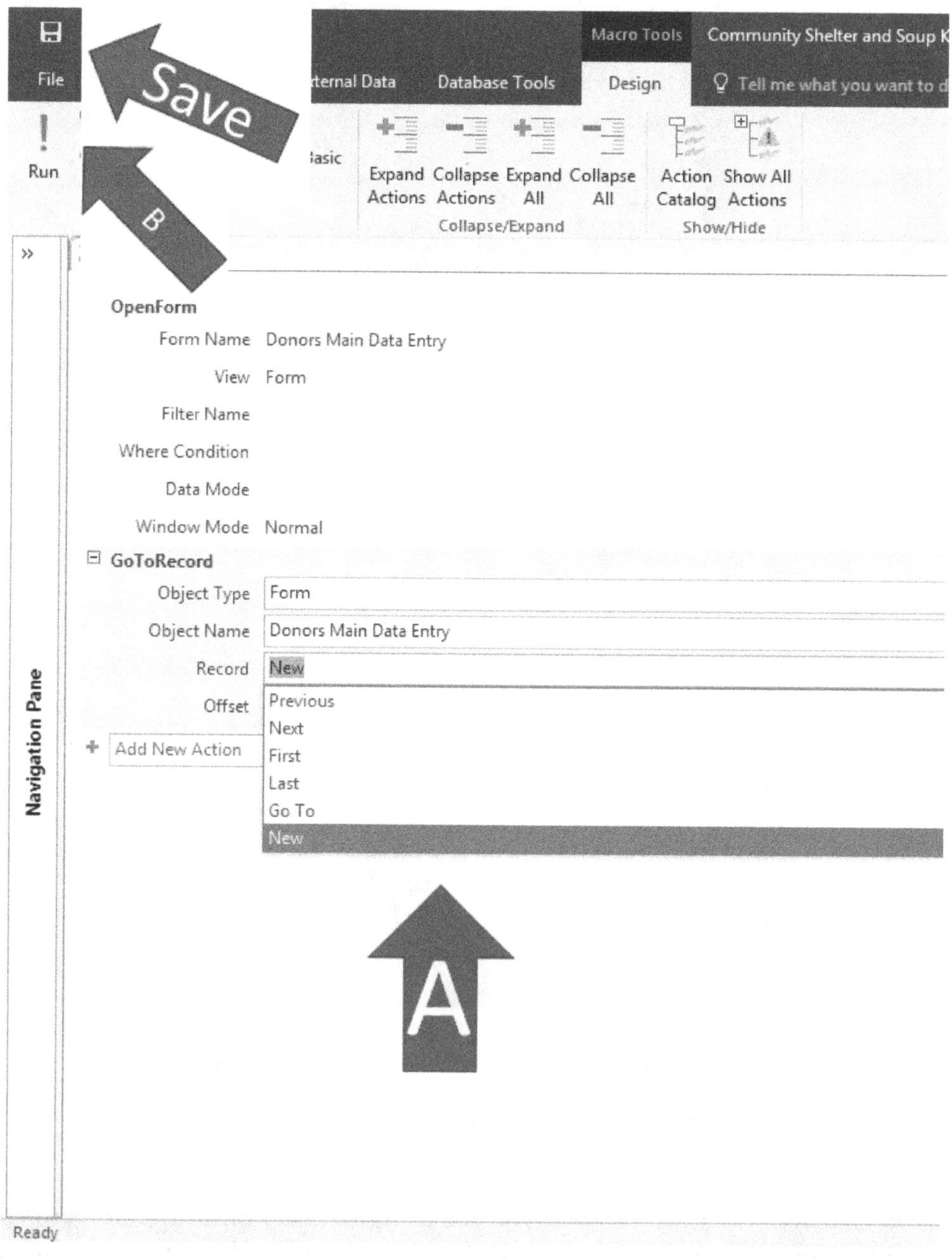

FIGURE 5.17   Creating a macro to open forms

Step 10: Run the **macro** by clicking the **Run button** in the **Design** ribbon of the **macro tools** (figure 5.17, arrow B). It will do exactly as you have commanded.

Step 11: Click **Save** (figure 5.17, arrow marked *Save*). Name the **macro** *Donors Main Data Entry*.

NOTE: You may want to invoke the **Gotorecord command** without going to the trouble of entering the **object type** or the **object name**. In many cases, Access will go to a new record of whatever **table**, **query**, or **form** is currently open. If you run the **macro** and Access tells you to enter them, then you will need to do so.

NOTE: When you save the **macro**, it will be listed among the other objects under **macros**, in the **navigation pane**.

NOTE: It is a good idea to name any **macros** the same name as the object that it is opening. If you need to change the **macro** later, it will make it easier to find the **macro** in the **navigation pane**.

Step 12: Close the **Macro**.

CREATING A MACRO TO OPEN REPORTS

Creating a **macro** to open a **report** is basically the same as it is for opening **forms**, except for three things:

1. Obviously, you must choose the **OpenReports** command in the **Add New Action** box rather than **OpenForm** command.
2. You must choose the **View** of **Print Preview** (figure 5.18, arrow). The **print preview** is the most common view used for reports unless you want the report to immediately print out upon running the **macro**. However, it is very uncommon for users to immediately print out a report upon running the **macro**, because users prefer to see a preview of a report before printing it to make sure that the report data and layout are correct.
3. You must **NOT** have the macro go to a new record, because you cannot add a record to a report. Thus, the **macro** will be a little simpler, because it will have only one step.

Therefore, when you create the **macro** to open the report named *Donors and Donations Group and Total*, the design view of the **macro** should look like what you see in figure 5.18. Please also name the **macro** *Donors and Donations Group and Total*.

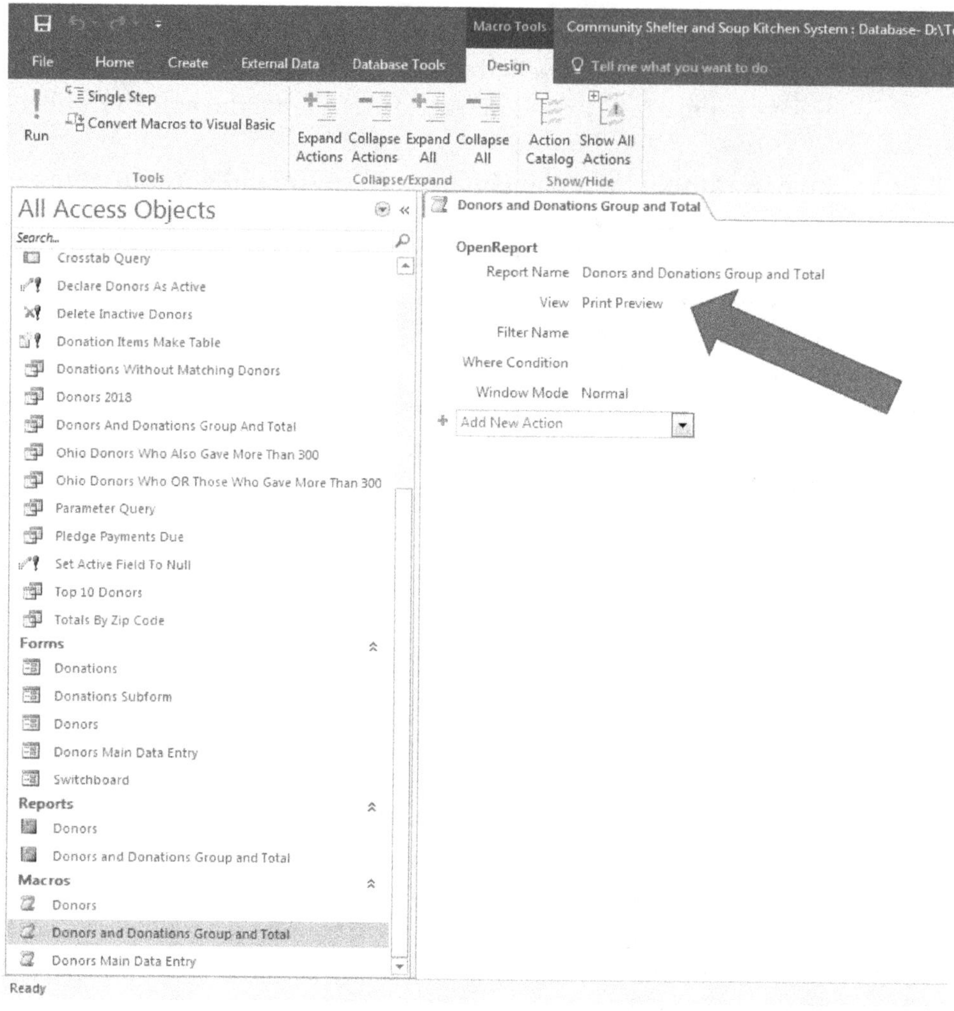

FIGURE 5.18   Creating a macro to open reports

CREATING A MACRO TO OPEN TABLES

Now let's create a **macro** that will open the *donors* table and take the user to a **new record**. Using the principles discussed earlier for creating a **macro**, you can now create it by applying the settings shown in figure 5.19. Therefore, make sure that the **macro** will do the following:

1. Open the **table** named *donors*.
2. Go to a new record in the *donors* table.

Name the **macro** *donors*.

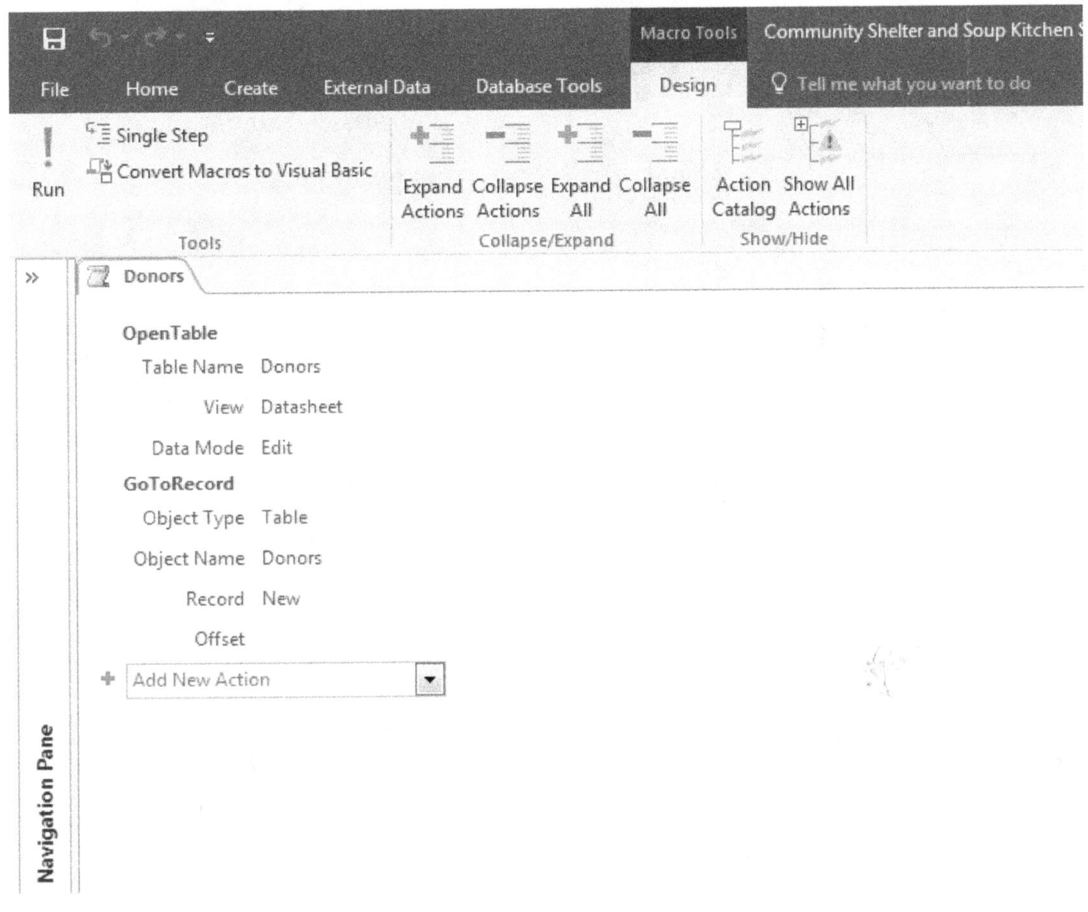

FIGURE 5.19   Creating a macro to open tables

CREATING A MACRO TO OPEN QUERIES

Now let's create a macro that will open the *calling list area code 513* **query**. Due to the fact that records will not likely be entered into the query (although queries **CAN** be used for updating records) there will be no need to add the **GoToRecord** command in the **macro**. Using the principles discussed earlier for creating a **macro**, you can now create it by applying the settings shown in figure 5.20. Therefore, make sure that the **macro** will open the **query** named *Calling List Area Code 513.* Also, name the **macro** *Calling List Area Code 513.*

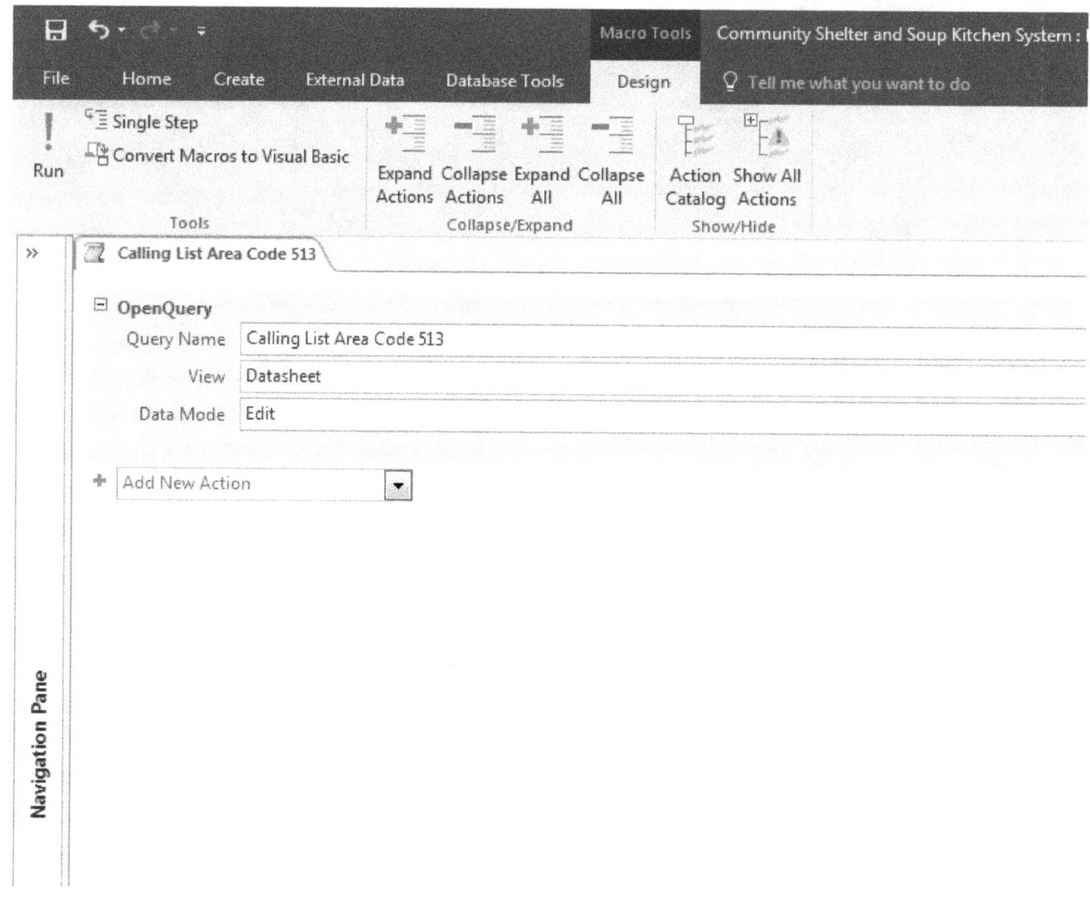

FIGURE 5.20   Creating a macro to open queries

CREATING A MACRO TO CLOSE ACCESS

One of the many things that **macros** can do is to close Access. To do so, do the following:

Step 1: Start a new **macro** as was shown in our previous discussion.

Step 2: In the **Add New Action** box, click the **QuitAccess** command (figure 5.21a, arrow).

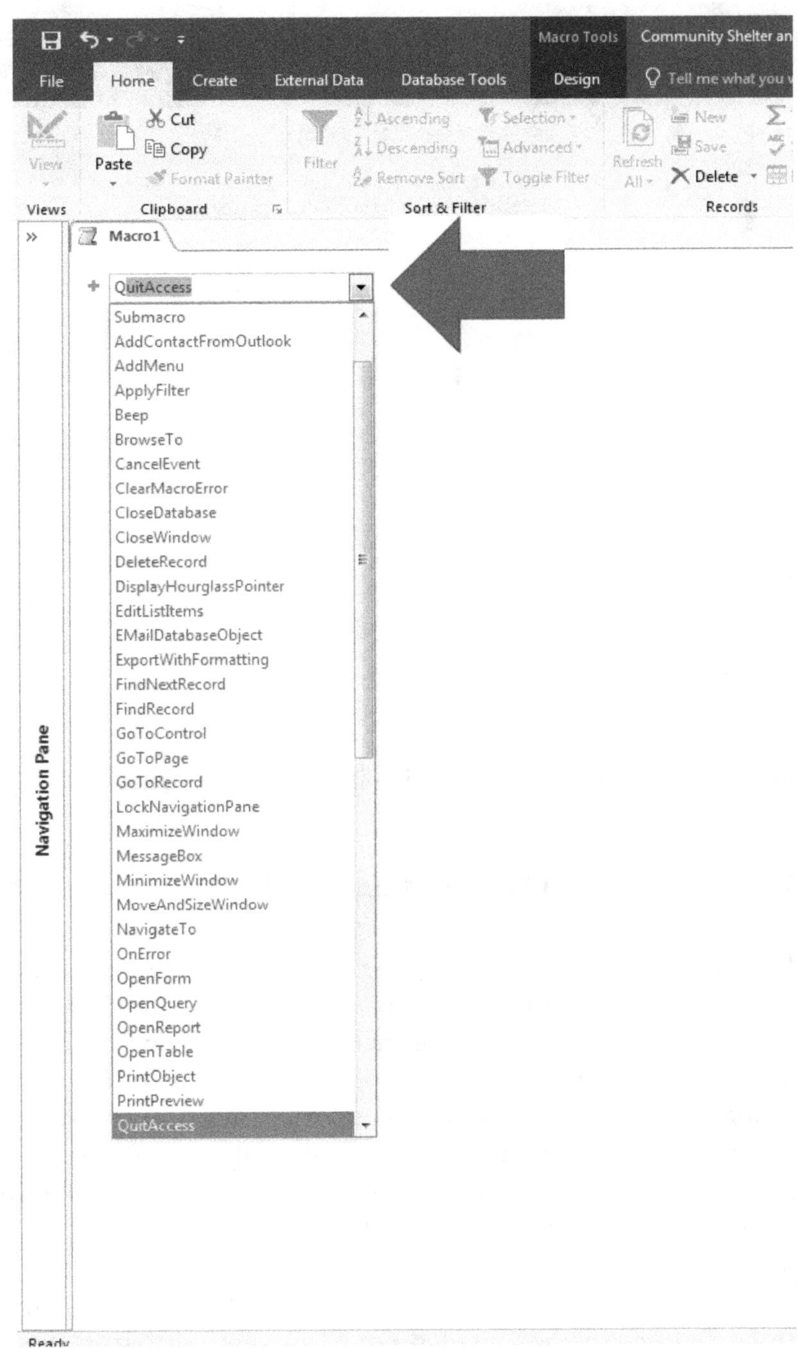

FIGURE 5.21a   Creating a macro to close access (step 1)

Step 3: Press **Enter.** When you do, your screen will look like figure 5.21b.

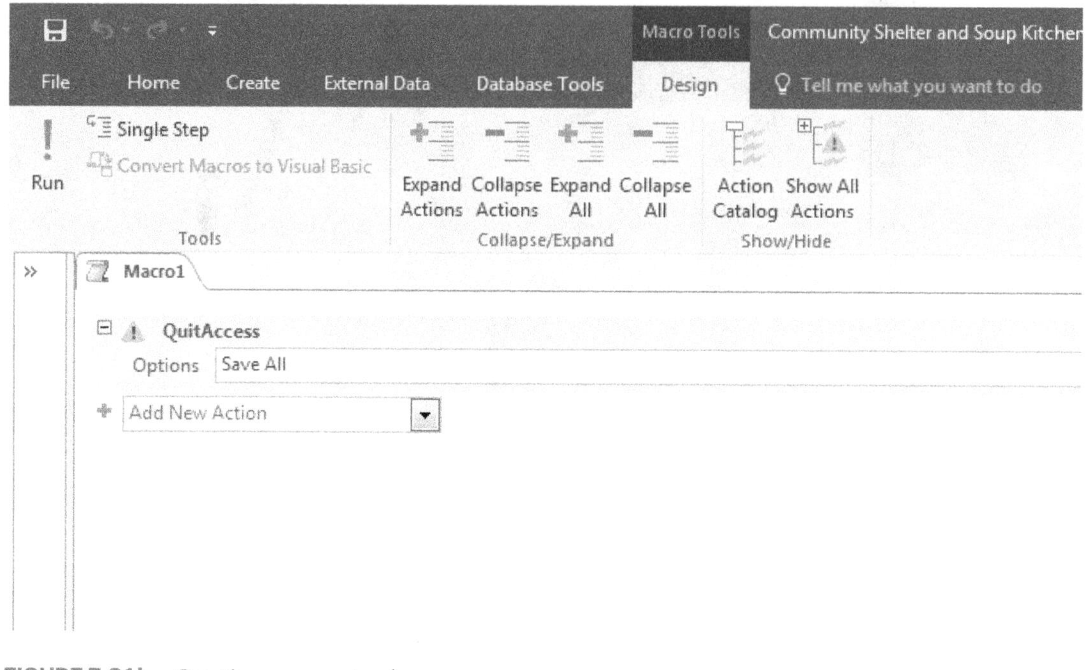

FIGURE 5.21b   Creating a macro to close access

Step 4: Close and save the **macro** and give it the name of **Quit**.

ATTACHING THE MACROS TO THE COMMAND BUTTONS

As stated before, the **command buttons** in our database are of no use unless a **macro** is attached to them. The **macros** of the database as they now stand are also of no use until the users will have a way of running them. We must now attach these **macros** to their prospective **command buttons**.

To attach a **macro** to a **Command button**, use the following steps:

Step 1: Return to the **design view** of the **switchboard** (figure 5.22).

Step 2: Click on the first button you placed at the top of the **switchboard** with the caption that reads *Donation Main Data Entry* (figure 5.22, arrow A).

Step 3: Click **Property Sheet** (figure 5.22, arrow B).

Step 4: Click on the **Event** tab of the **property sheet** (figure 5.22, arrow C).

Step 5: To the right of the box labeled **On Click** open the **drop-down list box**. Every **macro** that exists or has been created as an object in this database will be listed there. Choose the **macro** named *Donations Main Data Entry* that we created earlier (figure 5.22, arrow D). By doing this, in a sense, you are telling Access, *"In the event that you click on this button, run the Donors Main Data Entry macro."*

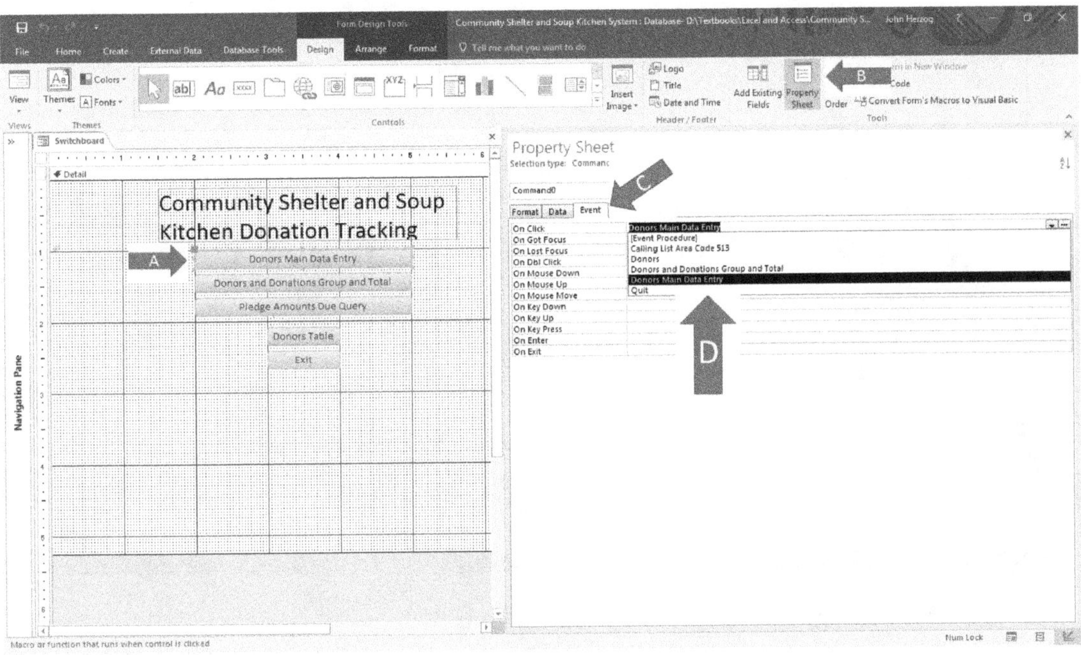

FIGURE 5.22 Attaching macros to command buttons

Step 6: Repeat this process for each of the remaining **command buttons** and **macros** by attaching the remaining **macros** with their respective buttons.

Step 7: Return to the **form view** of the switchboard and click the buttons to make sure that they do what you have set them to do.

NOTE: The **Command buttons** will not work in the form **design view**.

Step 8: Close and save the **switchboard**.

NOTE: As mentioned earlier, each time you click on a button to run a **macro**, it is called an **eevent**. In the **Event** tab, there are several other **events** such as **on Dbl-click**, on **Got Focus**, and many, many more. The most common **event** used is **On Click** and that is what we use, here.

ADDING A CLOSE BUTTON TO THE DONORS MAIN ENTRY DATA FORM

When the users of this database enter data into the *donors main data ntry* form, there are some who will not know how to close it. This is not a problem with **reports**, as Access will put a **Close Print Preview** button at the top of the screen when a report is in the **print preview**.

You can put a button on the **form** by taking the following steps:

Step 1: Create a new **macro** that will only do one step: It will close the existing window as seen in figure 5.23a.

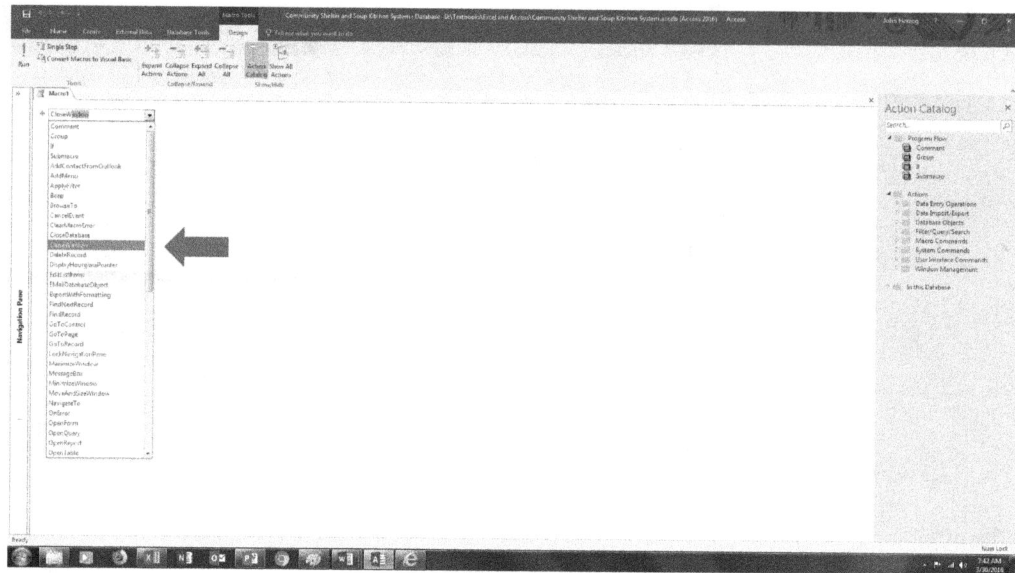

FIGURE 5.23a   Adding a close button to a data entry form

Step 2: Save the **macro** with the name of **Close**. It should look like what you see in figure 5.23b.

NOTE: It is best **NOT** to enter the **object type** or the **object name** (figure 5.23b). Access will close the form that is currently active. By leaving those boxes empty, you can use this macro for any form you want to close.

FIGURE 5.23b   Adding a close button to a data entry form

Step 3: Open the *donors main data entry* **form** in the **design view**.

Step 4: As shown in figure 5.24a and discussed before, put a **command button** in the **header** of the **form**. In this case, place it at the **8-inch** mark in the **Form Design grid** and give it a **caption** of **Main Menu** (5.24a, arrow A).

Step 5: Attach the *Close* macro to the **Command button** (5.24a, arrow B).

Step 6: Close and save the donors main data entry form.

Step 7: Let's test the *Main Menu* button now. Open the **switchboard** and click the *Donors Main Data Entry* form. After you have opened it, you should see that Access has taken you to a new record. When you click *Main Menu*, it will close the *donors main data entry* **form** and thus your **switchboard** (or **main menu**) will still be open. Therefore, in a sense, the **Main Menu** button, by closing the *donors main data entry* **form**, is displaying the **main menu**.

NOTE: An alternate way to create a **macro** for a **command button** is to create an **embedded macro**. To do so, complete the following steps:

Step 1: In the **form design view** select the button and click on the **three-dotted** box (called the **Expression Builder** button) in the **On Click Event** tab (figure 5.24b, arrow A).

Step 2: You would then see the **Choose Builder menu**. Click the **Macro Builder** option in (figure 5.24b, arrow B).

Step 3: That will take you to the **design view** of a **macro**. You would then do the same steps that were discussed earlier in creating a **macro**.

NOTE: THERE ARE TWO MAJOR PROBLEMS WITH USING THE EMBEDDED MACROS, AND THUS IT IS NOT RECOMMENDED THAT YOU USE THEM. First, if you attach an embedded macro to a Command button, the macro can only be applied to that Command button. If you wanted to open the object from another menu, you would have to recreate the macro all over again. Second, YOU SHOULD NOT PUT EMBEDDED MACROS IN DATABASES THAT WILL HAVE TWO OR MORE SIMULTANEOUS USERS. DOING SO WILL CREATE ERROR MESSAGES. IF YOU DO DECIDE TO USE THEM, THEY SHOULD ONLY BE PLACED INTO A DATABASE THAT WILL ONLY BE USED BY ONE PERSON AT A TIME.

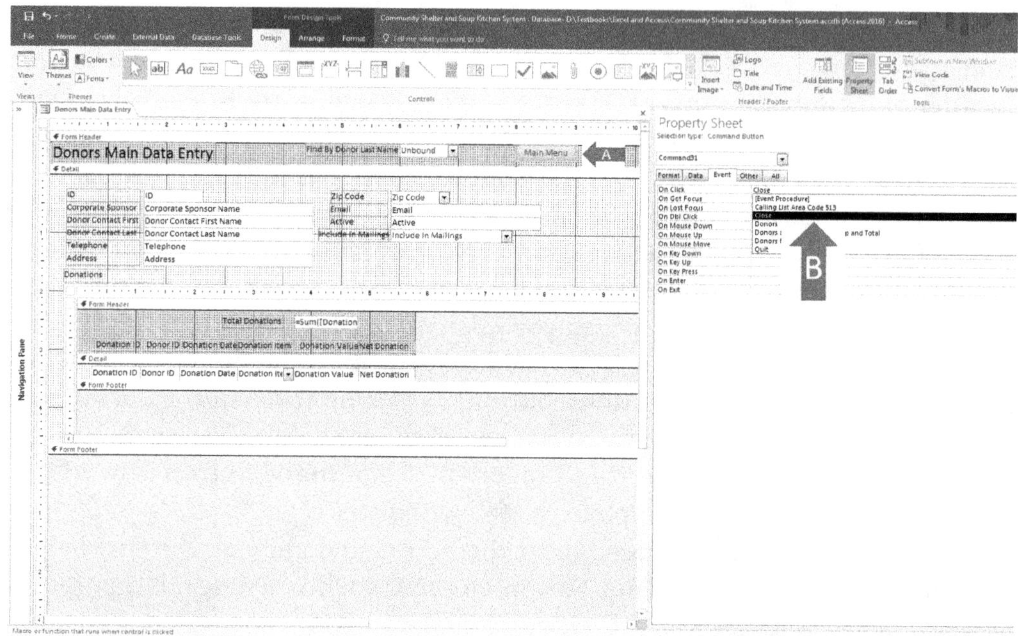

FIGURE 5.24a Adding a close button to a data entry form

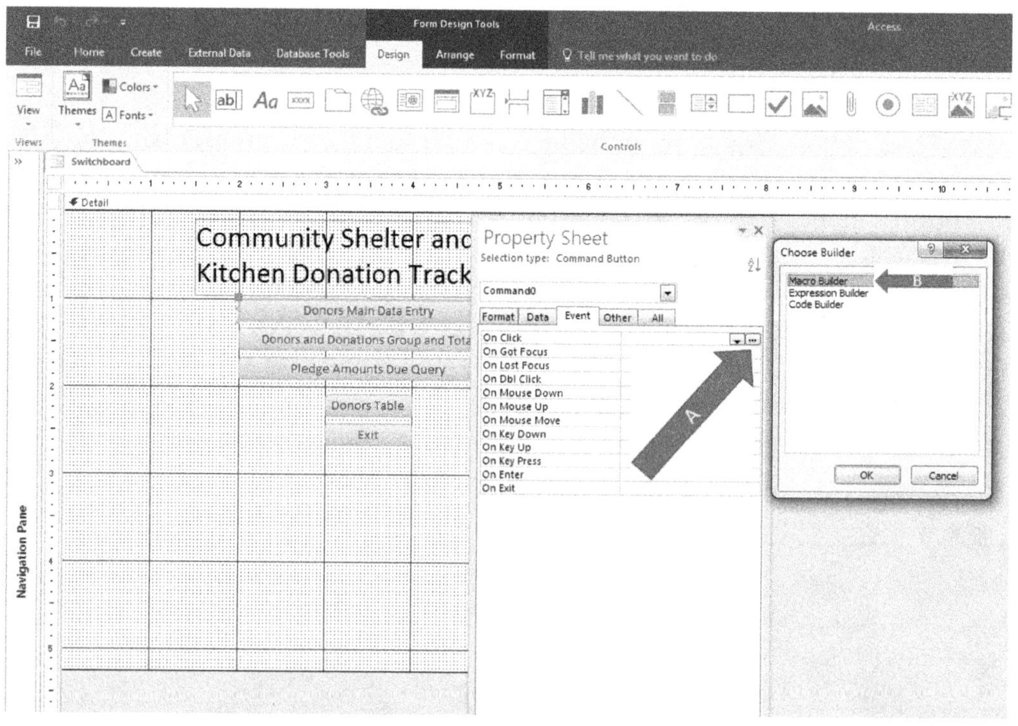

FIGURE 5.24b Creating embedded macros

MAKING THE DATABASE OPEN PROPERLY

As mentioned at the beginning of this chapter, it is vital that the users of this database know exactly what to do when they open it. If they only see what is shown in figure 5.25, they will be clueless. Therefore, we want our user-friendly **switchboard** to be the first thing the users see when they open the database. It is also important to protect the database so that users will not be able to change the design of the database objects and thus create serious damage to it.

To display the **switchboard** automatically when the database is opened, take the following steps:

Step 1: Click the **File** ribbon in the upper left of **Database window** (figure 5.25, arrow).

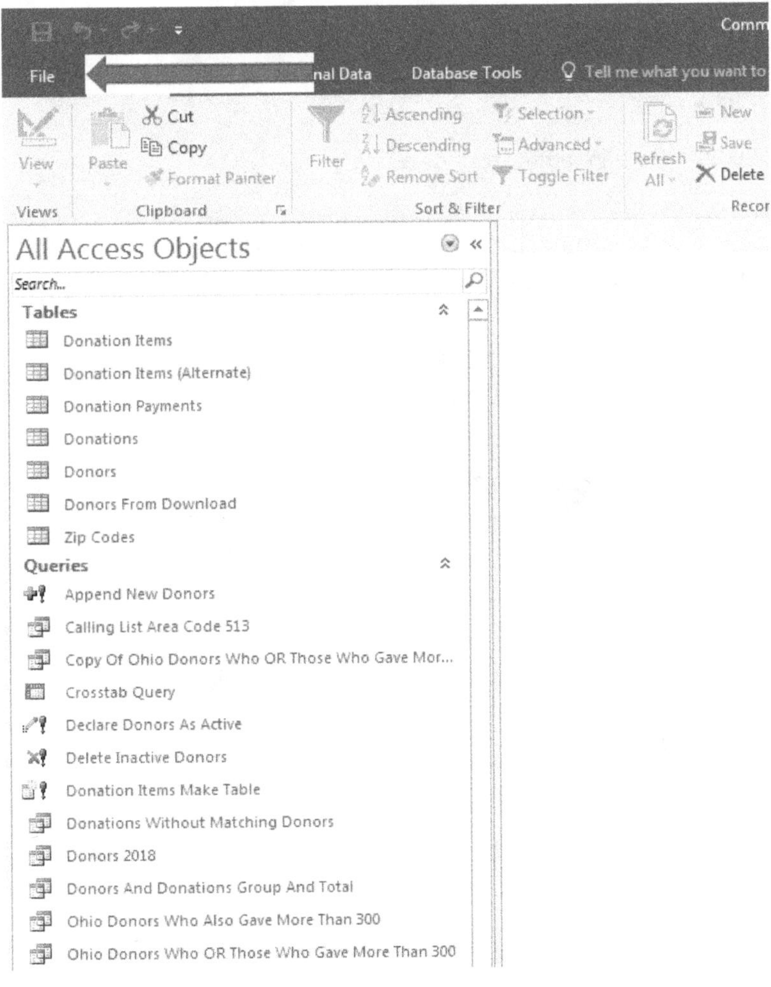

FIGURE 5.25   Start-up settings for a database

Step 2: That will take you to the menu shown in figure 5.26. Click **Options** (figure 5.26, arrow).

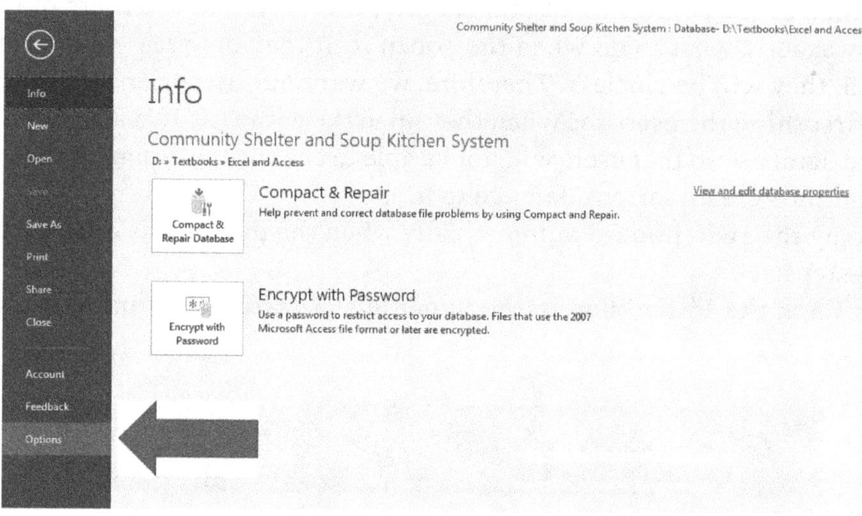

FIGURE 5.26   Start-up settings for a database

Step 3: That will take you to the **Access Options** menu shown in figure 5.27. Click **Current Database** (figure 5.27, arrow).

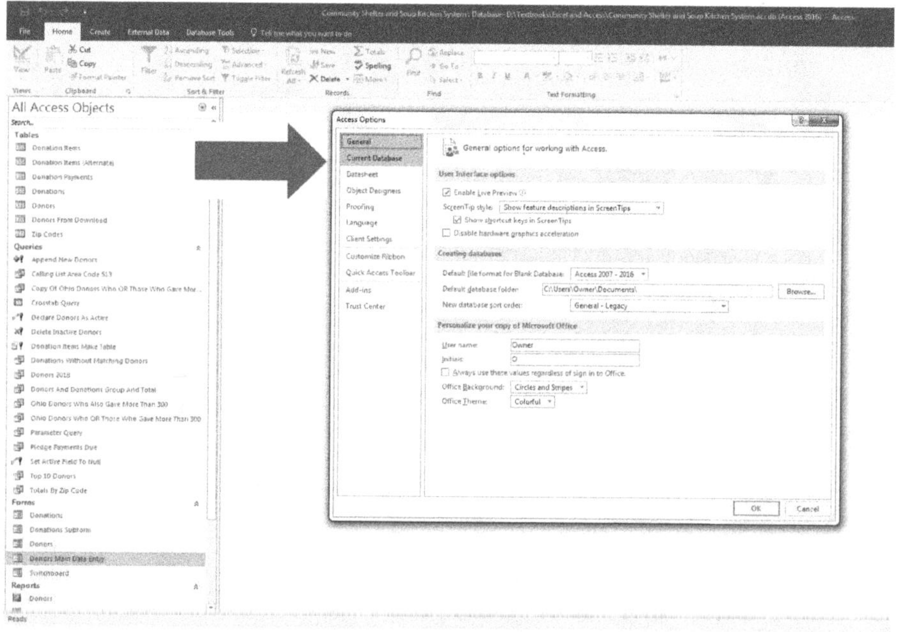

FIGURE 5.27   Start-up settings for a database

Step 4: That will take you to the menu shown in figure 5.28. In the **Display Form drop-down list box**, choose the **Switchboard form** (figure 5.28, arrow). By doing this, you are declaring what **form** is to be opened when the database is opened.

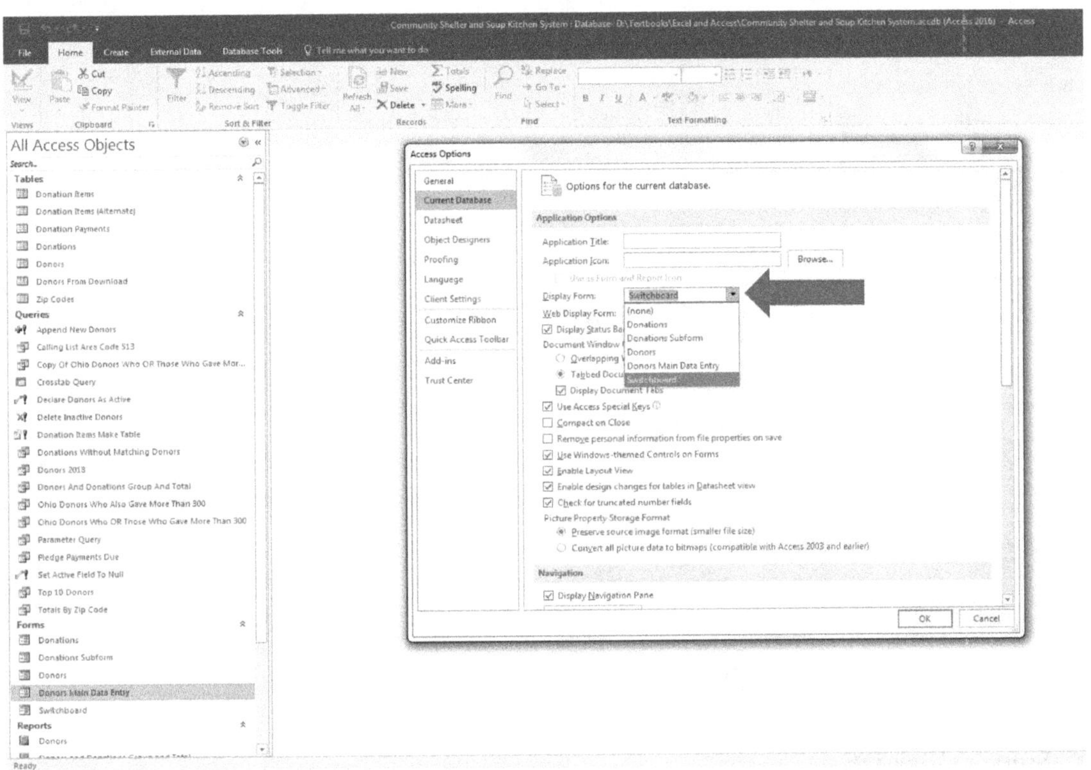

FIGURE 5.28   Start-up settings for a database

Step 5: Scroll down slightly in the **Access Options** menu until you can see the **Navigation** and **Ribbon and Toolbar option** areas (figure 5.29). Make sure that the **checkboxes** for **Display Navigation Pane** (figure 5.29, arrow A), **Allow Full Menus**, and **Allow Default Shortcut menus** (figure 5.29, arrow B) are **UNCHECKED**. By doing this, users will not be able to see the list of the database objects in the **navigation pane**; thus, they will not likely be able to damage or delete any of them. Hence, it will also prevent them from changing the design of any of the database objects.

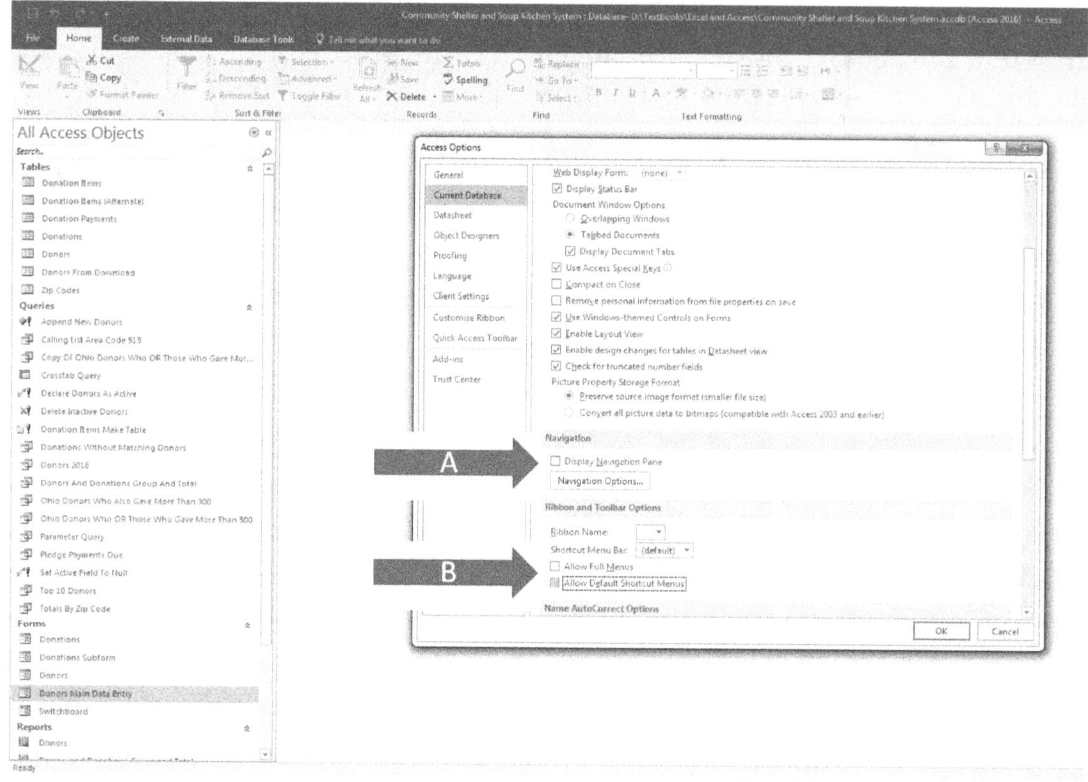

FIGURE 5.29   Start-up settings for a database

Step 6: Click **OK**.

Step 7: That will take you to the tiny menu you see in figure 5.30. It is simply informing you that many of the options you have just chosen will not be visible until you close and reopen the database. Click **OK** when you see that menu.

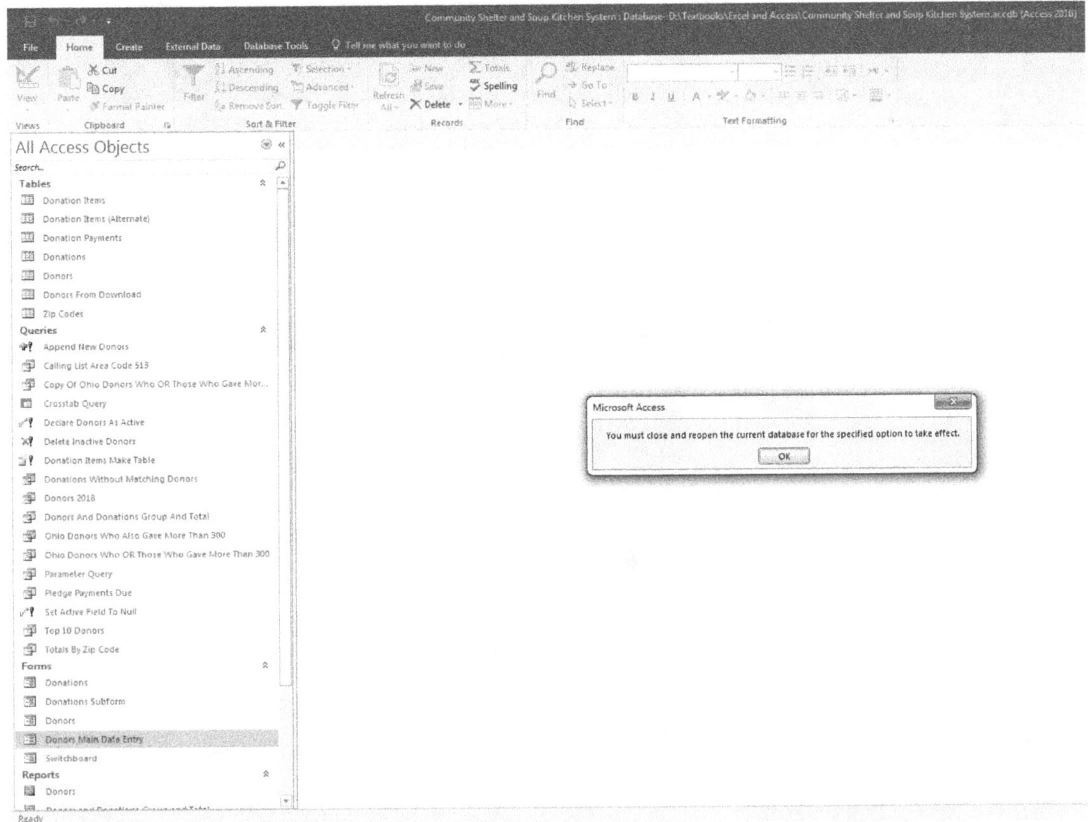

FIGURE 5.30   Start-up settings for a database

Step 8: Close the database and reopen it. Your screen will look like figure 5.31. Notice that the menus are only showing options that a user will need for entering, viewing, searching for, refreshing, and outputting data, but it will not allow a design change in the structure of the database.

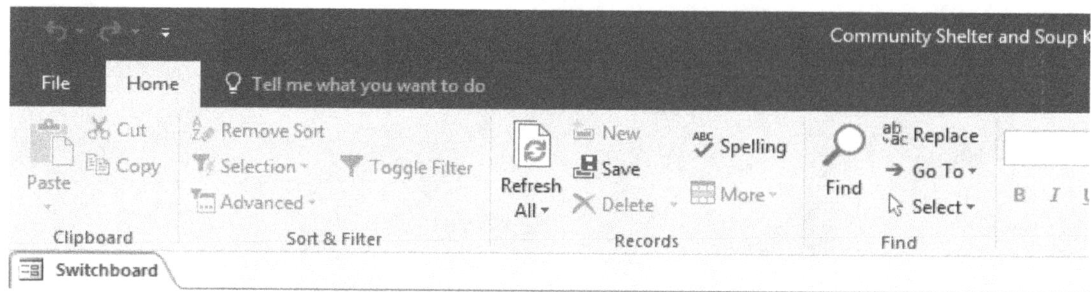

Community Shelter and Soup Kitchen Donation Tracking

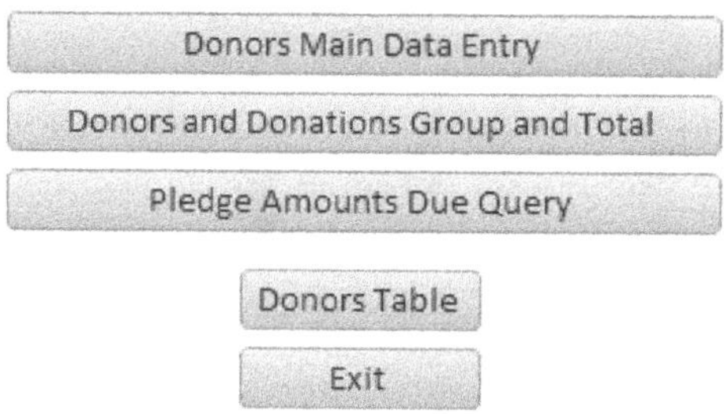

FIGURE 5.31   Start-up settings for a database

NOTE: From this point on, if **YOU**, the database creator and manager, want to make design changes, it is important for you to open the database **WHILE HOLDING DOWN THE SHIFT KEY**. That will open the database while bypassing the **startup settings** you have just now set. It will allow **YOU, AND YOU ALONE**, to make changes to the design of the database system. Data can still be entered, viewed, searched, and printed by all users.

Chapter 5: Assignment (Summary of Tasks)

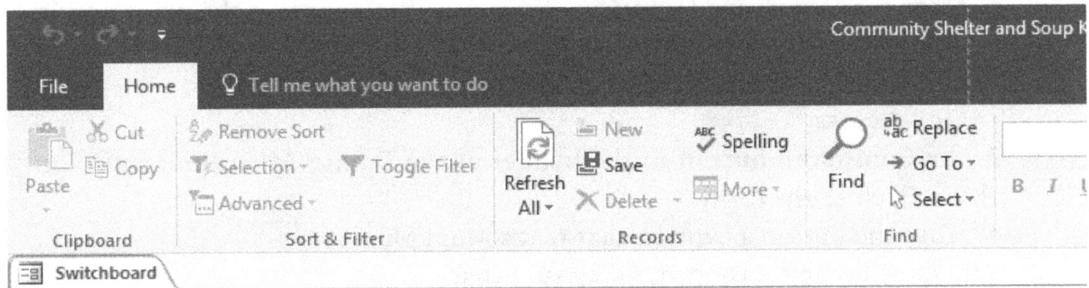

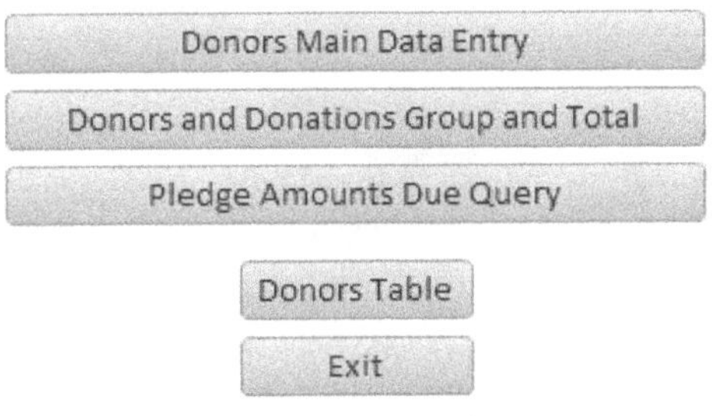

FIGURE 5.32 Switchboard assignment sample

1. Open the Community Shelter and Soup Kitchen System database.
2. Create a **switchboard** (or **main menu**) from scratch using a blank form.
   a. Remove the Record Selector and the Navigation buttons.
3. Add **Command buttons** to the **switchboard** as shown in figure 5.32 with the following captions:
   - Donors Main Data Entry
   - Donors and Donations Group and Total
   - Pledge Amounts Due
   - Donors Table
   - Exit

4. Create the following **macros** that will do the following:
   - Open the *Donors Main Data entry* form and then go to a new record.
   - Open the *Donors and Donations Group and Total* report in the **print preview**.
   - Open the pledge amounts due query.
   - Open the *donors* table and then go to a new record.
   - Quit Access.
5. Add a **Command button** in the **header** of the *Donors Main Data entry* form that will close the form.
   a. Give the button a caption that reads *Main Menu*.
6. When the database is open, make sure that
   a. the **switchboard** is the first thing that users will see;
   b. the navigation pane will be hidden; and
   c. the system will NOT allow full menus or allow default shortcut menus.

Chapter 5: Assignment (Alternate) (Summary of Tasks)

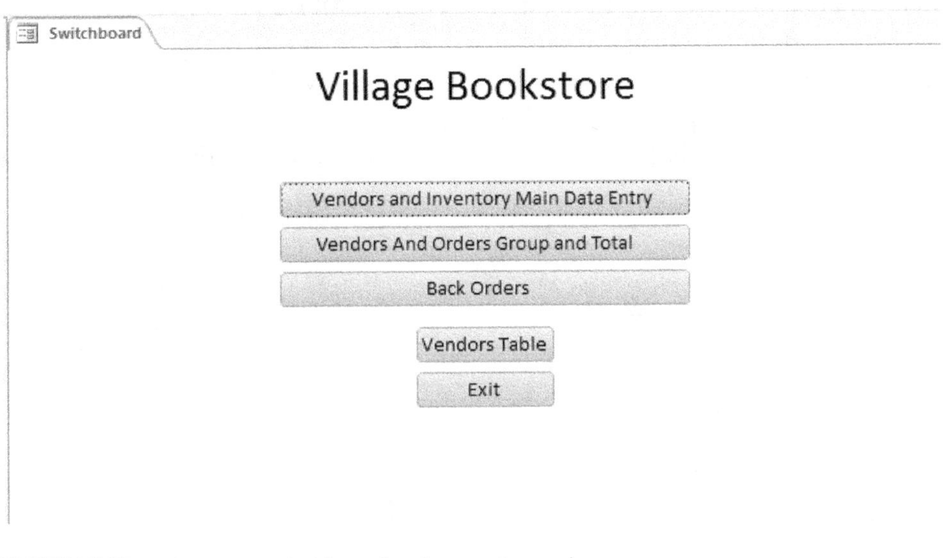

FIGURE 5.33   Alternate switchboard assignment sample

1. Open the village bookstore database system.
2. Create a **switchboard** (or **main menu**) from scratch using a blank form.
   a. Remove the Record Selector and the Navigation buttons.
3. Add **Command buttons** to the **switchboard** as shown in figure 5.33 with the following captions:
   - Vendors and Inventory Main Data Entry
   - Vendors and Orders Group and Total
   - Back Orders
   - Vendors Table
   - Exit
4. Create the following **macros** that will do the following:
   - Open the *vendors and inventory main data entry* form and then go to a new record.
   - Open the *vendors and orders group and total* report in the **print preview.**
   - Open the *back orders* query.
   - Open the *Vendors* table and then go to a new record.
   - Quit Access

5. Add a **Command button** in the **header** of the *vendors and inventory main data entry* form that will close the form.
   a. Give the button a caption that reads *Main Menu.*
6. When the database is open, make sure that
   a. the **switchboard** is the first thing that users will see;
   b. the navigation pane will be hidden; and
   c. the system will NOT allow full menus or allow default shortcut menus.

Viewing and Printing Data of a Given Record

IN THIS CHAPTER you will learn how to do the following:

1. Create a **query** of a given record.
2. Copy objects.
3. Change the **query** being read by an existing **report**.
4. Create a **macro** to run the **report** created that will first refresh the data.
5. Create a tab control.
6. **Compact and repair** an Access database.
7. Create Standard Action buttons.

CHAPTER 6 TASK: CREATING REPORTS OF DATA OF A GIVEN DONOR

From this point forward, if you close the *community shelter and soup kitchen system* database, you will need to reopen it while pressing and holding down the **Shift** key. Otherwise, you will not be able to make additional changes to the design of any of its objects.

You may remember back in chapter 4 that you created the **query** and **report** called *Donors and Donations Group and Total*. That **report** will be very helpful to those who want a summary of activity for each *donor*. However, suppose you wanted to print *donation* data of only one given *donor*. Many would make the mistake of going to the page in the **print preview** with that specific *donor* and then printing it out. That would be very time consuming, especially if you had thousands of *donors* in the report. It would also be a problem if the data about that *donor* was in the middle of data showing the *donations* of other *donors*. That data would have to be covered or removed for confidentiality purposes if the *donor* wanted to see that printout.

To resolve this problem, you must **first** do the following:

a. Make a copy of the *Donors and Donations Group And Total* **query** (giving it a name in this example of *Donors and Donations Group and Total (Given Donor))* and make the system run the **query** showing only the data of that donor, using what is called the **expression builder** (to be discussed shortly).

b. Make a copy of the *Donors and Donations Group and Total* **report** (also giving it a name of *Donors and Donations Group and Total (Given Donor).*

c. Make the *Donors and Donations Group and Total (Given Donor)* **report** read the new **query** (to be discussed shortly).

Making copies of **reports**, **queries**, or any **Access object** was covered in chapter 2. Now we can begin to make the new **query** you have just created to read only a given record (or *donor*). You can do so by taking the following steps:

Step 1: Open the *Donors Main Data Entry* **form IN THE FORM VIEW** (figure 6.1) and make sure that it is reading any one of the records. In this example, it is displaying *Record Number 7*, which is the record of the *donor* named *Finnigan's Pub.*

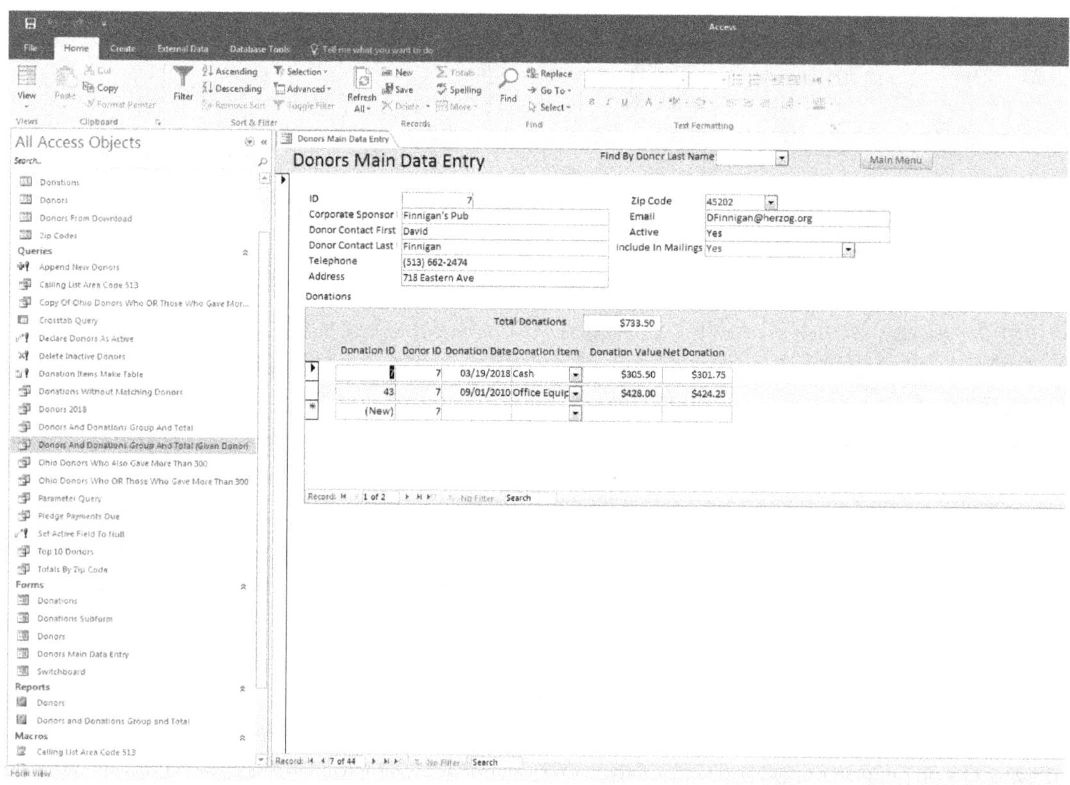

FIGURE 6.1 Creating a query of a given record

Step 2: Open the *Donors and Donations Group and Total (Given Donor)* **query** that you have just created in the **design view**.

Step 3: Your screen will look like figure 6.2. Click in the **Criteria box** below the *ID* field (6.2, arrow A).

Step 4: Click the **Builder** in the **Query Tools Design** ribbon (6.2, arrow B).

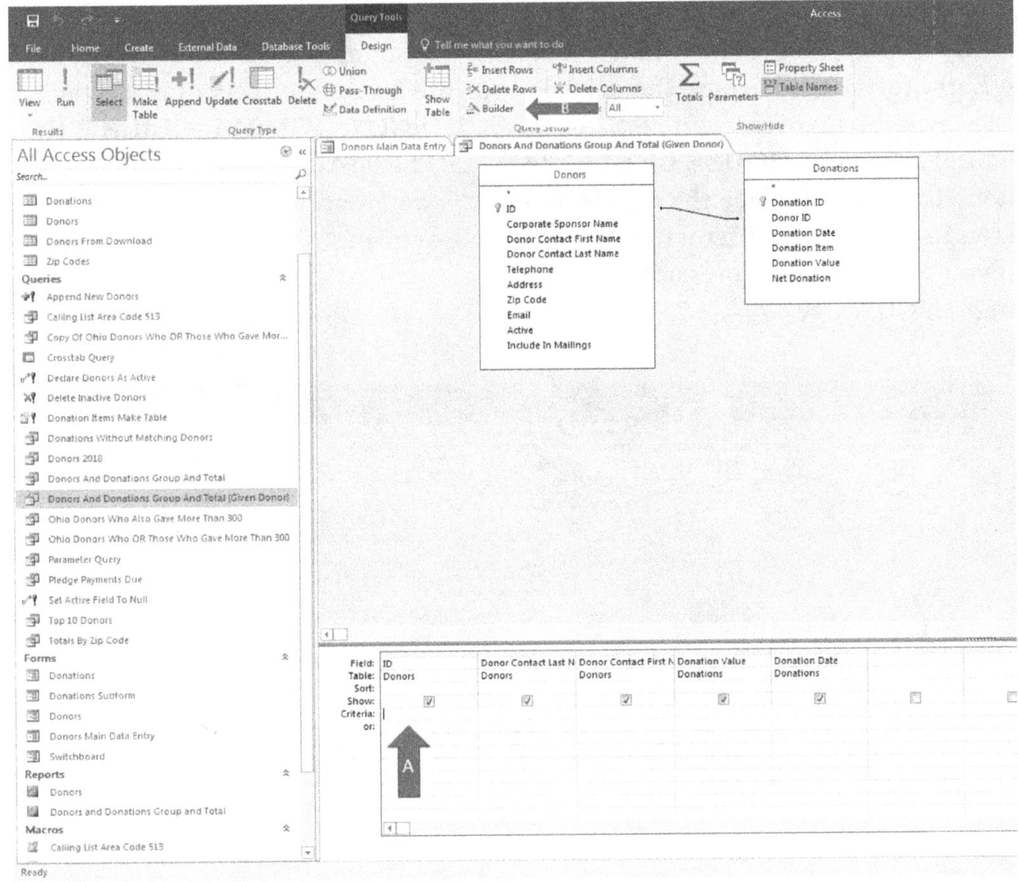

FIGURE 6.2   Creating a query of a given record

Step 5: That will give you the screen you see in figure 6.3. Double-click the name of database you are using, which of course in this case is the *Community Shelter and Soup Kitchen System* (figure 6.3, arrow A).

Step 6: Double-click the **Forms** option (figure 6.3, arrow B) beneath the name of the database.

Step 7: Double-click the **Loaded Forms** option (figure 6.3, arrow C).

NOTE: As you can see by this last step, we are looking for a record in only loaded forms. **THAT IS WHY IT IS VERY IMPORTANT THAT YOU OPEN THE FORM**

THAT YOU WANT TO USE FOR CRITERIA (the *Donors Main Data Entry* form in this example) **AS YOU DID BEFORE YOU BEGAN TO USE THE BUILDER!**

Step 8: Double-click the *Donors Main Data Entry* form among the **loaded forms** (figure 6.3, arrow D). It should be the only one there unless you have another form open. By doing this, it will give you a list of all of the **controls** in that form in the **Expression Categories** window (figure 6.3, arrow E).

Step 9: What you are trying to do here is to make sure that whatever *donor ID* is currently in the *donor ID* field that is currently being displayed in the *Donors Main Data Entry* **form** will become the criterion for the *ID* field of this **query**. Therefore, double-click the *ID* field (figure 6.3, arrow E) (**DO NOT DOUBLE-CLICK THE ID LABEL!!! NEVER CHOOSE LABLES FOR CRITERIA!**).

Step 10: That will place the expression that is referencing that field of that form into the **Expression Builder window** (figure 6.3, arrow F). The criteria have not been set if you don't see the expression there!

Step 11: Click **OK**.

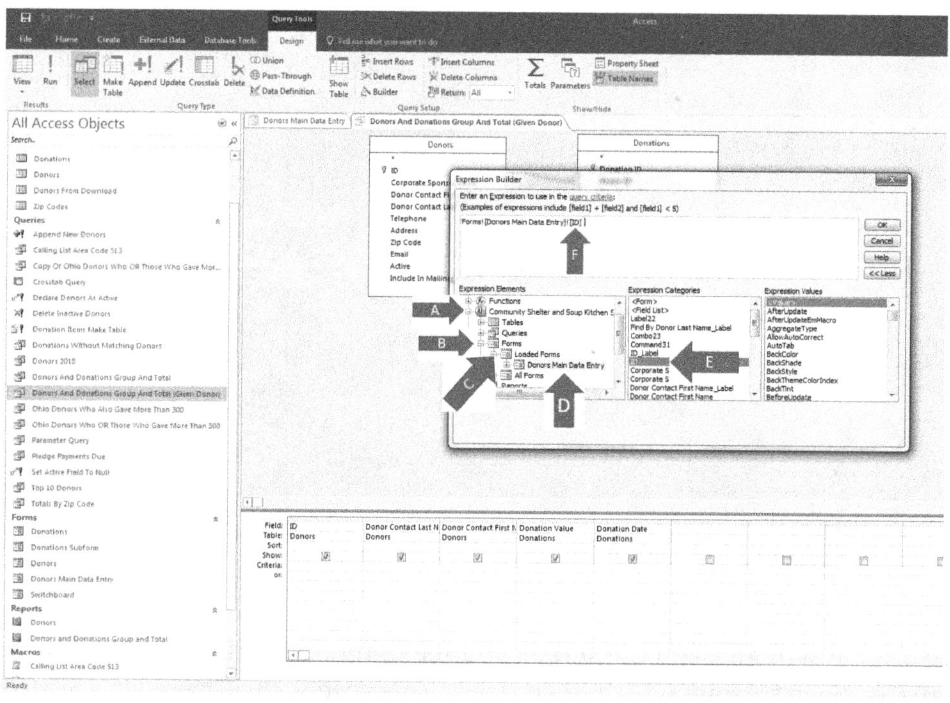

FIGURE 6.3 Creating a query of a given record

Step 12: That will take you to the screen you see in figure 6.4. Notice that the criteria box is now reading the what you placed in the **Expression Builder window**. Run the **query**.

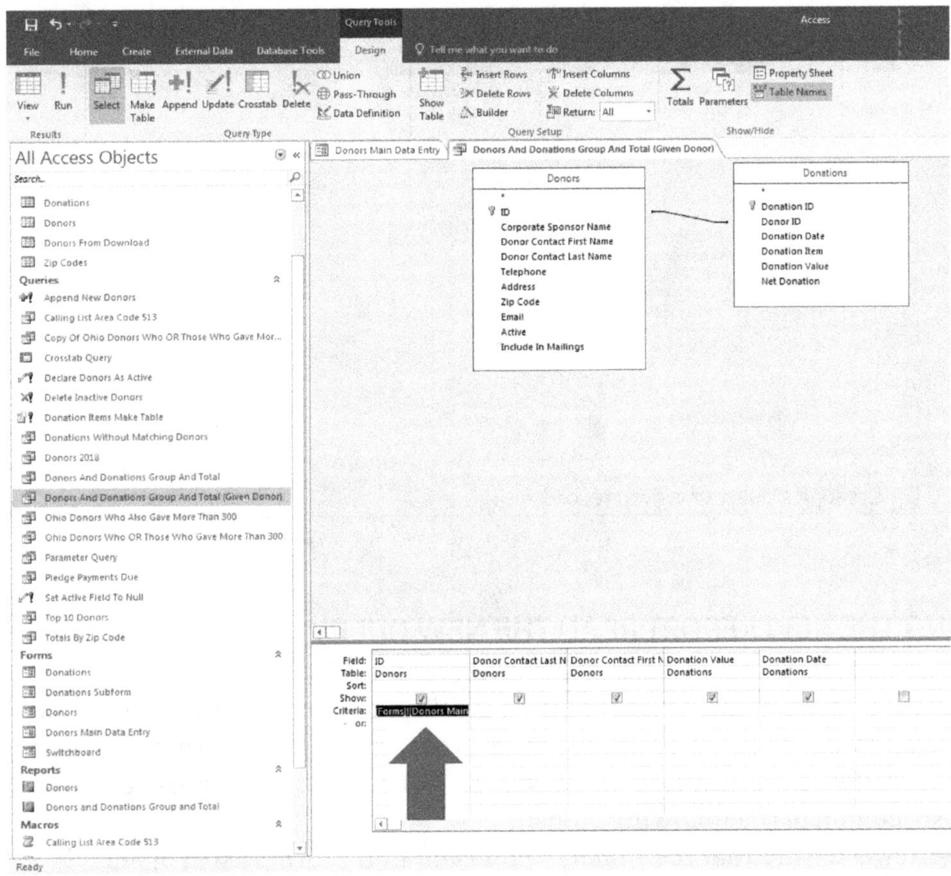

FIGURE 6.4   Creating a query of a given record

Step 13: That will take you to the screen you see in figure 6.5. Notice that it will show you only the first *donor* and *donations* that were displayed in the *Donors Main Data Entry* form (which in this example, as mentioned before, is the *donor* named *Finnigan's Pub*).

NOTE: If you move to a different record in the form and then rerun the query, it will display the record to which you have moved. Save the **query**.

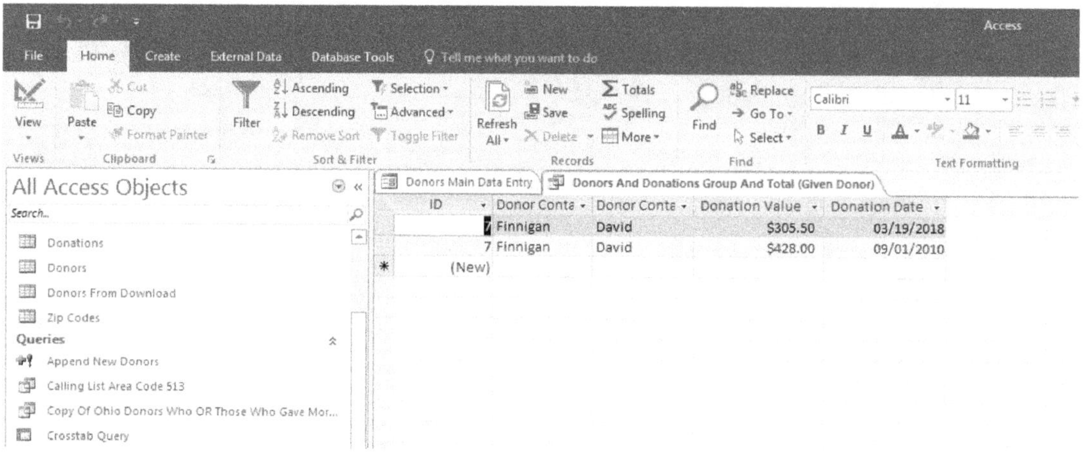

FIGURE 6.5 Creating a query of a given record

MAKING THE NEW REPORT READ THE NEW QUERY

Now that you have created and modified the *Donors and Donations Group and Total (Given Donor)* query that will only show a given, displayed record, you will need to have the *Donors and Donations Group and Total (Given Donor)* **report** read that query. To do so, complete the following steps:

Step 1: Make sure that the *Donors Main Data Entry* form is still open.

Step 2: Go to the **design view** of the newly created *Donors and Donations Group and Total (Given Donor)* **report** (figure 6.6).

Step 3: Double-click the **Form/Report Selection box** in the upper left of the **Design View grid** (figure 6.6, arrow A)

NOTE: As mentioned in chapter 5, just as you can select a control and view or edit the properties of it, you can also see the properties of an entire form (or report) by double-clicking the **Form/Report selection box**. When you do, it will open the **property sheet** for the entire **form** or **report** and it will also place a tiny black box in the **Form/Report selection** box.

Step 4: Change the **record source** in the **All** tab of the **property sheet** (figure 6.6, arrow B) so that it will now read the **query** *Donors and Donations Group and Total (Given Donor)*.

Step 5: Run the **report** in the **print preview**. Your screen will look similar to what you see in figure 6.7. Close and save the **report**.

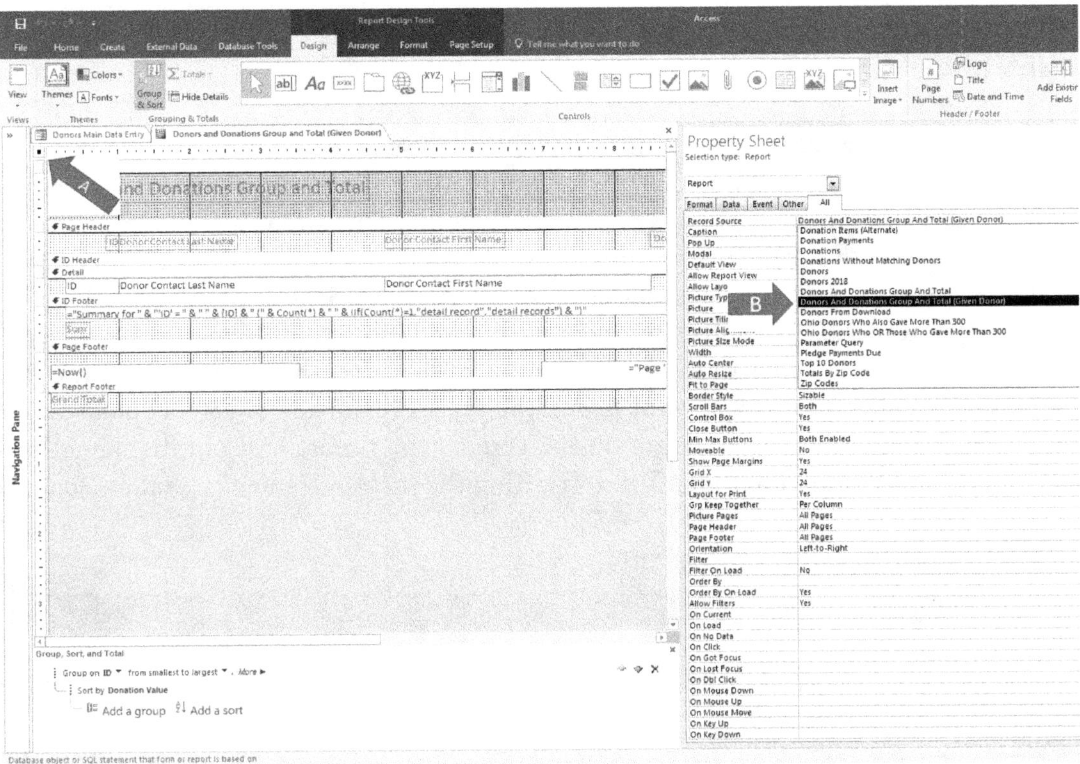

FIGURE 6.6 Changing a report record source

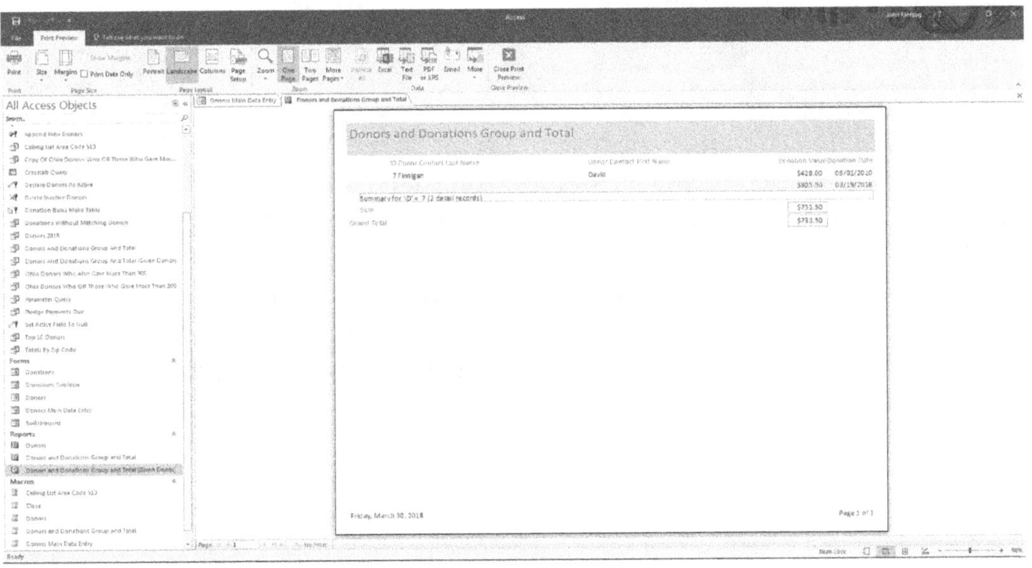

FIGURE 6.7 Changing a report record source results

CREATE A MACRO/COMMAND BUTTON TO RUN THE REPORTS OF DATA OF A GIVEN RECORD

Now let's make it easy to run the report that shows the data of a given *donor* by the use of a **Macro** and **Command button**. It would only make sense to run it from the same screen that is being read by the **query**. Create the **macro** by using the following steps:

Step 1: Using the concepts given earlier, create the **macro** with the specifications seen in figure 6.8. Please notice that the **MACRO IS SET TO REFRESH THE ACTIVE/CURRENT RECORD** before opening the *Donors and Donations Group and Total (Given Donor)* **report** in the **print preview**. The **refresh** step or **command** in the **macro** is important. If you want to run the report just after adding a *donation* or making a change to any of the data on the screen, the report will not display all of the updated information without first refreshing the record. Name the **macro** *Donors and Donations Group and Total (Given Donor)*.

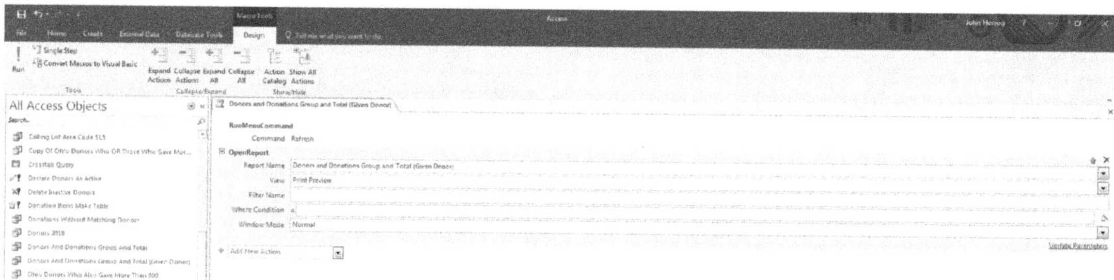

FIGURE 6.8   Creating a macro to run a report of a given record

Step 2: Also using the concepts given earlier, add a **Command button** in the **header** of the *Donors Main Data Entry* form to the right of the **Main Menu** button (figure 6.9) that will run the *Donors and Donations Group and Total (Given Donor)* **macro** (by attaching it as discussed earlier). Use *Print* as the caption for the **Command button** (figure 6.9).

NOTE: In figure 6.9 we have moved back to the first record, thus if you change the *Donation Value in a Donation of Maxwell Hawkes*, the changes should be reflected in the report when you click the **Command button** you have just created.

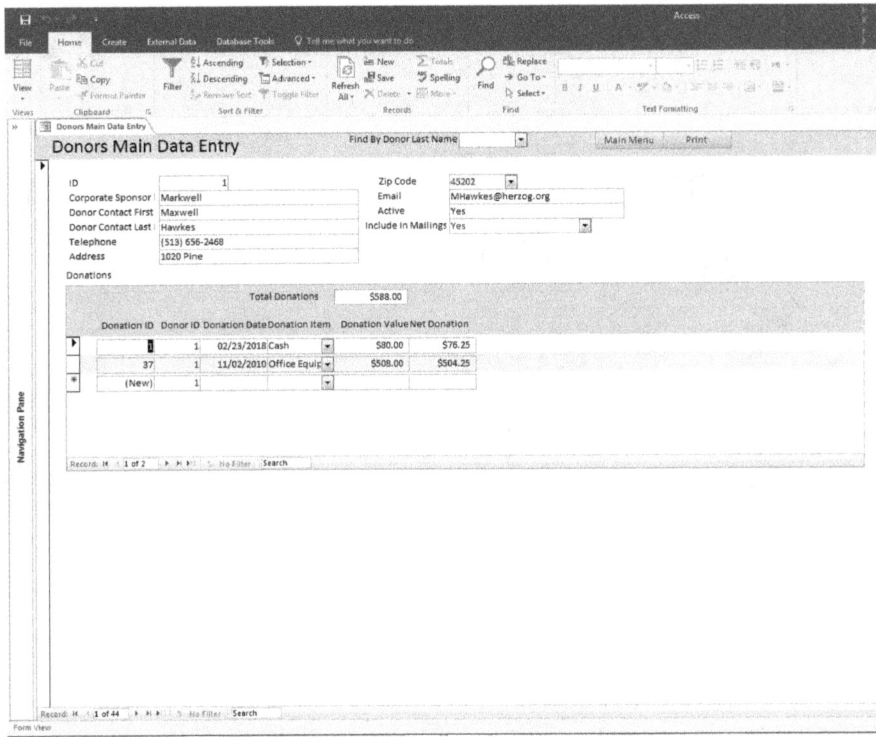

FIGURE 6.9   Creating a button to run a report of a given record

ADD A TAB CONTROL TO A FORM

There are some **forms** that require so many fields that it requires the user to have to use the scroll bar quite often. This is not good, because scrolling takes time, and if you have many users of the system, the minutes of wasted time will add up. Numerous fields in a form also make the screens look busy, convoluted, unorganized, and confusing. That's why **tab controls** are needed in forms. Even though your database doesn't have many fields, it is still important to show you how to create **tab controls**. In this example, you will only be putting the **tab control** in the **main form**. To do so, take the following steps:

Step 1: Open the *Donors Main Data Entry* form in the **design view** (figure 6.10).

Step 2: You must remove the controls in the **main form** in order to add the **tab control** in their place. An alternate way to select controls, discussed before, is to place your mouse on the **left ruler bar** in the **form design view** and click and drag downward (figure 6.10, arrow). It will draw lines of a box around the controls. It will select all controls that the lines touch and/or are surrounded by the lines, even if the lines don't completely engulf them.

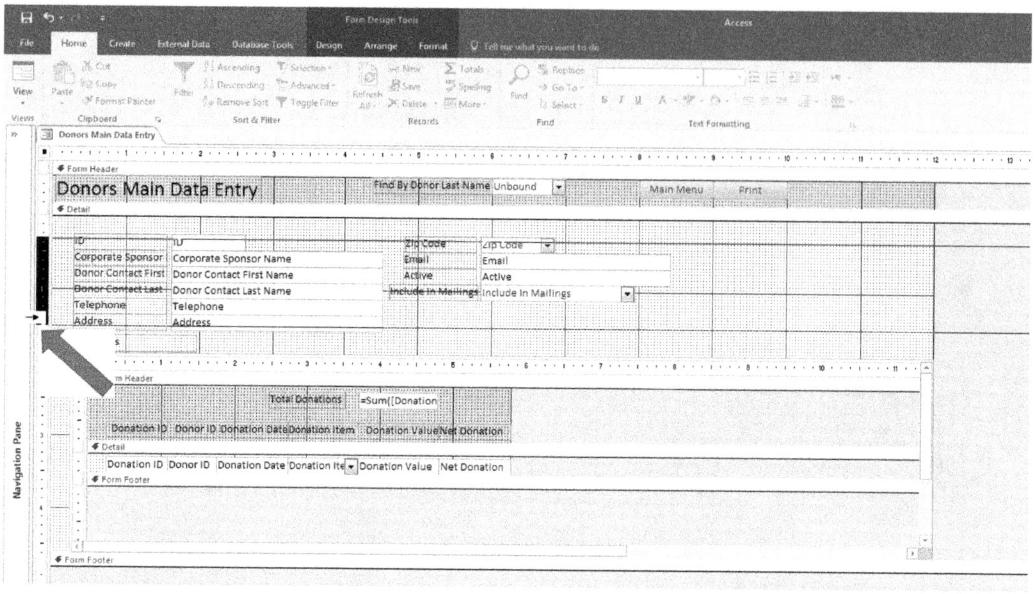

FIGURE 6.10　Adding a tab control

Step 3: After letting go of the mouse, your screen will look like figure 6.11. Click the **Cut** command in the **Home** ribbon (figure 6.11, arrow), or you can press and hold **Ctrl** and then tap the **X** on your keyboard.

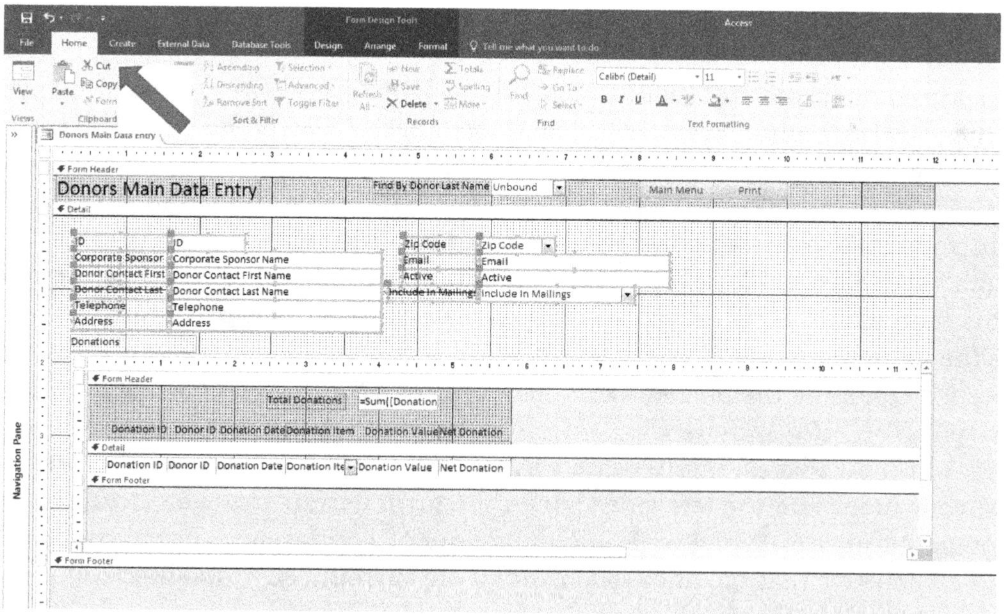

FIGURE 6.11　Adding a tab control

Step 4: That will temporarily remove all of your **controls** from the **main form**. In the **Controls toolbox** of the **Design** tab of the **form design tools**, click (do not click and drag) the Tab Control icon (figure 6.12, arrow A).

Step 5: Click once (do not click and drag) at the furthest point in the upper left of the detail area of the **Form Design grid**. The **tab control** will now be placed at that point (figure 6.12, arrow B).

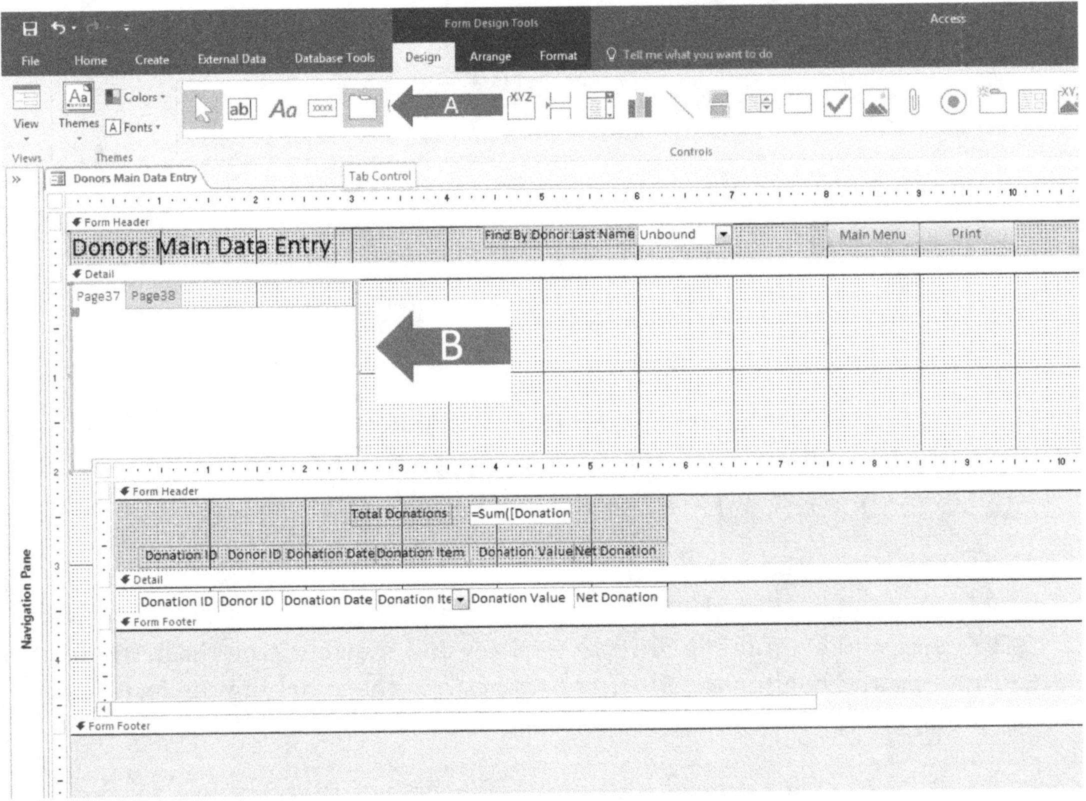

FIGURE 6.12 Adding a tab control

Step 6: Click **TWICE** on the **tab** in which you want to place the controls (in this case the tab is labeled as *Page37* as you can see in figure 6.13).

NOTE: In your case, the tab could be labeled something else. As mentioned earlier, the default numbers Access puts on **labels** and **controls** are automatic and are not relevant as we will change the them anyway.

Step 7: Click **Paste** in the **Home** ribbon (or you can press and hold **Ctrl** and then tap **V** on your keyboard). Your screen will look like figure 6.13.

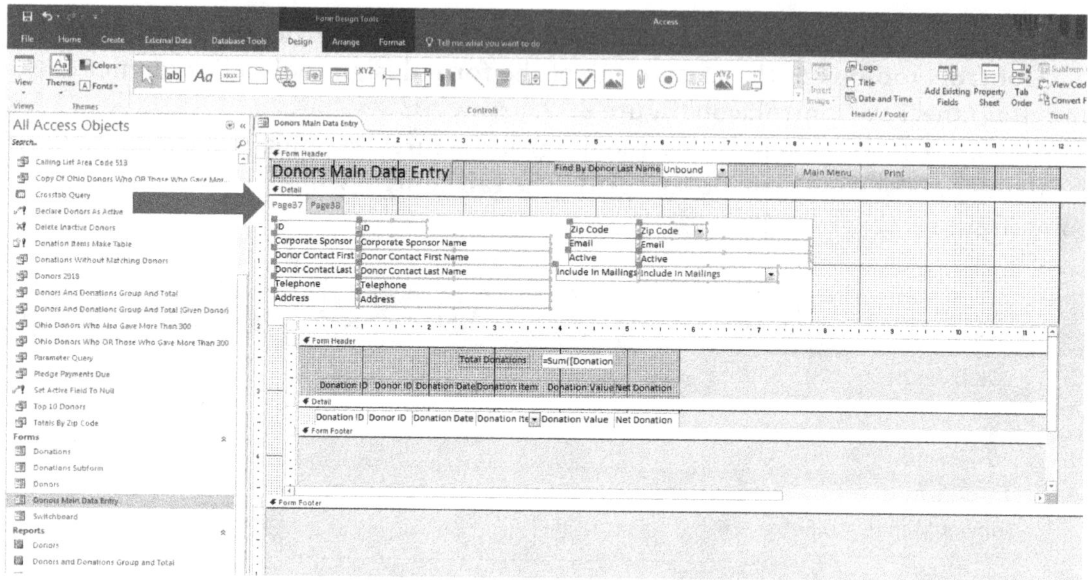

FIGURE 6.13 Adding a tab control

Now we need to move the *zip code, email, active,* and *include in mailings* fields into the other **tab**. To do so, complete the following steps:

Step 1: Select those fields **ONLY** by the following:

a. Clicking once on the **zip code** field
b. Pressing and holding the **Shift** or **Ctrl** key and then clicking the *email* field
c. Pressing and holding the **Shift** or **Ctrl** key and then clicking the *active* field
d. Pressing and holding the **Shift** or **Ctrl** key and then clicking the *include in mailings* field

Your screen should look like figure 6.14.

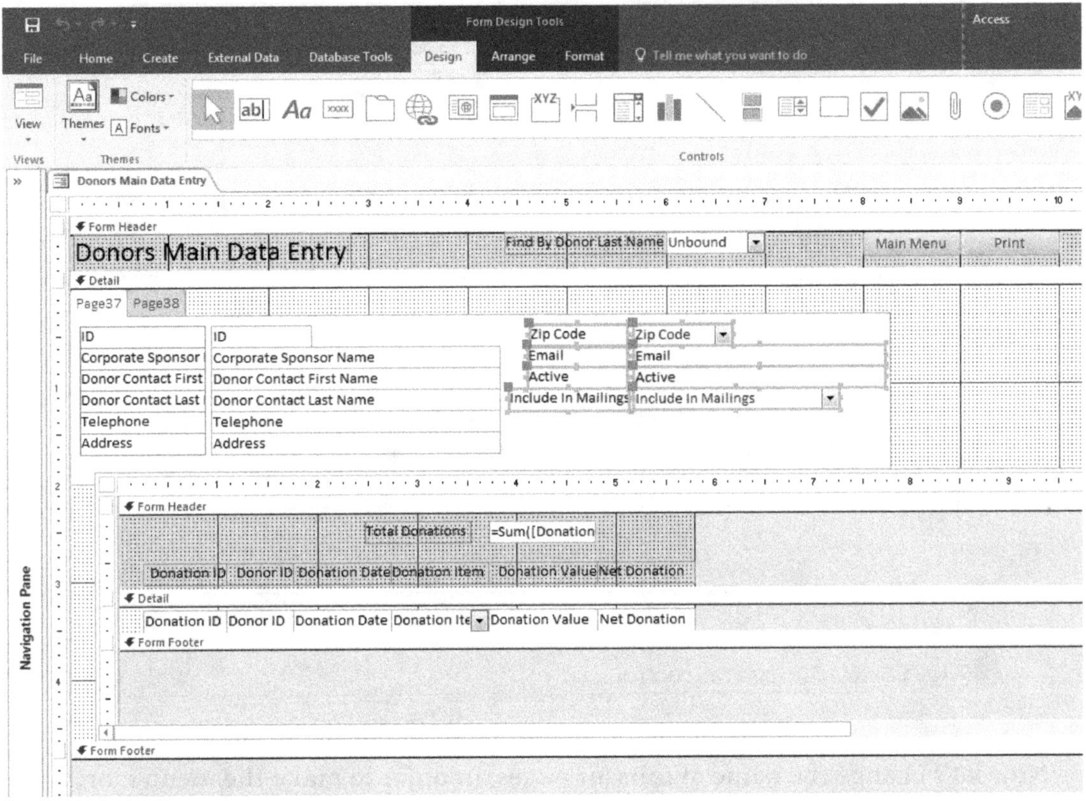

FIGURE 6.14   Adding a tab control

Step 2: Click **Cut** in the **Home** ribbon, or press **Ctrl** and then **X** on your keyboard (figure 6.15, arrow A).

Step 3: Click **TWICE** on the other **tab** (or **page,** which in this example is named *Page38)* as shown in Figure 6.15, arrow B.

Step 4: Click **Paste** in the **Home** ribbon (figure 6.15, arrow C). The fields will now be in a separate **tab** (or **page**) as seen in figure 6.15.

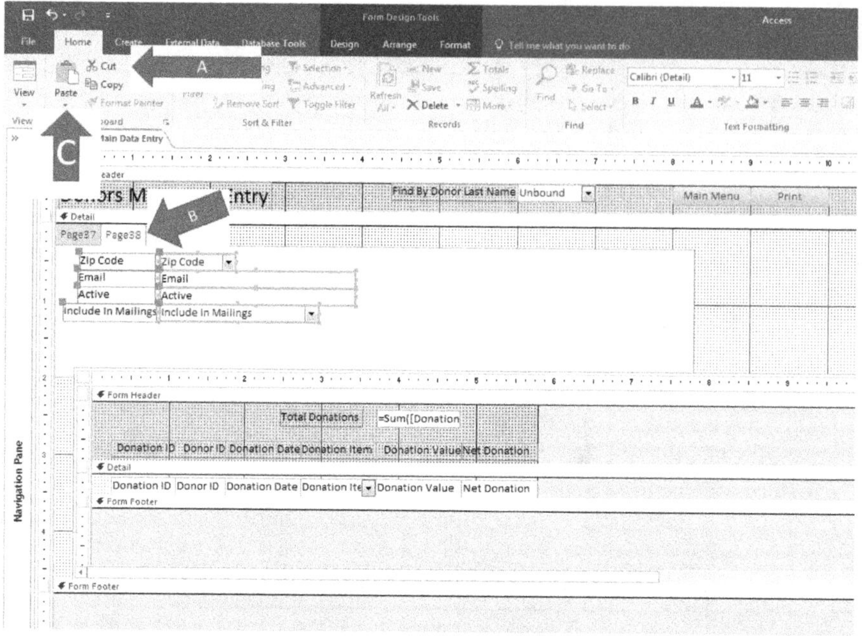

FIGURE 6.15   Adding a tab control

Now let's change the name of **tabs** (or **pages**) in order to make the menu more user friendly. To do so:

Step 1: Click on the **tab TWICE** to which you just sent the *zip code, email, active,* and *include in mailings* fields (as shown in Figure 6.16, arrow A).

Step 2: Click the **Property Sheet** (6-16, arrow B).

Step 3: In the **All tab** of the **property sheet**, in the box labeled **Name** type the words: *Zip Code Email and Misc.* (6-16, arrow C).

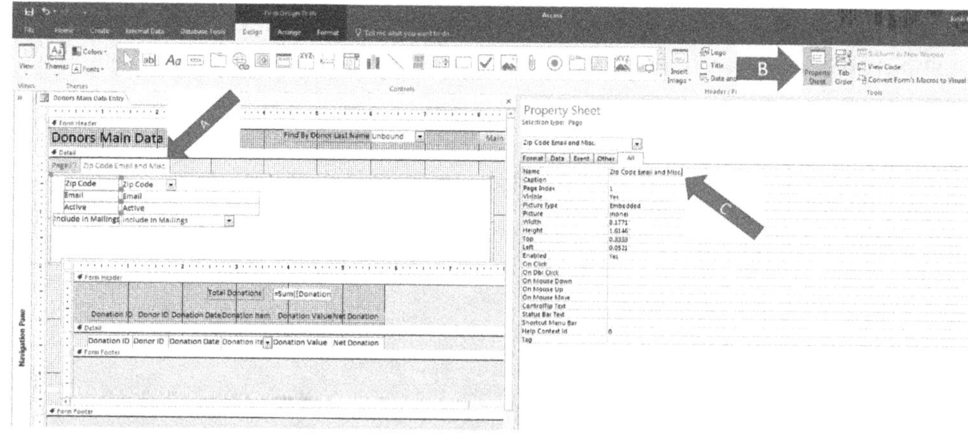

FIGURE 6.16   Adding a tab control

Step 4: Repeat this process with the other **tab**. Click on the other **tab** (in this case *Page 37*) and in the **Name box**, type the words: *Name Telephone Address*.

If the **tab control** overlaps with your **sub-form**, you will need to continue with the following steps:

Step 5: Click on the **sub-form** and tap the **Down Arrow Cursor Control key** (↓) on your keyboard until the label to the **sub-form** is completely visible (the **sub-form label** will automatically move with the **sub-form**). You may also want to make sure that the **sub-form** is completely to the left. To do so, tap the **Left Arrow Cursor Control key** (←) on your keyboard until the **sub-form** is as far to the left in the **grid** as possible. This will make it so there will be enough room to see the **tab control** you have just inserted.

Step 6: View the form in the **form view**. Your screen should look like what you see in figure 6.17.

Step 7: Save the **form** and close it.

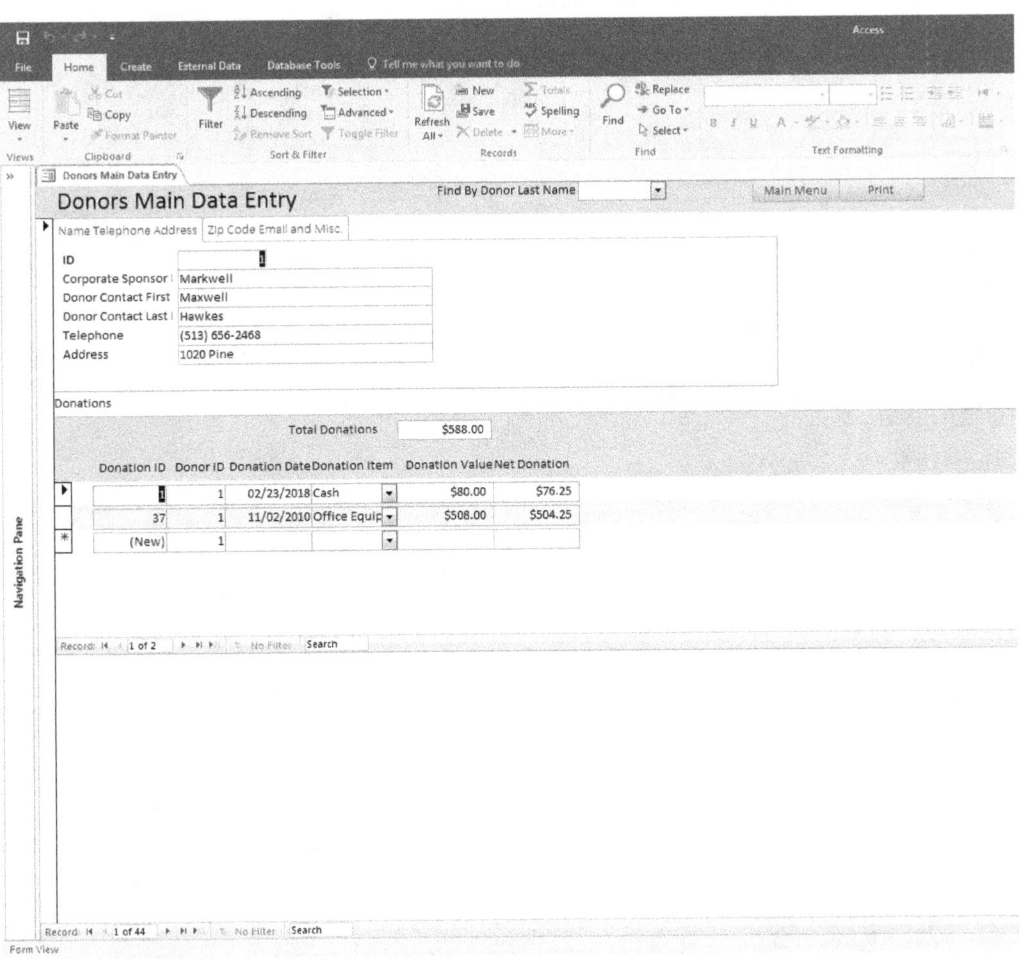

FIGURE 6.17   Adding a tab control final results

COMPACTING AND REPAIRING THE DATABASE

Every day that you use a database you should perform the **compact and repair** function of Access. Doing so will remove a lot of unused characters reserved for records. It will also remove any corruption to the database that can occur when it is passed around a network to its many users. Sometimes you will get error messages, or you may experience abnormal things about the database that cannot be easily explained. The **compact and repair** function can often remedy these things as well.

YOU CANNOT COMPACT AND REPAIR AN ACCESS DATABASE IF EVEN ONE PERSON BESIDES YOU IS WORKING ON IT (if your database is available on a network or if you have opened it twice).

To **compact and repair** the database, complete the following steps:

Step 1: Making sure that your database is still in a state where you can see all menus, click the **File** ribbon (figure 6.18, arrow).

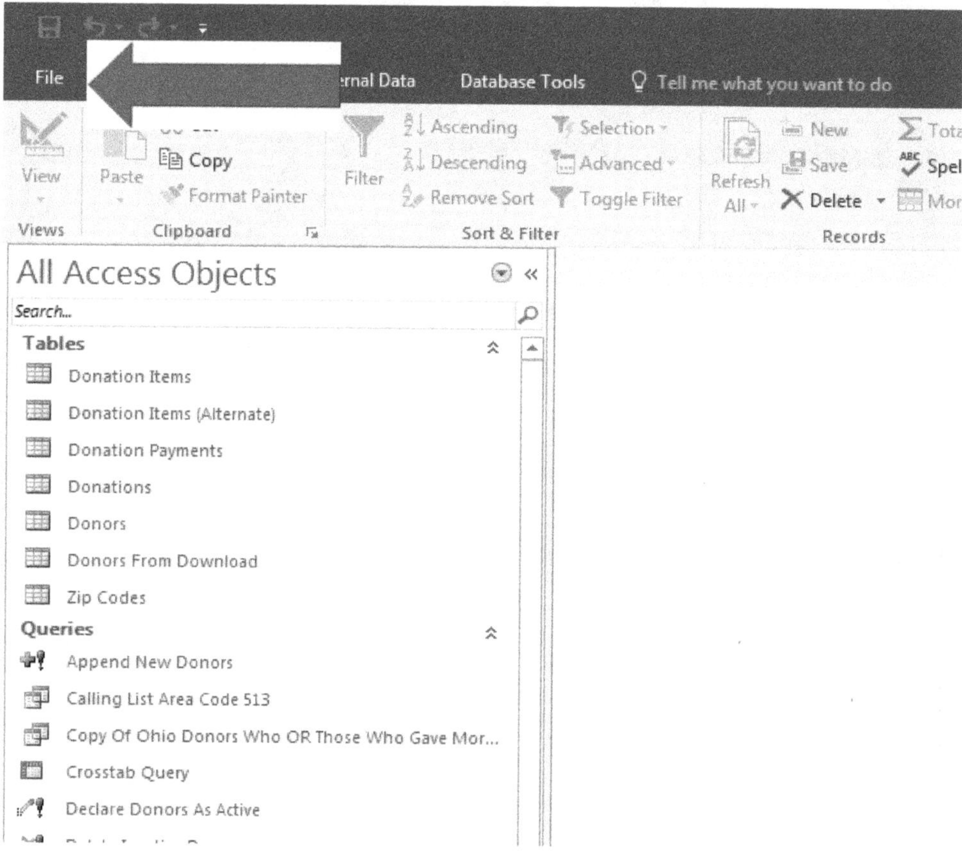

FIGURE 6.18 Compacting and repairing a database

Step 2: At the next menu, click **Compact and Repair** (figure 6.19, arrow).

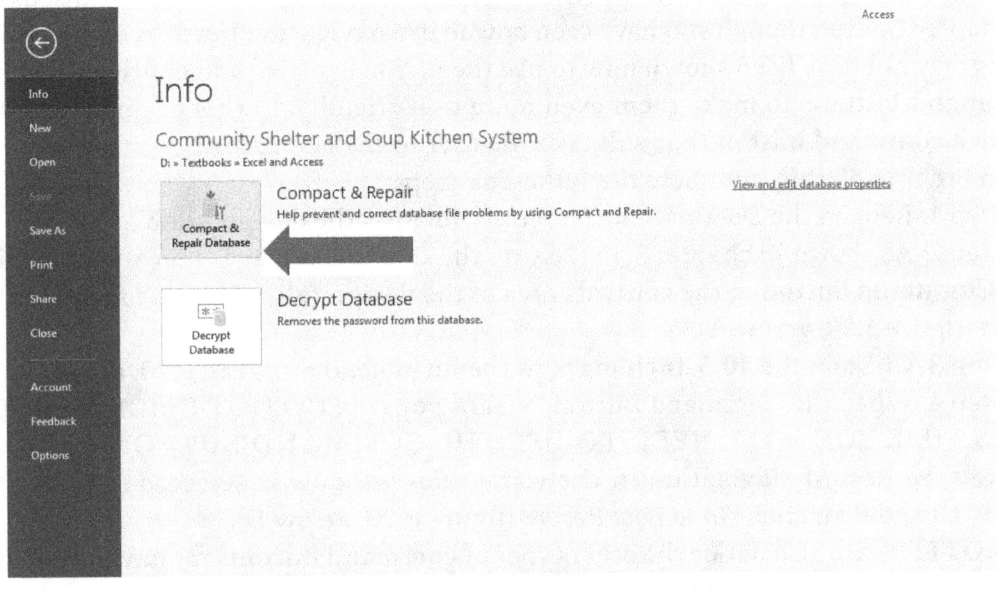

FIGURE 6.19   Compacting and repairing a database

When you do, it will compact and repair any corruptions and it will then reopen the database. If the database is quite large, it may take a few minutes.

NOTE: If your database has the **startup settings** in place that were discussed in chapter 5 (by which you hide the **navigation pane** and/or hide **full menus** and all **default shortcut menus**), these settings will once again be in place when your database reopens after performing the **compact and repair** function. You will have to close and then reopen the database while pressing and holding the **Shift** key in order to get back to state where you can continue to change the design of database objects.

CREATING AND USING STANDARD ACTION BUTTONS IN ACCESS

If you want to use some of the **command buttons** that Access offers without the trouble of having to create your own **macros** as described in chapter 5, you can use one of their **standard action buttons**. The problem is that when you do this **THESE BUTTONS WILL USE EMBEDDED MACROS AND** (as mentioned earlier in chapter 5) **EMBEDDED MACROS CAN CAUSE PROBLEMS WHEN YOU HAVE MORE THAN ONE SIMULTANEOUS USER ON A NETWORK.**

Assuming that you have only one user of the database, you can create these **command buttons** with no problem. So, let's try to do a few of these buttons. For example, there are those who prefer to add a **command button** to make a form more user friendly by having it **add a new record** or **go to a new record**. This is often done because users are often unfamiliar with how to use the **navigation buttons** (discussed in chapter 1). Even though you have seen how to use **navigation buttons** earlier, other users may not have been shown how to use them. You can also add **captions** to these **command buttons** to make them even more user friendly. In this example, we will make a **command button** that will take the user to the **last record**.

In order to do this, complete the following steps:

Step 1: Reopen the *Donor Main Data Entry* **form** in the **design view**.

Step 2: As shown in chapter 5, make sure the **Control Wizard** is on and then click the **Command button** in the **controls** area of the **design tab** within the **form design tools** (figure 6.20, arrow A).

Step 3: Click on the **10.5-inch** mark in the **form header** (figure 6.20, arrow B).

Step 4: When the Command Button Wizard pops up, DO NOT CLICK CANCEL. THIS TIME YOU WILL NEED TO USE THE COMMAND BUTTON WIZARD. Make sure Record Navigation in the Categories window is selected (figure 6.20, arrow C) and then click *Go to Last Record* (figure 6.20, arrow D).

NOTE: Please also notice the other types of **command buttons** for **navigation** that you can have put into your **form** that are listed in the **actions** list:

- Go to First Record
- Go to Next Record
- Go to Previous Record

These are not the same as the **Find Next** or **Find Record** commands. **Find Next** or **Find Record** will give **pop-up menus** that will give you a way to search for records of given criteria.

Step 5: Click **Next** (figure 6.20, arrow E).

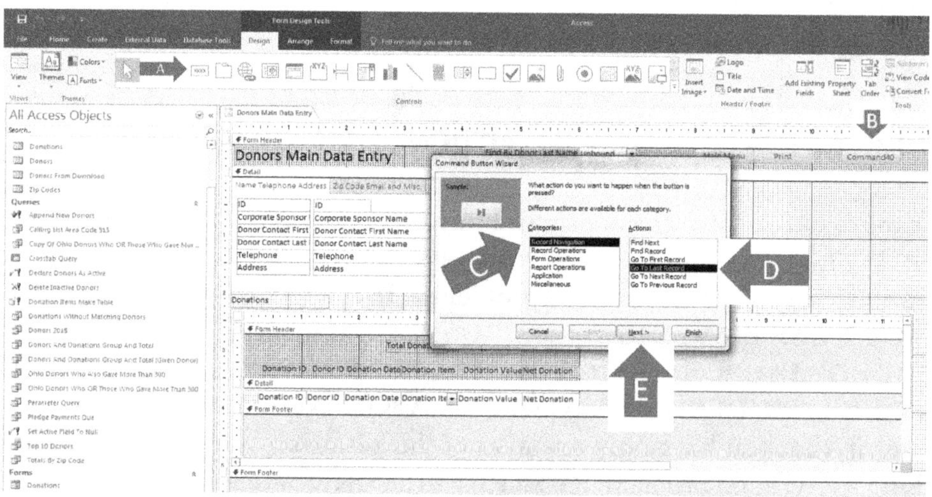

FIGURE 6.20 Creating standard action buttons in forms

Step 6: At the next menu, type *Go to Last Donor* as the **caption** in the **text box** (figure 6.21, arrow A). Do **NOT** use the **picture** option, because it will give you the picture of a **navigation button** and that is what the average user doesn't understand.

Step 7: Click **Finish** (figure 6.21, arrow B).

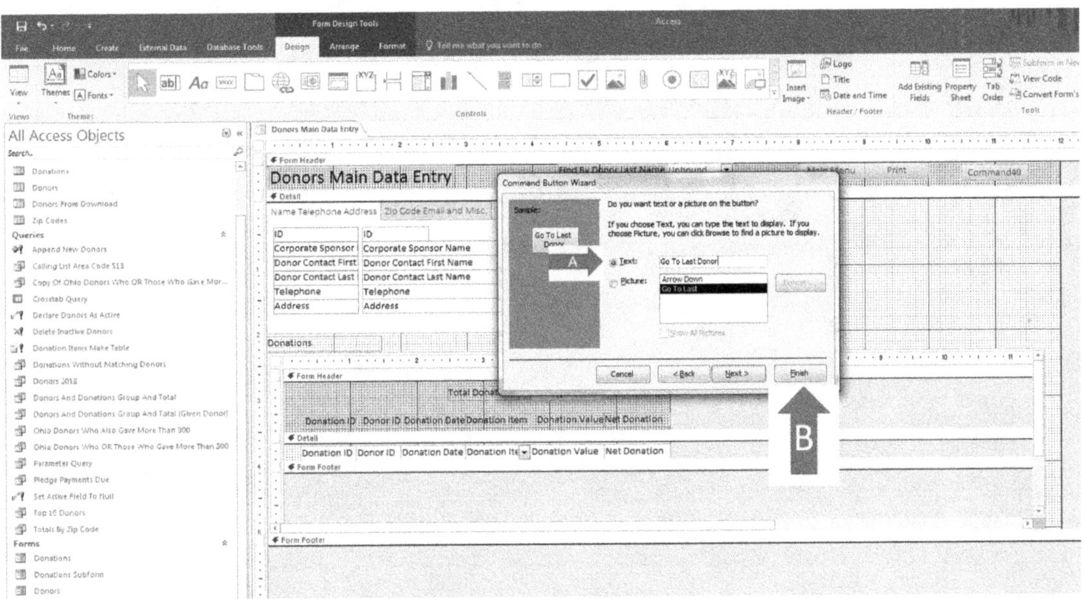

FIGURE 6.21   Creating standard action buttons in forms

Step 8: Go to the **form view**. Your screen should look like figure 6.22. Try the button out. It will take you to the last record.

NOTE: It is highly recommended that instead of putting captions on these buttons that read *Go to Last Record or Go to First Record*, that you should have them indicate what a given record holds. That is why in this example you have been asked to caption the button as *Go to Last Donor*.

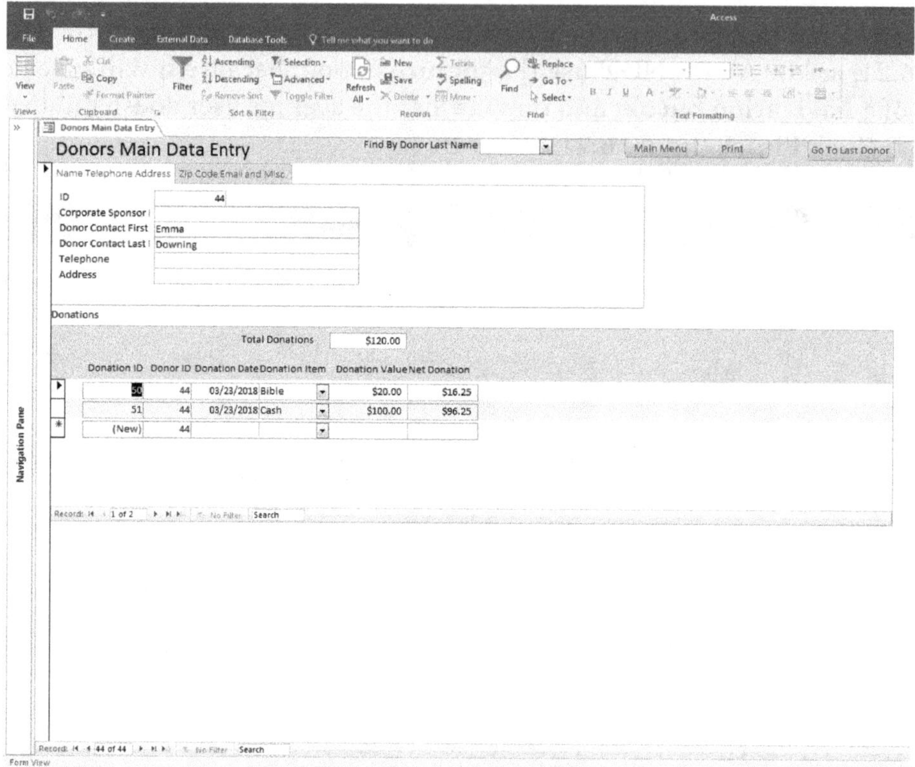

FIGURE 6.22   Creating standard action buttons in forms

Repeat this process for all of the other **command buttons**.

NOTE: When you want to add a **command button** that will take you to a **new record** or to **delete a record**, you can repeat the same process except that at the first step of the **Command Button Wizard**, choose **Record Operations** (figure 6.23, arrow A) and then choose either **Add New Record** or **Delete Record**, whichever task you are trying to accomplish (figure 6.23, arrow B). Please also notice the other **command buttons** that you can put in your forms that are listed in the Actions list:

- Duplicate a Record
- Print Record
- Save Record
- Undo Record

The other categories in the **Command Button Wizard** will complete tasks we also did with **macros** we created from scratch. **Form operations** will do the following:

- Open form
- Close form

- Apply form filter
- Print a form
- Print current form
- Refresh form data

The report operations will do the following:

- Open report
- Mail report
- Preview report
- Print report
- Send report to file

The **application category** will only invoke the **Quit command** that we discussed earlier.

The miscellaneous category will do the following:

- Print table
- Invoke the auto dialer
- Run macro
- Run query

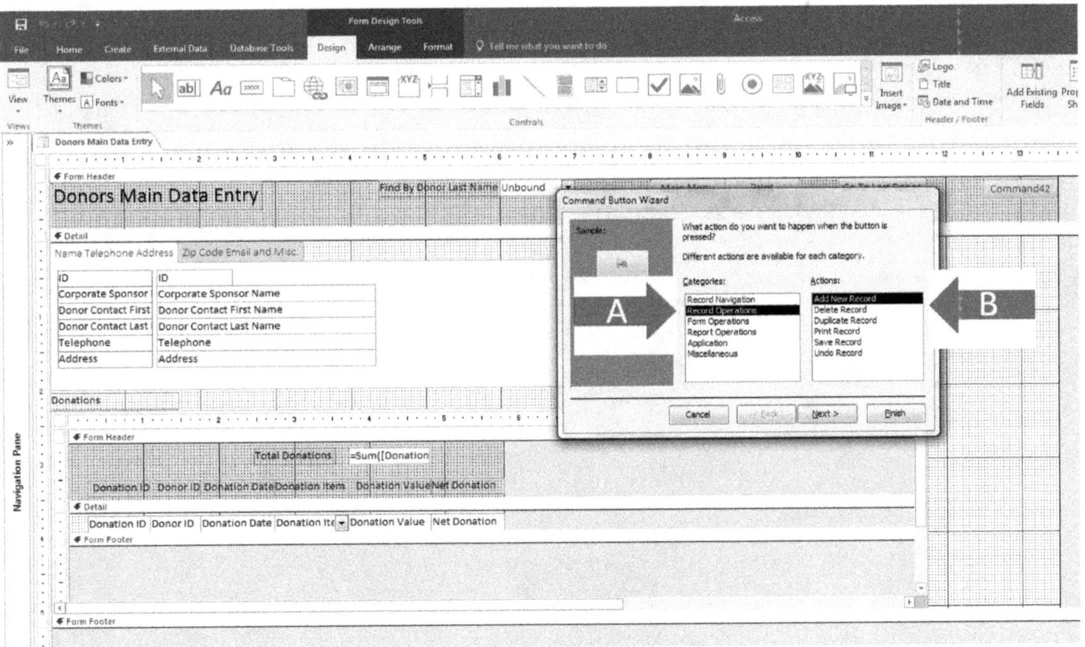

FIGURE 6.23 Creating standard action buttons in forms

After you have added buttons to *Go to First Donor, Go to Last Donor, Go to Next Donor, Go to Previous Donor, Add a New Donor* and *Delete a Donor*, using these principles, and then using the skills you learned resizing and aligning controls, your screen could look a lot like figure 6.24.

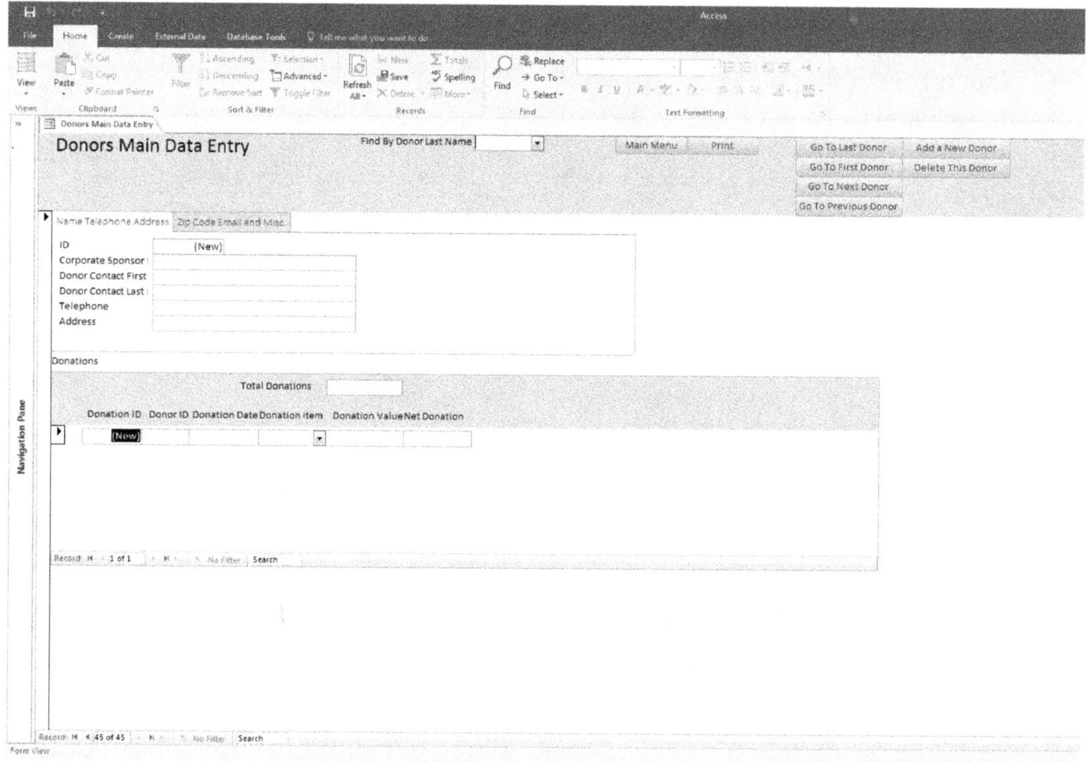

FIGURE 6.24 Creating standard action buttons in forms

Chapter 6: Assignment (Summary of Tasks)

1. In the Community Shelter and Soup Kitchen System database copy the *Donors and Donations Group and Total* query and give it the name of *Donors and Donations Group and Total (Given Donor)*.
2. Open the *Donors Main Data Entry* form in the **form view** and make sure that it is reading any one of the records. **KEEP THIS FORM OPEN UNTIL YOU FINISH ITEM 7 OF THIS ASSIGNMENT**.
3. Open the *Donors and Donations Group and Total (Given Donor)* query in the **design view** and use the **expression builder** to make sure it will display only the record that is currently displayed in the *Donors Main Data Entry* form (using the *ID* Field for the criteria).
4. Make a copy of the *Donors and Donations Group and Total* report and name it *Donors and Donations Group and Total (Given Donor)*.
   a. Change the **record source** of that new report to now become the **query** *Donors and Donations Group and Total (Given Donor)*.
5. Run the report in the **print preview** to make sure it only shows the *donor* and *donations* of the record displayed in the *Donors Main Data Entry* **form**.
6. Create a **macro** that will refresh the current record and then open the *Donors and Donations Group and Total (Given Donor)* **report** (that you just created) in **print preview**. Name the **macro** *Donors and Donations Group and Total (Given Donor)*.
   a. Place a **Command button** in the header of the *Donors Main Data Entry* **form** to run the **macro**.
   b. Make the caption on the button read *Print*.
7. Add a **tab control** to the *Donors Main Data Entry* form.
   a. In the first tab, display the ID, corporate sponsor, donor contact first name, donor contact last name, telephone, and address fields. Name the tab *Name Telephone Address*.
   b. In the second tab, display all of the remaining fields (the *zip code, email, active,* and *include in mailings* fields). Name the tab *zip code email and misc*.
8. Compact and repair the database.
9. Create **command buttons** in the **header** of the *Donors Main Data Entry* form that, when clicked, will do the following:
   a. Take the user to the **last** donor (with the caption of *Go to Last Donor*).
   b. Take the user to the **first** donor (with the caption of *Go to First Donor*).
   c. Take the user to the **next** donor (with the caption of *Go to Next Donor*).
   d. Take the user to the **previous** donor (with the caption of *Go to Previous Donor*).
   e. Add a **new** donor (with the caption of *Add New Donor*).
   f. **Delete** the current *donor* (with the caption of *Delete This Donor*).

Make your screen look like figure 6.24.

Chapter 6: Assignment (Alternate) (Summary of Tasks)

1. In the Village Bookstore database system, copy the *Vendors and Orders Group and Total* query and give it the name of *Vendors and Orders Group and Total (Given Vendor)*.

2. Open the *Vendors and Inventory Main Data Entry* form in the **form view** and make sure that it is reading any one of the records **KEEP THIS FORM OPEN UNTIL YOU FINISH ITEM 7 OF THIS ASSIGNMENT.**

3. Open the *Vendors and Orders Group and Total (Given Vendor)* query in the **design view** and use the **expression builder** to make sure it will display only the record that is currently displayed in the *Vendors and Inventory Main Data Entry* form (using the *ID* field for the criteria).

4. Make a copy of the *Vendors and Orders Group and Total* report and name it *Vendors and Orders Group and Total (Given Vendor)*.
   a. Change the **record source** of that new report to now become the **query** *Vendors and Orders Group and Total (Given Vendor)*.

5. Run the report in the **print preview** to make sure it only shows the *vendors* and *orders* of the record displayed in the *Vendors and Orders Main Data Entry* **form**.

6. Create a **macro** that will refresh the current record and then open the *Vendors and Orders Group and Total (Given Vendor)* **report** (that you just created) in **print preview**. Name the **macro** *Vendors and Orders Group and Total (Given Vendor)*.
   a. Place a **command button** in the header of the *Vendors and Inventory Main Data Entry* **form** to run the **macro**.
   b. Make the caption on the button read *Print*.

7. Add a **tab control** to the *Vendors and Inventory Main Data Entry* form.
   a. In the first tab, display the ID, vendor name, contact first name, contact last name, contact email, and address fields. Name the tab *Name and Contact Data*.
   b. In the second tab, display all of the remaining *vendor* fields. Name the tab *Zip Code Telephone and Misc*.

8. Compact and repair the database.

9. Create **command buttons** in the **header** of the *Vendors and Inventory Main Data Entry* form that, when clicked, will do the following:
   a. Take the user to the **last** *vendor* (with the caption of *Go to Last Vendor*).
   b. Take the user to the **first** *vendor* (with the caption of *Go to First Vendor*).
   c. Take the user to the **next** *vendor* (with the caption of *Go to Next Vendor*).
   d. Take the user to the **previous** *vendor* (with the caption of *Go to Previous Vendor*).
   e. Add a **new** *vendor* (with the caption of *Add New Vendor*).
   f. **Delete** the current *vendor* (with the caption of *Delete This Vendor*).

Creating Show-All Queries, IIF Expressions, Pop-Ups, Option Groups, and Conditional Macros

IN THIS CHAPTER you will learn how to do the following:

1. Create queries that show **ALL** records in the **parent table** even if they have no matches in the **child table**.
2. Use IIF Expressions in a query.
3. Create **pop-up menus** and know why they are needed.
4. Create an **option group** and know why they are needed.
5. Use **conditional macros** and know why they are needed.
6. Create union queries.

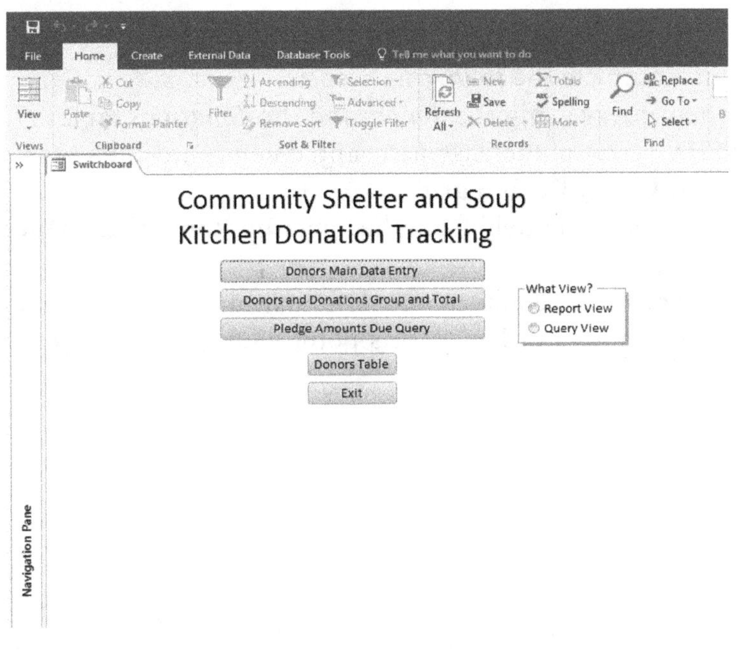

FIGURE 7.1   Union queries

317

CHAPTER 7 TASK: SHOWING ALL DONORS, CREATING OPTION GROUPS, CONDITIONAL MACROS, AND ACTION BUTTONS

To Show All Donors with or without Donations

When you ran the *Donors* and *Donations Group and Total* query, you may have noticed that the *donors* shown in figure 7.2 were not in the query. That's because they did not make a donation, thus **THEY HAD NO MATCHING *DONATIONS* IN THE *DONATIONS* TABLE**. These records are often forgotten. It is important to many, therefore, to show them in reports in order to inform the **report** users that there are those who are not donating. Perhaps in this example they were entered as prospective *donors*. Whatever the case, many will want to know they are there.

Donor Contact Last Name	Donor Contact First Name
Bruner	Nancy
Kreidler	Samuel
McIntosh	Michael
Salvador	Luciana
Ferris	Eileen
Martinson	Margaret
Montana	Marie

FIGURE 7.2   Missing donors from queries and reports

In order to do this, complete the following steps:

Step 1: Create a new **query** using the *donors* and the *donations* tables.

Step 2: Send the *ID, donor contact last name* and *donor contact first name* (all from the *donors* **table**) to the **Query Design grid**.

Step 3: Send the *donation item* field from the *donations* **table** to the **Query Design grid**.

Step 4: Sort the **query** by the *donor contact last name* and *donor contact first name* fields, from left to right, in that order. At this point your **query design** should look like figure 7.3.

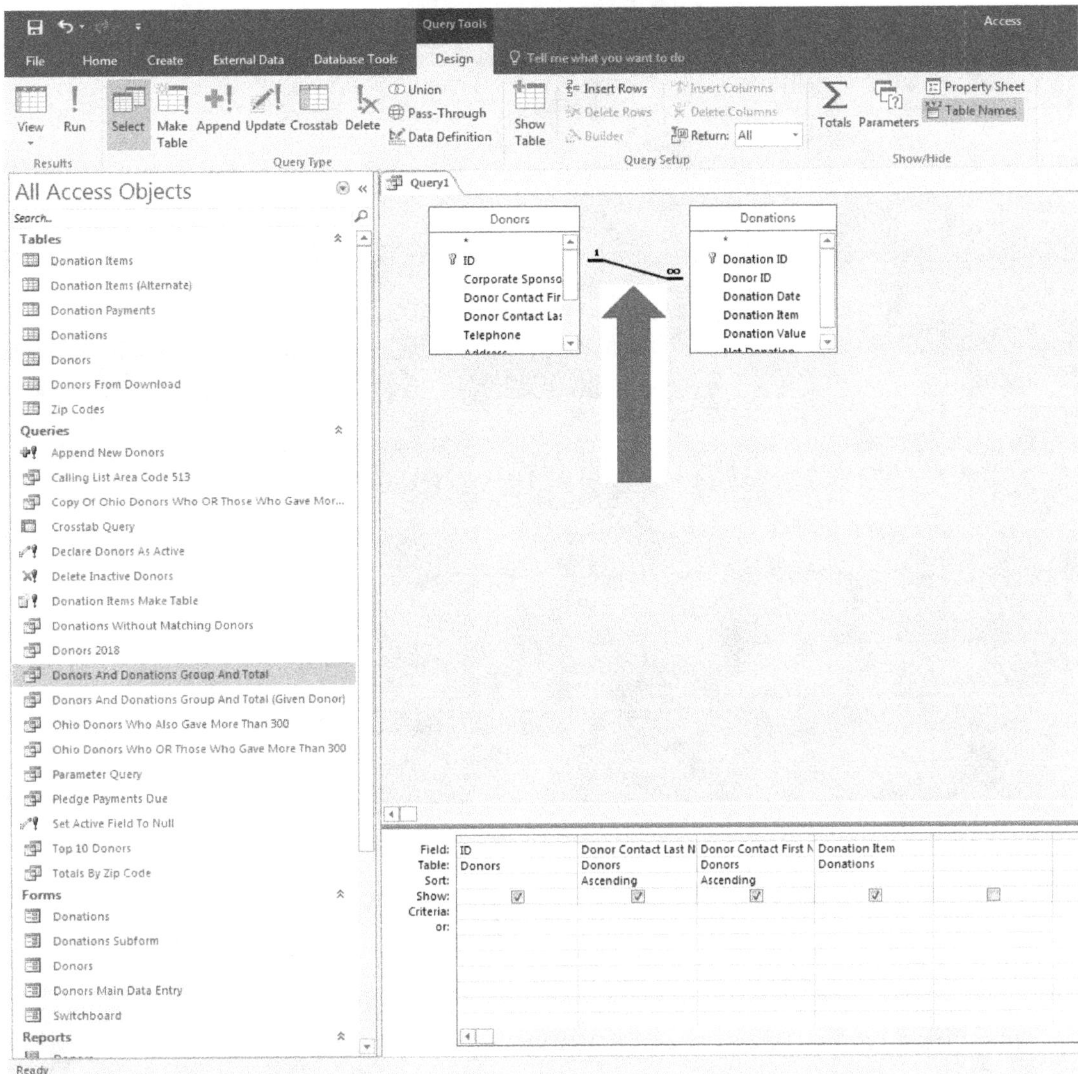

FIGURE 7.3 Changing join properties in a query

Step 5: Double-click the **slanted relationship Line** between the two tables that you have chosen for the **query** (**NOT** the parts where there is a **1** or **infinity** (∞) **symbol**). When you double-click the **slanted relationship line**, it will display the **Join Properties pop-up menu** (figure 7.4, arrow).

Step 6: Check the **2nd Option button** in that menu that would select the option that reads **Include All Records From Donors.**

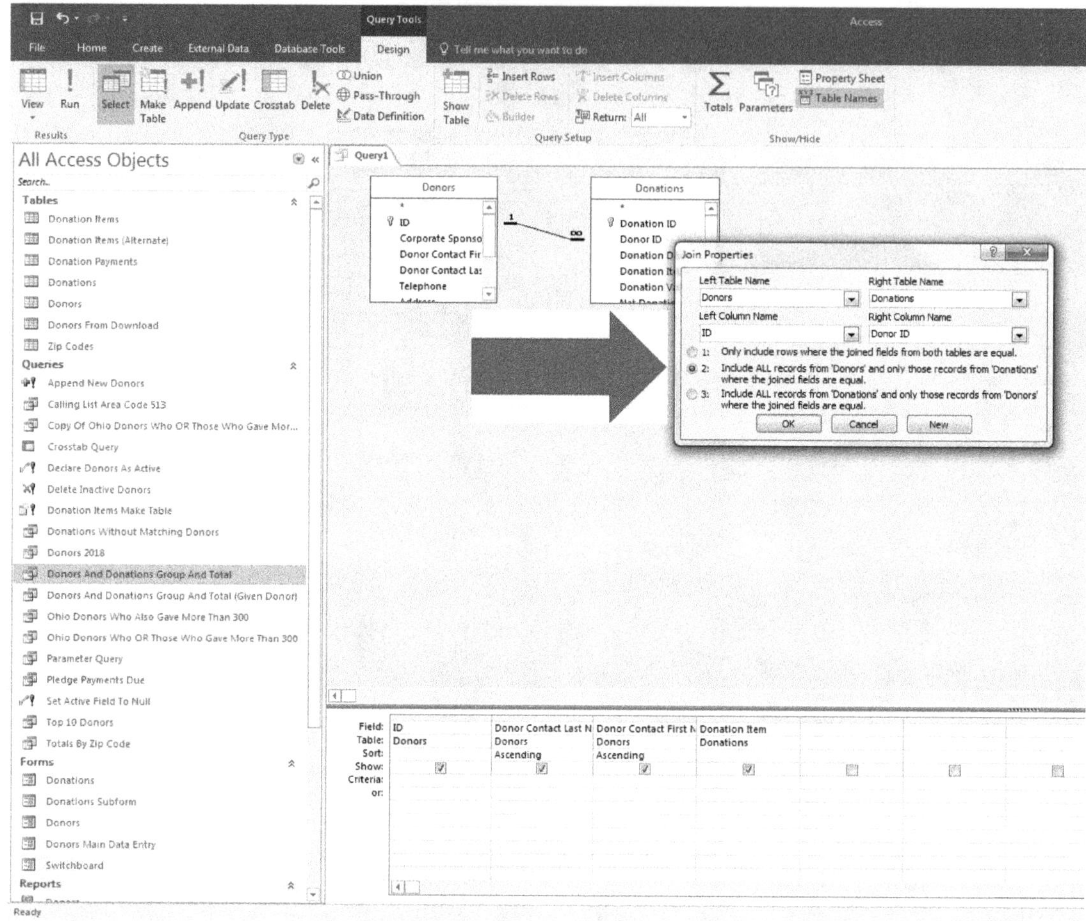

FIGURE 7.4   Changing join properties in a query

Step 7: Click **OK**.

Step 8: That will take you to what you see in figure 7.5. Please notice that Access has placed an **arrow** on that **slanted relationship line** that points to the **child** table. That arrow is indicating which of these two **tables** will be displaying **null** data if there is no matching **child** records between the *donor* and *donations* **table**. Run the **query**.

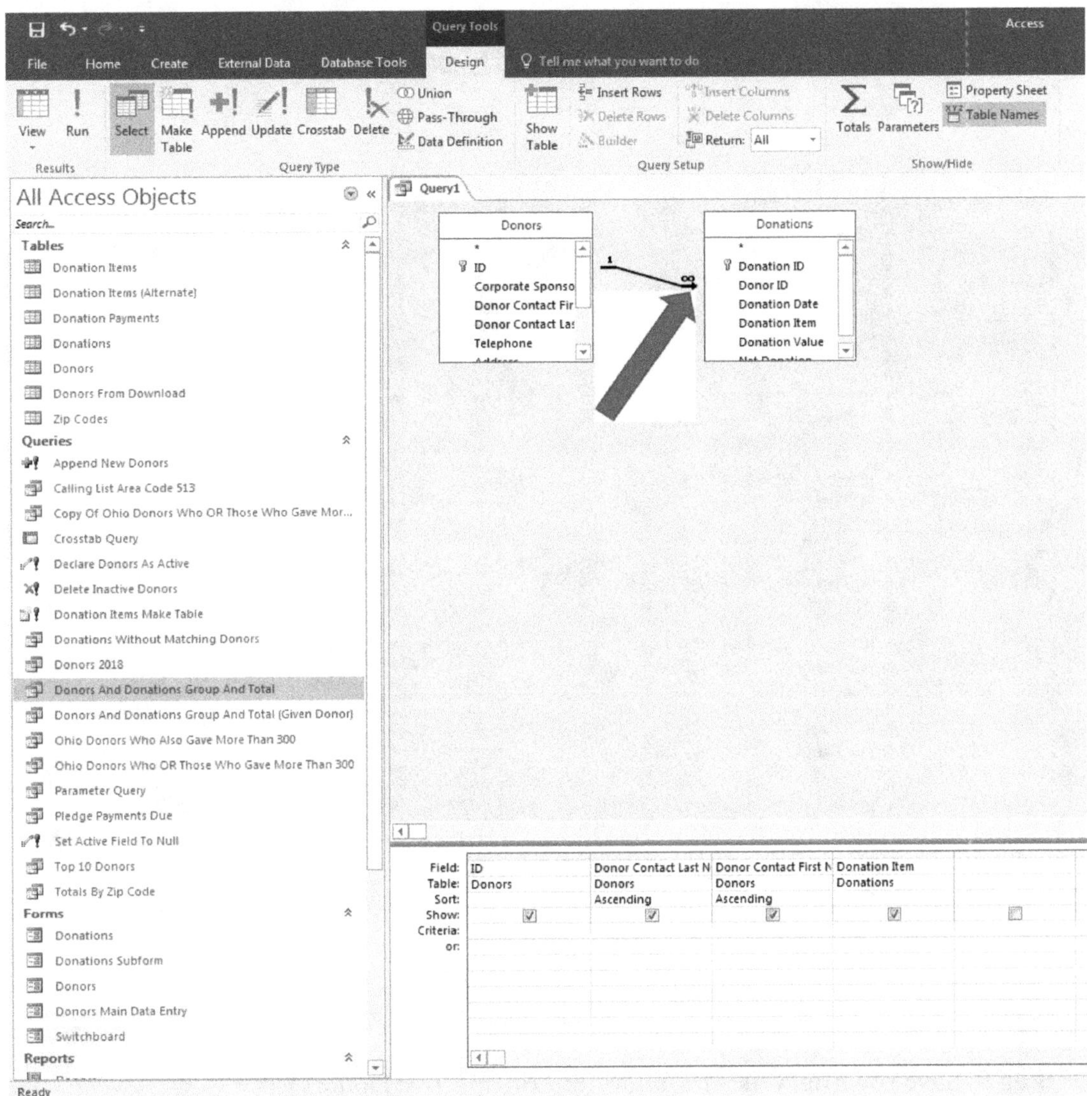

FIGURE 7.5 Changing join properties in a query

The results can be seen in figure 7.6. Please notice that the people in the *donors* table who have had no *donations* have a **null value** to the right of their names. Again, that is because they have made no *donations* and therefore, there is no match between the *donations* and the *donors* table for those people.

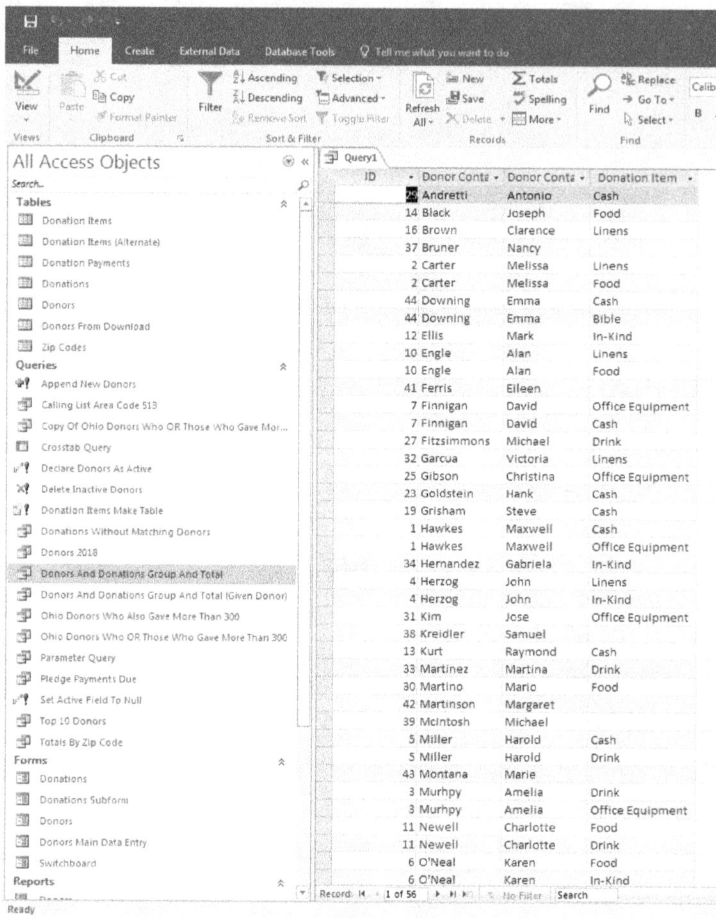

FIGURE 7.6   Changing join properties in a query

Step 9: Save the **query** as *All Donors and Prospective Donors.*

This concept can be used in numerous ways. You can even look for matches and nonmatches between tables and queries that are not even formally related with referential integrity.

The problem with our query is that it is confusing to the average user, who might see all of those null cells in the *donation item* field and think that there was a data entry error. In order to remove the confusion, you can use the **IIF function** to indicate to the user that these people are not yet *donors* and are currently only *prospective donors.* Please notice that the **IIF function** in Access has two I's. If you have used Microsoft Excel you will find that the **IIF function** in Access is very, very similar to Excel's **IF function**.

THE IIF FUNCTION IN ACCESS CAN BE USED ANYTIME THAT YOU WANT TO DISPLAY DATA UNDER GIVEN CIRCUMSTANCES. THIS IS

A VERY POWERFUL TOOL IN DATABASE MANAGEMENT. THIS EXAMPLE IS ONLY ONE OF THOUSANDS WITH WHICH YOU COULD USE IT. To apply it to our example, complete the following steps:

Step 1: Return to the **design view** of the *All Donors and Prospective Donors* **query**.

Step 2: Delete the *donation item* field from the **Query grid**.

Step 3: In place of the *donation item* field, enter the following expression (in the top box of the first blank column):

Donated Item: IIF([Donation Item] is null, "NO Donations",[Donation Item])

If you want to **zoom** in on what you are typing, press the **Shift** key and then tap the **F2** key on your keyboard (as discussed in chapter 3) to see better what you are writing, and it would be well advised to do so (figure 7.7a).

Step 4: If you are in the **Zoom** box, click **OK** after you have written the expression.

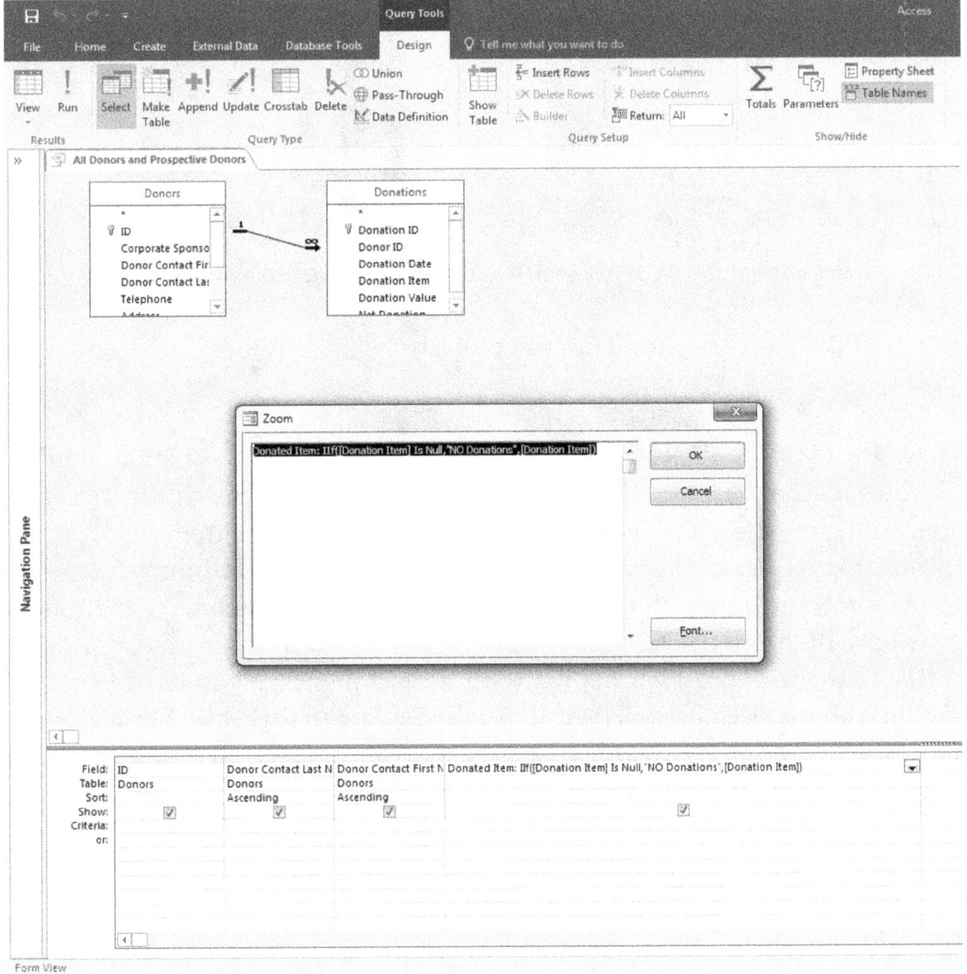

FIGURE 7.7a Using the IIF function in a query

Please try to understand the logic of the **IIF function**. Just as in Excel, the **IIF function** has three **arguments** and they are separated by a **comma** (,). The **first argument** makes a statement. If the statement is true, the query will display whatever you have written in the **second argument** (which is between the two commas). If the statement in the **first argument** is false, the **query** will display whatever you have written in the **third argument** (figure 7.7b).

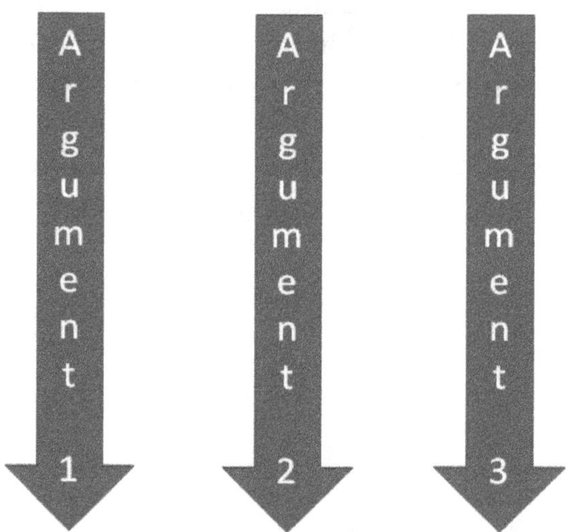

FIGURE 7.7b   Using the IIF function in a query

IN LAYPERSON'S TERMS this expression you have just written would thus read, "If the donation item field has a **null value**, then display the words that read NO donations; otherwise, display whatever data is in the donation Item field."

Step 5: Run the **query**. The results should be what you see in figure 7.8.

When you run the **query**, those who previously had a **null value** for their *donation items* now have the words *NO Donations* in place of a *donation item* (figure 7.8). Please also notice that the expression begins with a field name as you used in an earlier chapter. You cannot use *donation item* as the field name in this case, because you can't use the same field name in a **table** or **query** more than once. It would also cause a **circular reference** in the **expression**, which means that your field name has referred to itself as a field. **Circular references** in an **expression** therefore are not permitted. Hence, we have changed the field name to one that is similar to *donation item* so the users can still know what data they are viewing.

FIGURE 7.8   Using the IIF function in a query

CREATING A POP-UP MENU

Anyone who has ever used the internet, or any computer application has seen **pop-up menus**. They are often annoying, but when using a database, they can be very helpful. Suppose you are entering a new *donor* and you find that his or her *zip code* is not in our *zip code* **drop-down list box** for that field. You would thus have to add it to the *zip code* **table**. If this happens regularly, rather than having to close the *donor* **form** and then open and update the *zip code* **table**, you can create a **pop-up menu** that will allow you to enter the *zip code* without leaving the *donors* screen. This will speed up your data entry and remove confusion.

To create a **pop-up menu**, use the following steps:

Step 1: Open the *zip code* table.

Step 2: Click the **Create** ribbon (figure 7.9, arrow A)

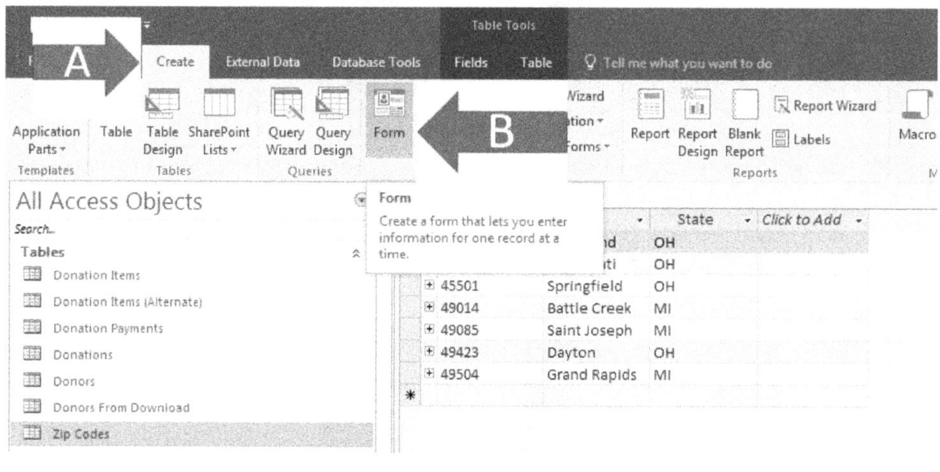

FIGURE 7.9 Creating a pop up form

Step 3: Click **Form** (figure 7.9, arrow B).

Step 4: That will create the **Form** that you see in figure 7.10. Click **Save** in the **quick access toolbar** in the upper left of your screen (figure 7.10, arrow A). When the **Save As Prompt** appears, save the form as *zip code* by clicking **OK** (figure 7.10, arrow B).

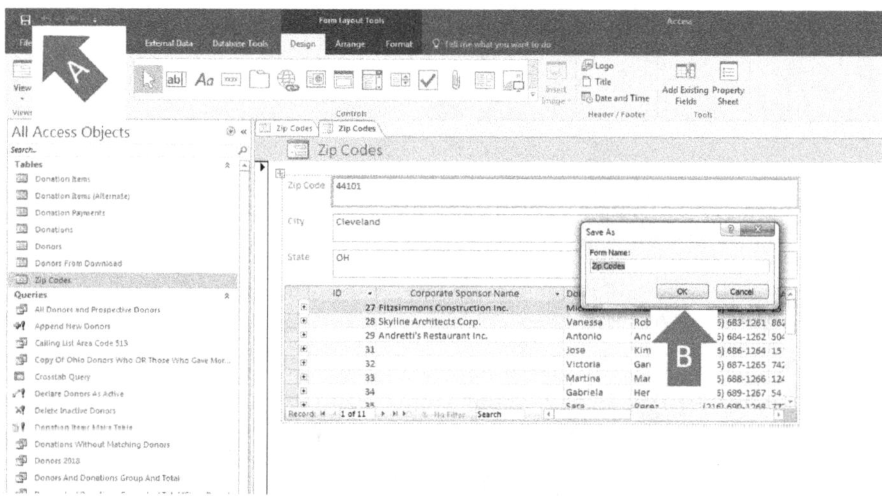

FIGURE 7.10 Creating a pop up form

Step 5: As you have often done in the past, go to the **form design view** of the newly created *zip code* **form**.

Step 6: If Access includes the *table.donors* **sub-form** in this **form**, click once on it and then press the **Delete** key on your keyboard. In this instance, having the *donors* **table** as a **sub-form** in this **form** would not be useful or practical.

Step 7: As discussed in chapter 4, resize the **field controls** and shore up the dead space in the **form**. After doing so, it should look similar to figure 7.11.

Step 8: Double-click the **Form/Report selection box** in the upper left of the **Design View grid** (figure 7.11, arrow A). When you do, it will open the **property sheet** for the form and it will also place a tiny black box in the **Form/Report selection box**.

Step 9: In the **All** tab of the **property sheet**, you will see the **pop-up property**. Change the **No** to a **Yes** (figure 7.11, arrow B).

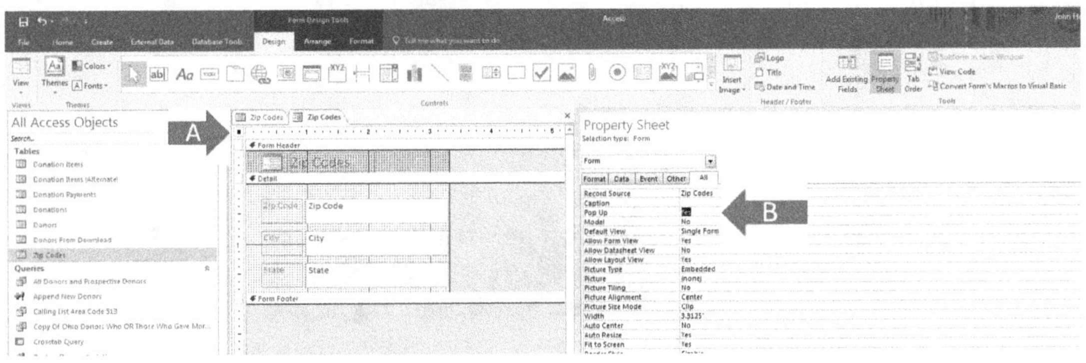

FIGURE 7.11 Creating a pop up form

Step 10: Go to the **form view** of the **form** and notice that it is now a **pop-up form** (figure 7.12).

Step 11: Save the **form** again and close it.

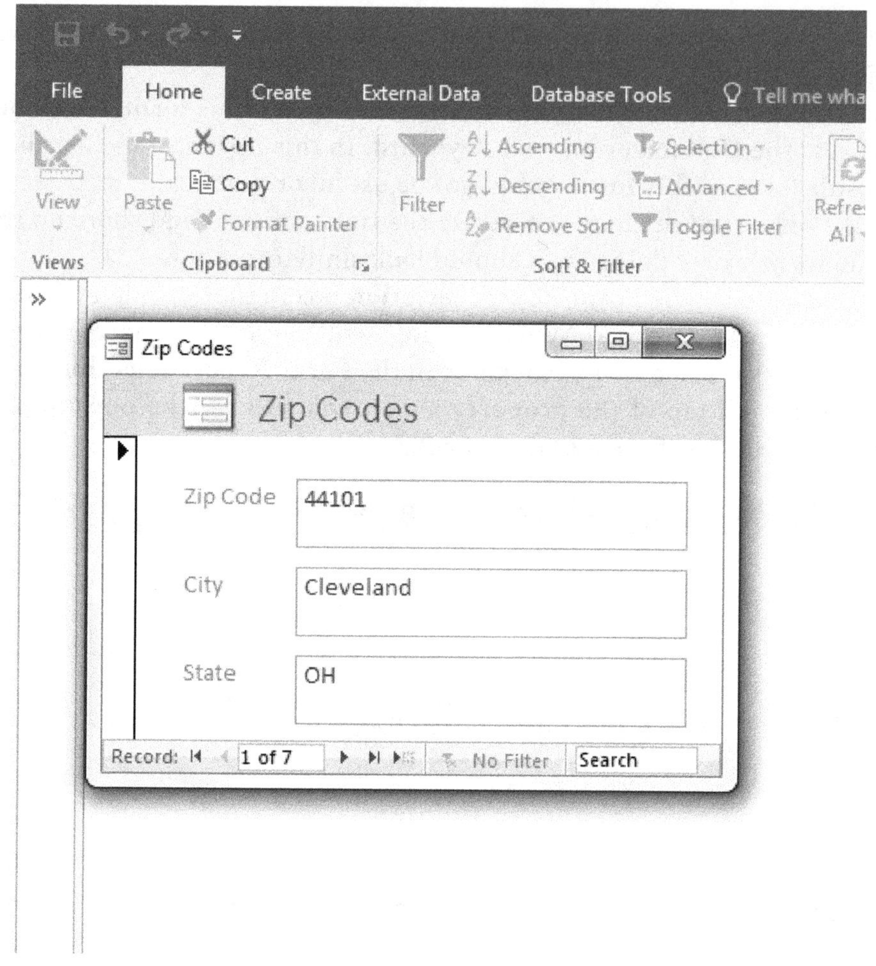

FIGURE 7.12   Creating a pop up form

MAKING THE POP UP MORE USEFUL

If you want to open the *zip code* **pop-up menu** you have just created, while entering a *donor*, you will need a **command button** to do so and to also close it. Those buttons can be created by using the procedures already discussed earlier. The *close* **macro** that was created earlier can also be used as the one that will be attached to a *Close and Save* button that you will need to place in the **header** of the *zip code* **pop-up form**.

Figure 7.13 shows that the *Zip Code* button has been added to the header of the *Donors Main Data Entry* **form** and that the *zip code* **pop-up menu** has a *Close and Save* **command button** applied. If you click and drag the *zip code* **pop-up menu** to where you want it while using it, it will open there from now on if you click **Save** (in the **quick access toolbar**) once you have positioned it.

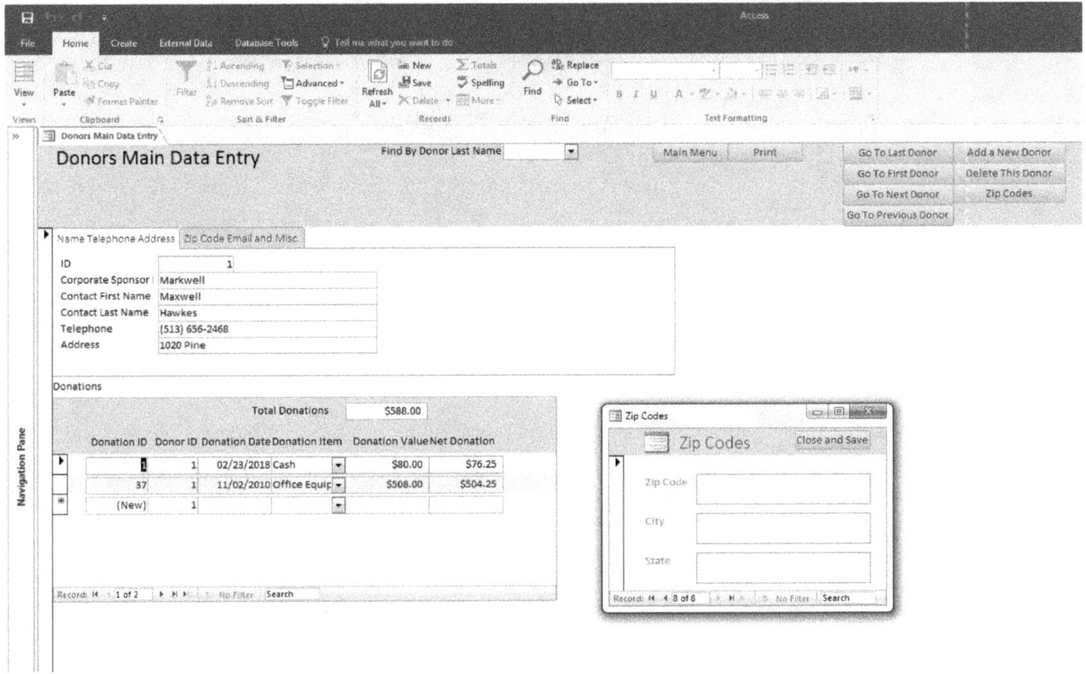

FIGURE 7.13 Making a pop up more useful

There is just one problem. If you enter a new *zip code* into the *zip codes* **pop-up menu** and then close it, the newly added *zip code* will not show in the *zip code* **combo box** (or **drop-down list box**) in the *Donors Main Data Entry* **form** unless the user would click **Refresh All** in the **Home** ribbon. But, what if the user forgets to do so or doesn't know that it is necessary? To prevent this problem, you will need to create a **macro** that will refresh the data automatically once the *zip code* **combo box** is selected, or as Access would say when the **combo box** has the **focus**.

To do this, use the following steps:

Step 1: Create a new **macro** with only one command or **action** in it, which would be the **Refresh** command (figure 7.14).

FIGURE 7.14   Making a pop up more useful

Please notice that if you type **Refresh** in the **new command box**, Access will automatically declare it a **RunMenuCommand**.

Step 2: Name the **macro** *Refresh*.

Step 3: Open the *Donors Main Data Entry* **form** in the **design view** (figure 7.15).

Step 4: Click on the *zip code* field.

Step 5: Open the **property sheet** (remember, you can also open the **property sheet** by double-clicking a control or by right-clicking on a control and choosing **properties**).

Step 6: Go to the **Event** tab (figure 7.15, arrow B).

Step 7: Set the **Got Focus Event** in that tab to run the *Refresh* **macro** when the *zip code* field has the focus of the user (figure 7.15, arrow C).

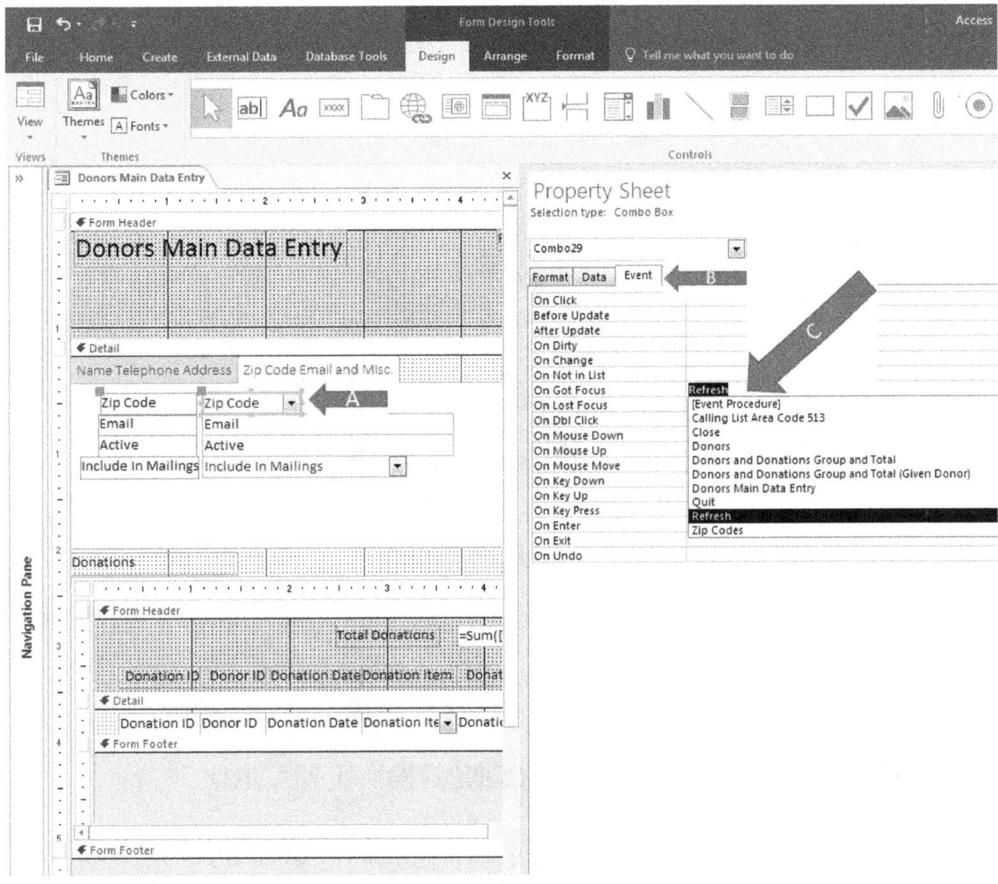

FIGURE 7.15   Making a pop up more useful

Let's try it out. If you return to the **form view** of the *Donors Main Data Entry* **form**, click the *zip code* **command button**. When the *zip code* **pop-up menu** is displayed, enter *11111* as the *zip code* and leave the *city* and *state* fields blank. After you close the *zip code* **pop-up menu**, open the *zip code* **drop-down list box** in the *Donors Main Data Entry* form. It will automatically show the newly added *zip code* (figure 7.16). Since the *11111 zip code* was only entered to demonstrate this concept, it would be best to now remove it from the *zip code* **pop-up menu**.

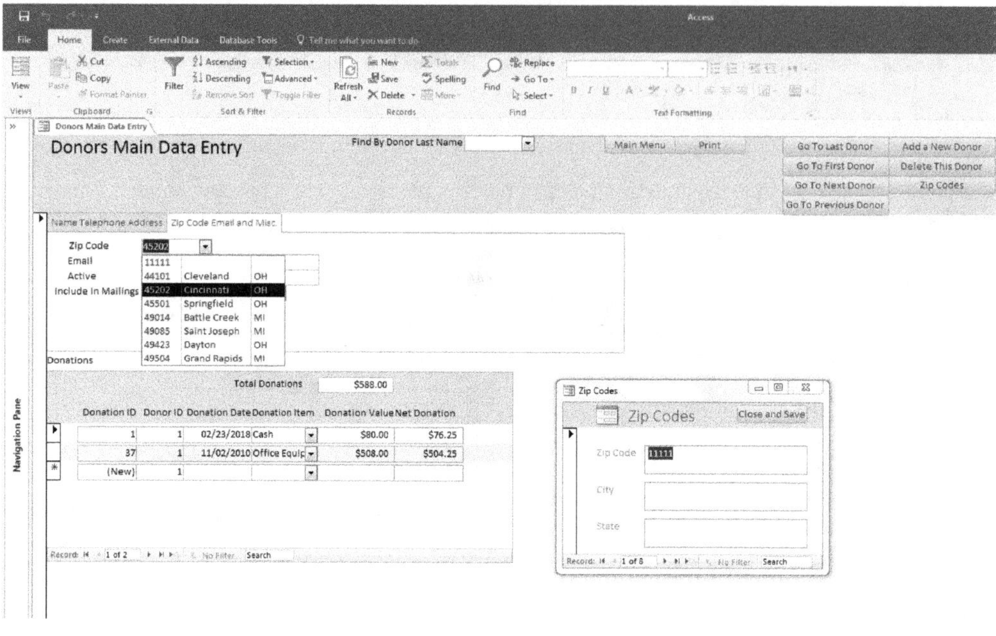

FIGURE 7.16 Making a pop up more useful

CREATING AN OPTION GROUP AND A CONDITIONAL MACRO

Sometimes you will want to view **forms**, **queries**, and **reports** based on a given condition or preference. Suppose for example that you would like to have the user decide whether the *Donors and Donations Group and Total* information will be displayed in a **query** or a **report**. You can give them a choice by using what are known as **option groups**. **Option groups** are also called **option boxes** and **frames** by **access** and users. To avoid confusion, in this book we will always refer to them as **option groups**.

An **option group** will store a numeric value within itself depending on what option you choose. For example, in the **option group** seen on the **switchboard** in figure 7.17, you can see that a user chose the *Report View* button within the **option group**, which is labeled as (and later we will see is named as) *What View?* As a result of that choice, Access has made it so that the user has declared that the *What View?* **option group** would then have a value of 1. Conversely, in this example, if the person chose the *Query View* button, that would have given the *What View?* **option group** a value of 2.

You will then tell **Access** to carry out the command of opening the *Donors and Donations Group and Total* **report** if the *What View?* **option group** is equal to 1. You will also tell Access to carry out the command of opening the *Donors and Donations Group and Total* **query** if the *What View?* **option group** is equal to 2.

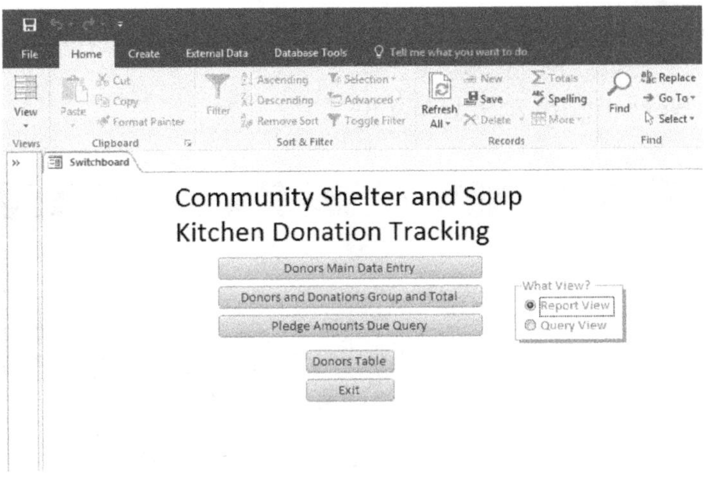

FIGURE 7.17   Creating an option group

Having decided this, let's now create the **option group** by doing the following steps:

Step 1: Open the **switchboard** in the **design view** (figure 7.18).

Step 2: Make sure, as discussed in chapter 4, that the **Control Wizard** is on.

Step 3: Click the **Option Group** icon in the **form design tools** (figure 7.18, arrow A) and do **NOT** get it mixed up with the **bound object frame** (which is similar in appearance but is completely different).

Step 4: Click the mouse to place the **option group** to the right of the *Donors and Donations Group and Total* **command button** (figure 7.18, arrow B).

Step 5: That will display the **Option Group Wizard**. Type the **label names** in the boxes as indicated by figure 7.18, arrow C, with the first in this example being *Report View* and the second being *Query View*.

Step 6: Click **Next** (figure 7.18, arrow D).

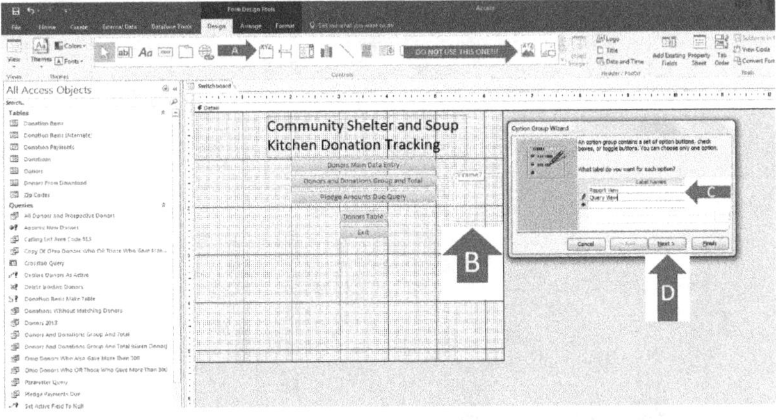

FIGURE 7.18   Creating an option group

Step 7: In the next menu, indicate that you do **NOT** want a default option (figure 7.19, arrow A).

Step 8: Click **Next** (figure 7.19, arrow B).

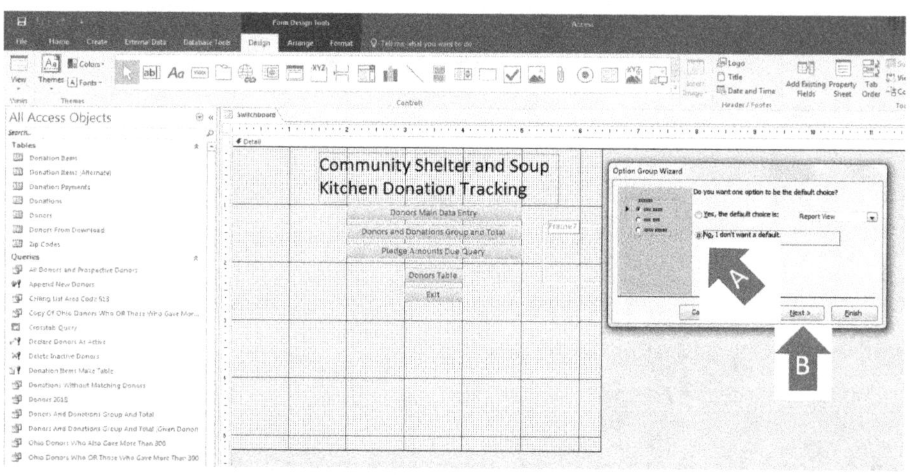

FIGURE 7.19　Creating an option group

Step 9: At the next menu (figure 7.20), notice that the values we want for each view are set to the number values we discussed earlier. As shown, *Report View* will have a value of 1 and *Query View* will have a value of 2. Therefore, all you need to do here is click **Next** (figure 7.20, arrow).

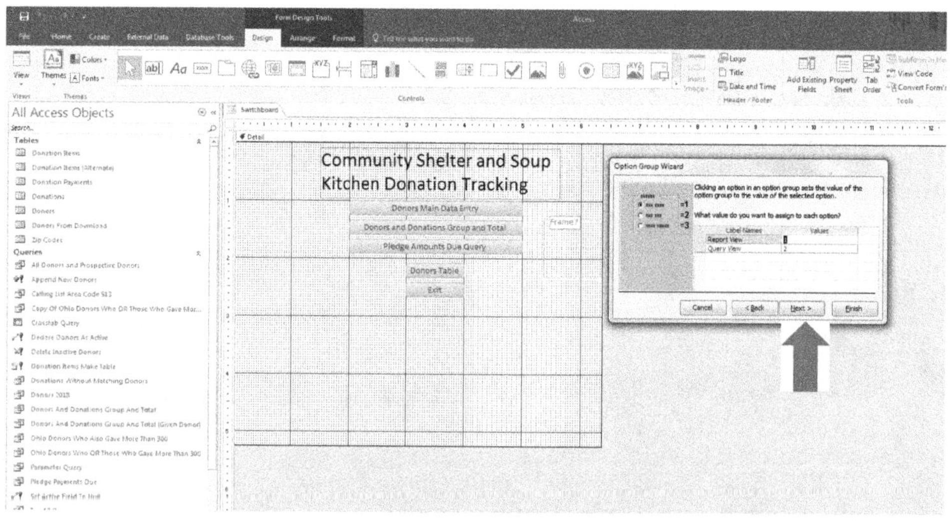

FIGURE 7.20　Creating an option group

Step 10: In the next menu it will ask for what appearance you want for the **option group**. In this example, use the default **option button** choice and choose the **shadowed** style (figure 7.21, arrow A). Then click **Next** (figure 7.21, arrow B).

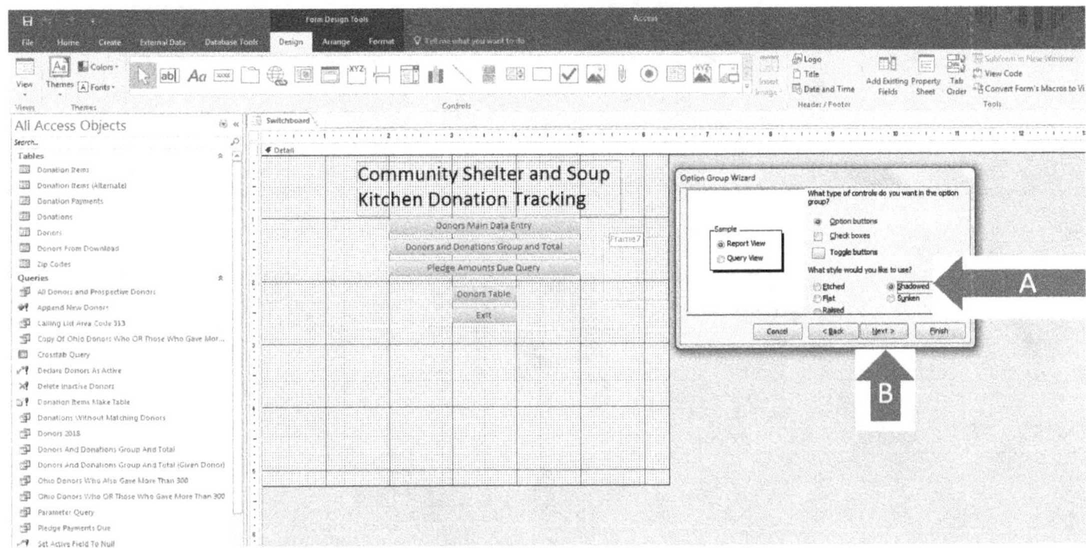

FIGURE 7.21   Creating an option group

Step 11: In the last step of the **Option Group Wizard**, enter the **caption** for the **option group,** which will be *What View?* (figure 7.22, arrow A).

NOTE: THIS CAPTION IS NOT THE SAME AS NAMING THE OPTION GROUP. YOU MUST ALSO NAME THE OPTION GROUP (OR FRAME) BY CLICKING ON ITS BORDER LATER AND GIVING IT THAT NAME IN THE NAME OPTION OF THE PROPERTIES SHEET IN THE FORM DESIGN VIEW! FAILURE TO DO SO WILL CREATE ERRORS IN THE CONDITIONAL MACRO WE WILL CREATE LATER.

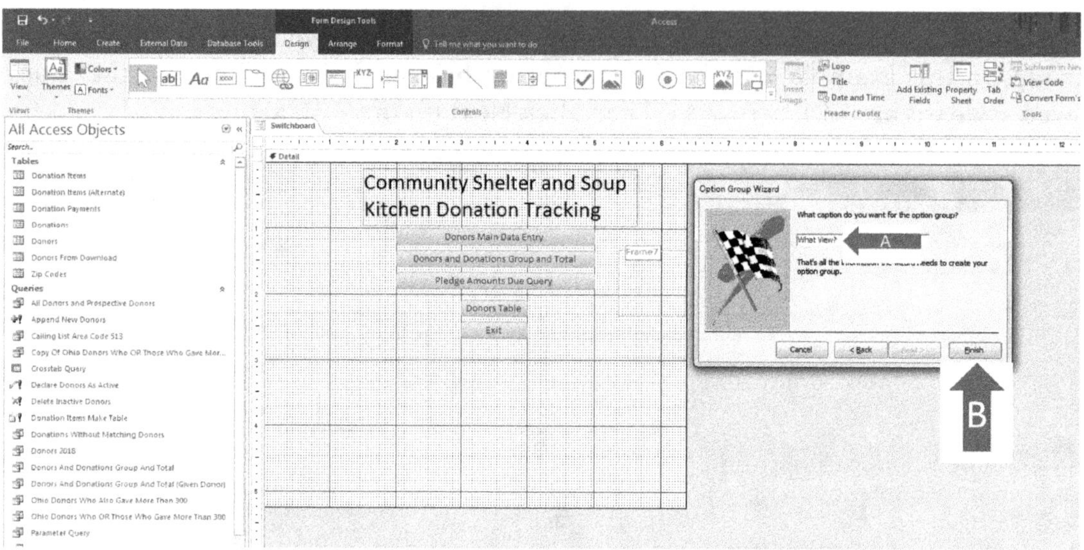

FIGURE 7.22 Creating an option group

Step 12: Click **Finish** (figure 7.22, arrow B).

Go to the **form view** to see what you have so far. Your screen will now appear as figure 7.23.

NOTE: A decision you may make at this point would be to select the entire **switchboard** and change the font color to black. Remember, the default font color is usually unacceptable to most users as it is usually too light for most to see and read.

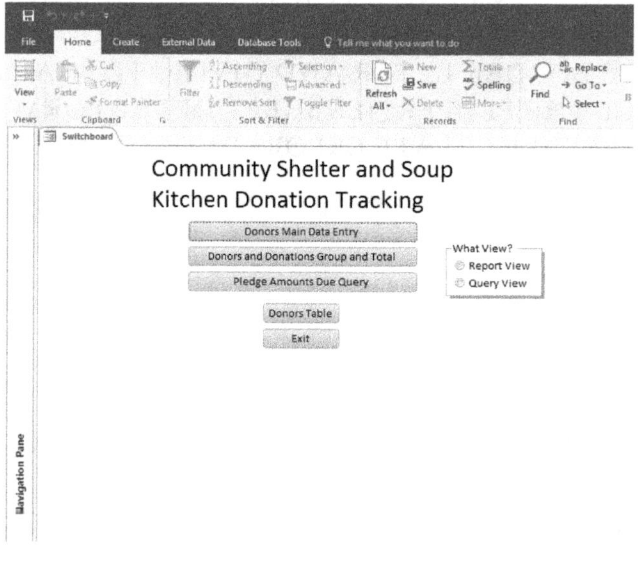

FIGURE 7.23 Creating an option group

Now we must make a **macro** and designate a **command button** that will utilize the **option group** we have just created.

Step 13: Return to the **design view** of the **switchboard** and click on the **BORDER** of the **frame** of the **option group** and **NOT** the *What View?* **caption/label** as shown in figure 7.24, arrow A.

Step 14: Open the **property sheet** and click the **All** tab (figure 7.24, arrow B).

Step 15: In this example, make sure that *What View?* is the name of the **frame** (also displayed in figure 7.24, at arrow B).

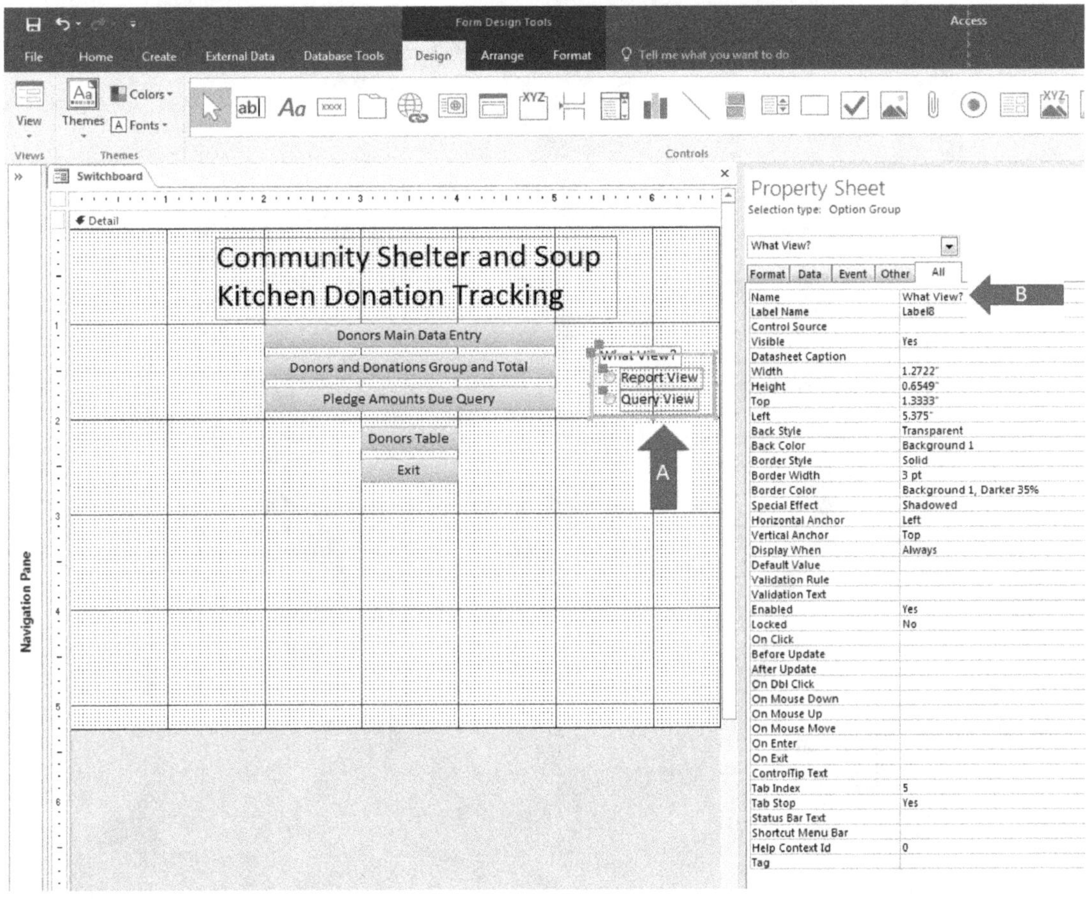

FIGURE 7.24   Creating an option group

Step 16: Close and save the **switchboard**.

Step 17: Create a new **macro** and choose the **IF command** from the **commands drop-down list box** (figure 7.25, arrow).

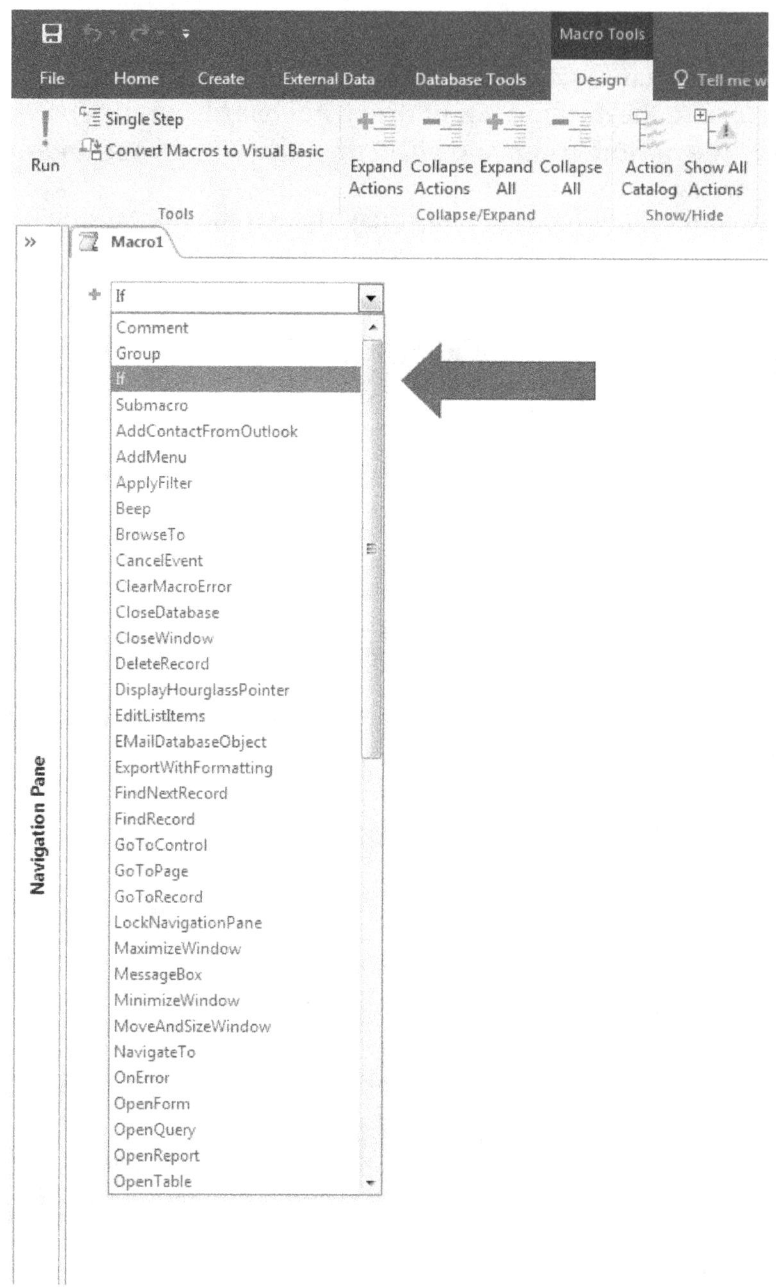

FIGURE 7.25 Creating a conditional macro

Step 18: Once you choose the **IF command**, you will get a box to its right (figure 7.26, arrow A). Type the following in the box:

[What View?]=1

NOTE: You must have **square brackets** [] surrounding the words *What View?* and **NOT** parentheses or set signs! Access will treat the frame of the **option group** as a field, and fields (as discussed in previous chapters) are referenced by **square brackets** [].

Step 19: When you press enter, an **Add New Action** box will appear. In that box, choose the **OpenReport command** from the **drop-down list box** (figure 7.26, arrow B).

Step 20: Make the *Donors and Donations Group and Total* report (figure 7.26, arrow C) the one that will open if the *What View?* **option group** button that is chosen is the *Report View* which, will be equal to 1. That will make the *What View?* **option group/frame** equal to 1.

Step 21: Make sure that when the **report** opens that it will open in the **print preview** (figure 7.26, arrow D).

Step 22: Click on the **Add New Action** box that is **BELOW** the words **End If** (figure 7.26, arrow E) and enter another **IF command** again. Once again, you will get a box to the right of it (figure 7.26, arrow E). Type the following in the box:

[What View?]=2

Step 23: When you press enter, once again, an **Add New Action** box will appear. In that box, choose the **OpenQuery command** from the **drop-down list** box (figure 7.26, arrow F).

Step 24: Make the *Donors and Donations Group and Total* **query** (figure 7.26, arrow G) the one that will run if the *What View?* **option group** is equal to 2.

Step 25: Make sure that when the **query** opens it will open in the **datasheet view** (figure 7.26, arrow H).

Step 26: Save the **macro** naming it *Conditional Macro*. Your screen should look like figure 7.26.

Now step back a moment and think about the logic of this macro in the order in which you wrote it. Remember, **macros** will do the commands in the order that you place them, from top to bottom, within the **macro**. It will check first to see if your **option group** (named *What View?*) is equal to 1. That is the value it will be if the user chooses the *Report View* option. If it is, then it will open the **report** named *Donors and Donations Group and Total* in the **print preview** as you commanded. If it's not equal to 1, the **macro** will do nothing. It will then go to the second condition. It will now check to see if your **option group** (named *What View?*) is equal to 2. That is the value it will be if the user chooses the *Query View* option. If it is, it will open the **query** named *Donors and Donations Group and Total* in the **datasheet view** as you commanded.

Step 27: Close the **macro**.

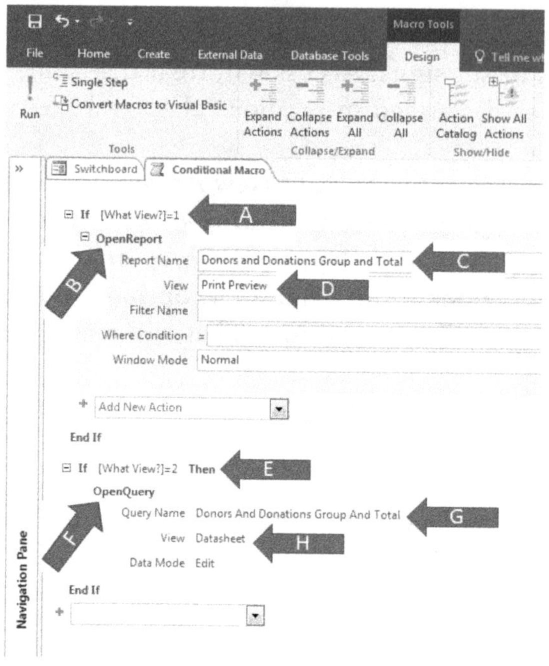

FIGURE 7.26 Creating a conditional macro

Step 28: Return to the **design view** of the **switchboard** and select the **Command button** with the caption that reads *Donors and Donations Group and Total* (figure 7.27, arrow A).

Step 29: Open the **property sheet** and in the **On Click Event** box in the **Event** tab, make sure that the **macro** named *Conditional Macro* that you just created will run when the **Command button** is clicked (figure 7.27, arrow B).

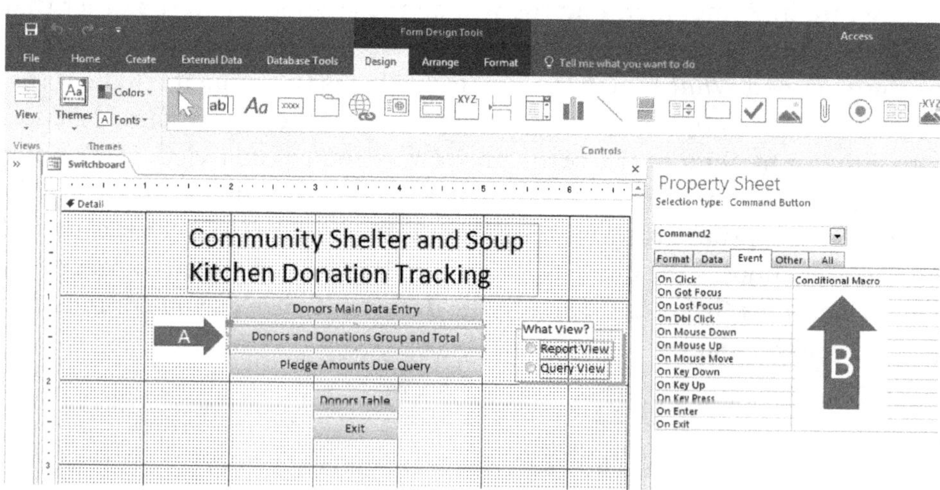

FIGURE 7.27 Creating a conditional macro

Step 30: Let's try out the **option group**. Return to the **form view** of the **switchboard** and try out the **option group** buttons. Click on the *Report View* option button and then click the **command button** marked *Donors and Donations Group and Total*. It should run the *Donors and Donations Group and Total* **report** in the **print preview**. Close the **report** and then click the *Query View* **option button**, and then click the **command button** marked *Donors and Donations Group and Total*. It should run the *Donors and Donations Group and Total* **query**. You will see that the **report** or **query** will run as you have commanded them to in each **condition** or **option group** selection.

Step 31. Close and save the **switchboard**.

CREATING A UNION QUERY

There are often times when you will need to put data together from tables or queries that are from different data sources. For example, many accounting systems draw data from numerous areas of an organization. They will often have to be stacked into one **query** or **report** for reporting and viewing purposes. Why then would we not just enter all of the data into one table? Sometimes you can't. Sometimes the data is being entered by many different departments within organizations and it isn't practical to put the data into one table.

Please remember that all **queries** in Access are created by **structured query language** (**SQL**). It is written behind the scenes for you as you create **queries** the way you have when you click and drag fields from **tables** to where you want them in the **Query Design View grid**. To see the **SQL** language of a given **query**, you can view it in the **query design**. To do so, use the following steps:

Step 1: Start a new **query** that is using the *donors* table with all of its fields (except the *ID* field). Figure 7.28 shows what the **design view** of that **query** would look like. To see the **SQL** statement that was created as this **query** was created, click the **View** button in the upper left of your screen and then click **SQL View** (figure 7.28, arrow).

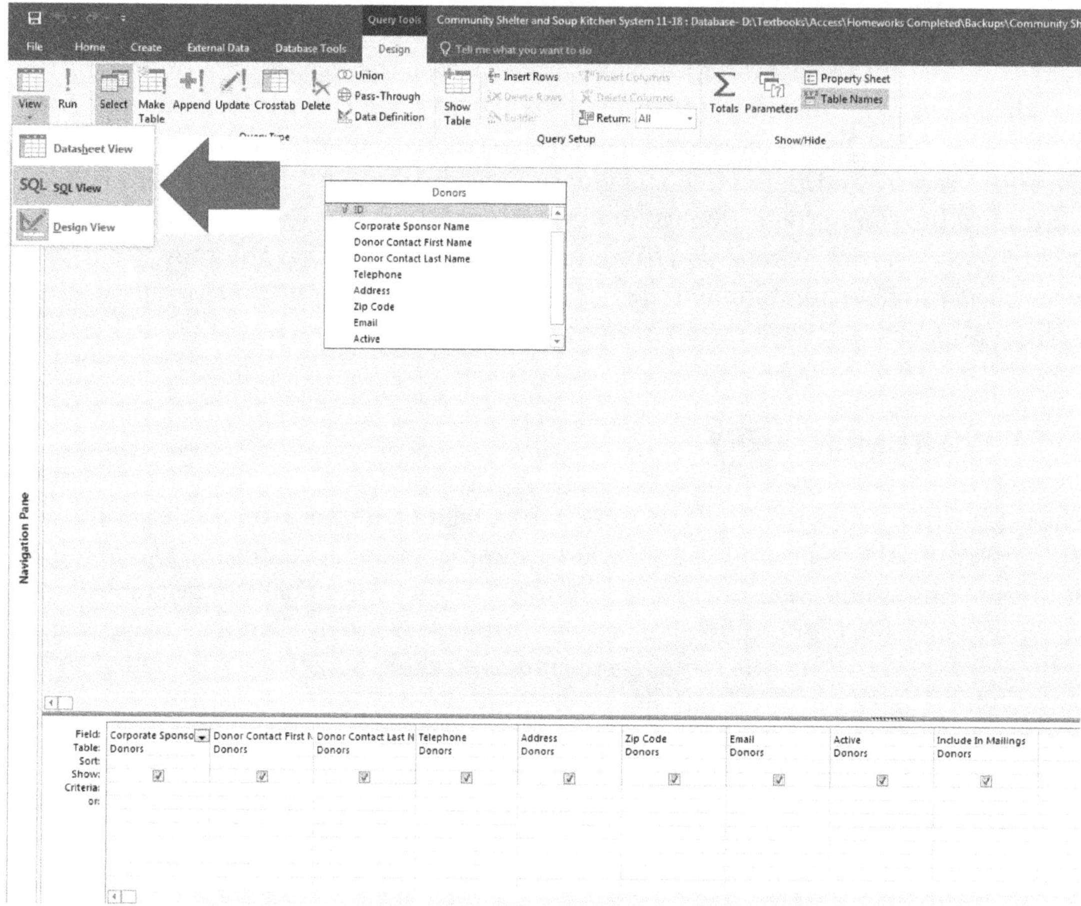

FIGURE 7.28   Creating a union query

Step 2: That will take you to the **SQL statement** (figure 7.29, arrow).

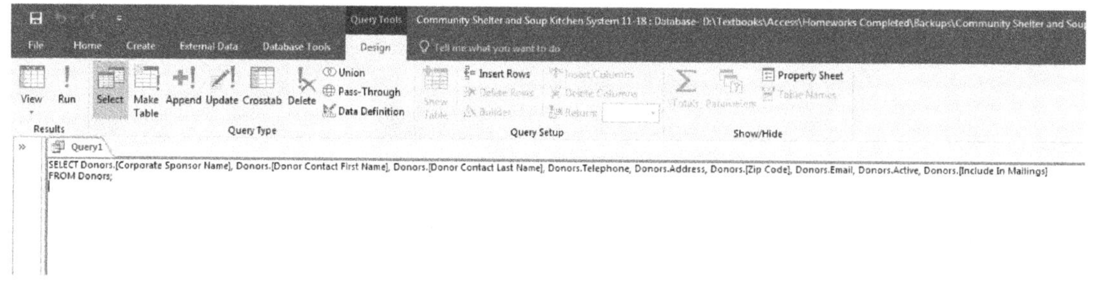

FIGURE 7.29   Creating a union query

Please notice that Access is instructed to select each of those fields in the order that you see from the *donors* table. This is called a **select statement**.

NOTE: It is also permitted to write the same **SQL statement** without preceding each field with the table name (figure 7.30). Doing so would make it faster to write the **SQL statement** if you decided to write it from scratch.

NOTE: If you type the **SQL statement** from scratch, notice that each field must be separated by a **comma** (,) and that the **ENTIRE** statement must always end with one **semicolon** (;).

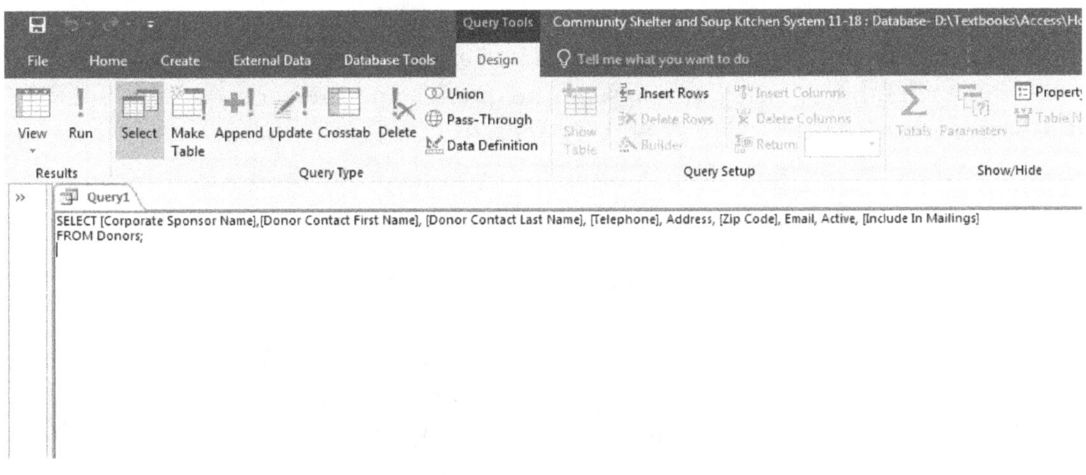

FIGURE 7.30 Creating a union query

To convert this **query** into a **union query**, you must first make sure that

1. the fields in the table or query with which you unite this table have the same number of fields, and
2. the data types and the type of content must match exactly from both tables (or queries).

For example, if you wanted to unite the *donors* table with the *donors from download* table that you imported in an earlier chapter, the **SQL statement** would look like what you see next:

```
SELECT Donors.[Corporate Sponsor Name], Donors.[Donor Contact First
Name], Donors.[Donor Contact Last Name], Donors.Telephone, Donors.Address,
Donors.[Zip Code], Donors.Email, Donors.Active, Donors.[Include In Mailings]
FROM Donors
```

UNION SELECT [Donors From Download].[Corporate Sponsor Name], [Donors From Download].[Donor Contact First Name], [Donors From Download].[Donor Contact Last Name], [Donors From Download].Telephone, [Donors From Download].Address, [Donors From Download].[Zip Code], [Donors From Download].Email, [Donors From Download].Active, [Donors From Download].[Include In Mailings] FROM [Donors From Download];

The *corporate sponsor name* of the *donors* **table** is first in the **select statement**; therefore, the *corporate sponsor name* of the *donors from download* **table**, or a field with the same data and data type must also be first in the **union select statement**.

NOTE: Rather than typing all of that out by hand, to save time, it is best to create a query of each table or query in two separate queries and then copy each of **THEIR SQL statements** into **ONE**, **NEW SQL statement** in a new query. At that point, you would only need to make the following changes:

1. Remove the semicolon from the first **select statement**.
2. At the beginning of the second **select statement**, add the word *Union*.

If you run the query, you will see the results shown in figure 7.31. Save the query as *Union Query*. Notice that the icon for the **query** in the **navigation pane** will be wedding rings, symbolizing that the query is the marriage between two or more **tables** and/or **queries** (or data sources) (figure 7.31, arrow).

NOTE: You can add numerous **union select statements** to unite numerous **tables** and **queries** in the same **union query**. You are not limited to two **tables** or **queries**. You can also use a mixture of **queries** or **tables**.

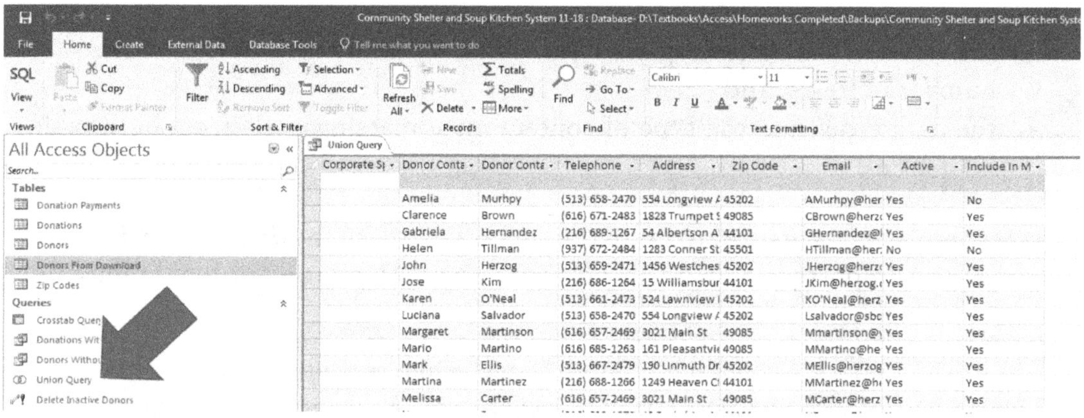

FIGURE 7.31 Creating a union query

Chapter 7: Assignment (Summary of Tasks)

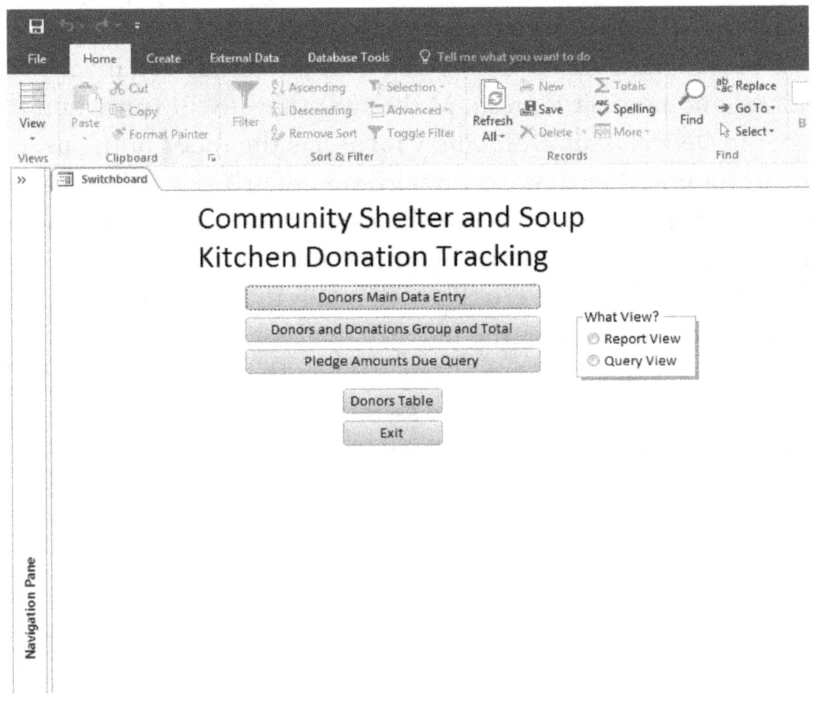

FIGURE 7.32   Homework sample of finished product

1. In the *Community Shelter and Soup Kitchen System* database, create a **query** using the *donors* and the *donations* **tables**. Name the **query** *All Donors and Prospective Donors*. Use the following specifications:
   a. Send the *ID, donor contact last name,* and *donor contact first name* fields (all from the *donors* table) to the **query design grid**.
   b. Sort the **query** by the *donor contact last name* and *donor contact first name* fields, from left to right in that order.
   c. Change the **join properties** of the **query** so that all records in the *donors* table will display even if they have no match in the *donations* table.
   d. Create a new field in the first blank column, top box (at the right of the *donor contact first name* field) with the expression:

   Donated Item: IIF([Donation Item] is null,"NO Donations",[Donation Item])

2. Create a **pop-up** form of the *zip code* table.
3. Remove the *table.donors* **sub-form** from the form if it appears.

4. Add a **command button** to the *Donors Main Data Entry* form that will open the *zip code pop-up* **form**. Give the **command button** a caption that reads *zip codes*.

5. Add a **command button** in the header of the *zip code* **pop-up form** that will close it. Give it a **caption** that reads *Close and Save*.

6. Make a **macro** that will have only one command that will invoke the **Refresh** command and name it *Refresh*. Make sure that it will run when the *zip code* field in the *Donors Main Data Entry* **form** has the **focus** of the user.

7. Create a **shadowed option group** with **no default** that will give the user the ability to choose between seeing the *Donors* and *Donations Group and Total* **report** in the **print preview** or the **query datasheet view**.

   a. Make a **conditional macro** that will honor the choice the user makes in the **option group** and name the macro *Conditional Macro*.

8. Make that be the **macro** that now runs when the *Donors and Donations Group and Total* **command button** in the **switchboard** is clicked.

9. Create a **union query** between the *donors* table and the *donors from download* table.

Chapter 7: Assignment (Alternate) (Summary of Tasks)

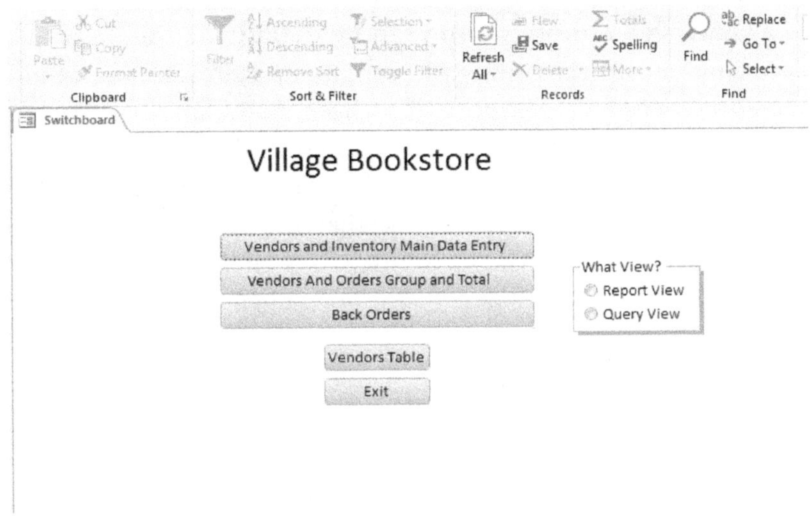

FIGURE 7.33   Alternate homework sample of finished product

1. In the *Village Bookstore Database System* database, create a **query** using the *vendor* and the *inventory orders* tables. Name the **query** *All Vendors (Active or Inactive)*. Use the following specifications:
   a. Send the *ID* and *vendor name* from the *vendors* **table** to the **Query Design grid**.
   b. Sort the **query** by the *vendor name* field.
   c. Change the **join properties** of the **query** so that all records in the *vendors* table will display even if they have no match in the *Inventory Orders* table.
   d. Create a new field in the first blank column, top box (at the right of the *vendor name* field) with the expression:

   Vendor Activity: IIF([Book Title] is null,"NO Activity",[Book Title])

2. Create a **pop-up** form of the *zip code* table.
3. Remove the *table.vendors* **sub-form** from the form if it appears.
4. Add a **command button** to the *Vendors and Orders Main Data Entry* form that will open the *zip code pop-Up* form. Give the **command button** a caption that reads *zip codes*.
5. Add a **command button** in the header of the *zip code* **pop-up** form that will close it. Give it a **caption** that reads *Close and Save*.

6. Make a **macro** that will have only one command that will invoke the **Refresh** command and name it *Refresh*. Make sure that it will run when the *zip code* field in the *Vendors and Orders Main Data Entry* **form** has the **focus** of the user.

7. Create a **shadowed option group** with **no default** that will give the user the ability to choose between seeing the *Vendors and Orders Group And Total* **form** in the **print preview** or the **query datasheet view**.

   a. Make a **conditional macro** that will honor the choice the user makes in the **option group** and name the macro *Conditional Macro*.

8. Make that be the **macro** that now runs when the *Vendors and Orders Group and Total* **command button** in the **switchboard** is clicked.

9. Create a **union query** between the *vendors* table and the *append vend* table.

Document Management

IN THIS CHAPTER you will learn how to do the following:

1. Create a **report** that will appear in a word processing document format
2. **Concatenate** data (with formatting) so that your text will mingle with **field data**

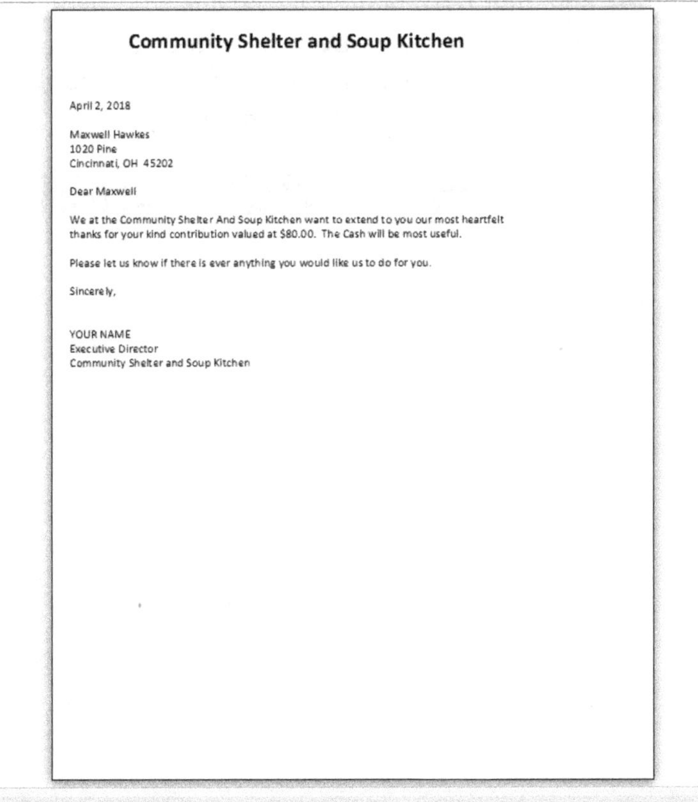

FIGURE 8.1 Document management sample

CHAPTER 8 TASK: PERFORMING DOCUMENT MANAGEMENT

As mentioned in the introduction chapter of this book, **document management** makes it possible for your data to be shared for any and all uses. As mentioned before, many argue **mail merging** will take care of this, but a database will merge data routinely by the click of the mouse and therefore much, much faster than a **mail merge**. In this example, we will use *donor* data to create thank-you notes for said *donors* without having to type the data all over again in a word processing document.

This process will show you a more in-depth use of the **concatenate** coding in Access, discussed briefly in chapter 3. This coding will allow you to mingle data fields within your documents, whether they be legal documents, letters, or anything in which data usage can change.

The first thing you must do is create a **query** that your document will read. Using the skills you have learned so far, complete the following steps:

Step 1: Start a new **query** using the *donors, donations,* and *zip code* **tables**. Send the *ID, donor contact first name, donor contact last name, address,* and *zip code* fields from the *donors* table into the **Query Design grid**. Also send the *city* and *state* fields from the *zip code* table, and the *donation ID, donation value, donation item,* and *donation date* fields from the *donations* **table** into the **Query Design grid**.

Step 2: Name the **query** *Donations Thank You Letters*. When you do, the **design view** of your **query** should look like figure 8.2. Keep the **query** open.

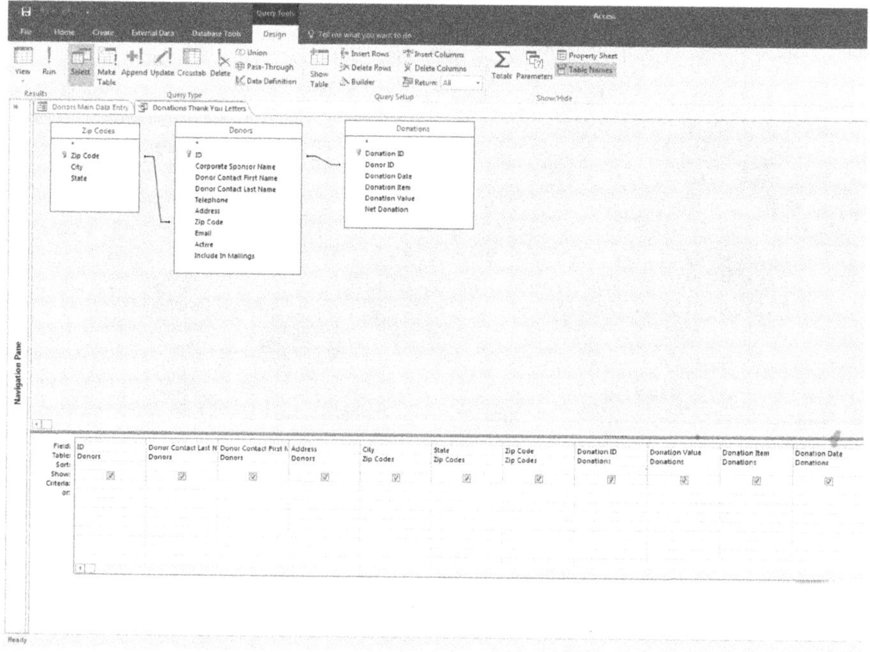

FIGURE 8.2   Document management query

Step 3: Open the *Donors Main Data Entry Menu* and then return to the **query design** of your new **query**. When you do, click in the **Criteria** box, below the *donation ID* field. Also drawing on the skills you learned in this book, click the **Expression Builder** in the **query design** (figure 8.3, arrow A) to make sure that the current *donation* in the **sub-form** of the *Donors Main Data Entry Menu* is the one that will be selected in this **query**. Your expression should look like what you see at arrow B in figure 8.3.

Note: You will have to double-click the *Donors Main Data Entry* **form** and the *Donations* **sub-form** in the **Expressions Elements** window (figure 8.3, arrow C) in order to create the expression.

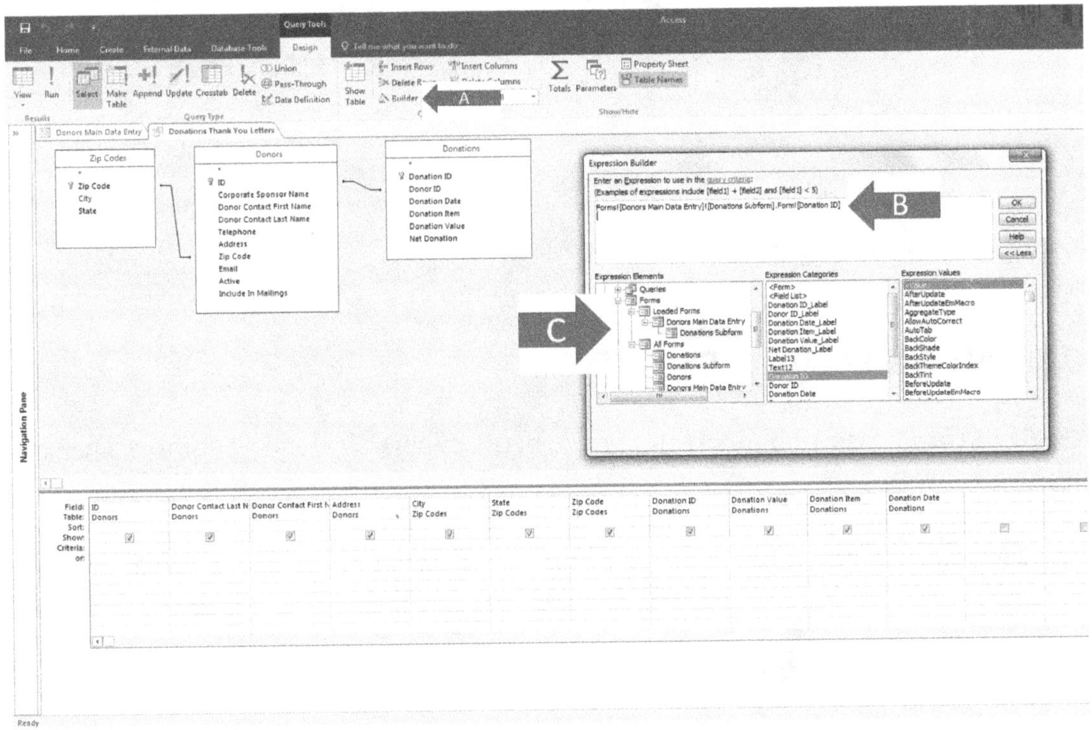

FIGURE 8.3   Document management query expression

Step 4: Click **OK** and your **query design view** should now look like figure 8.4. If you run the **query** it should only display the data from the *donation* that is currently displayed in the *Donors Main Data Entry* **form**.

Step 5: Close and save the **query**.

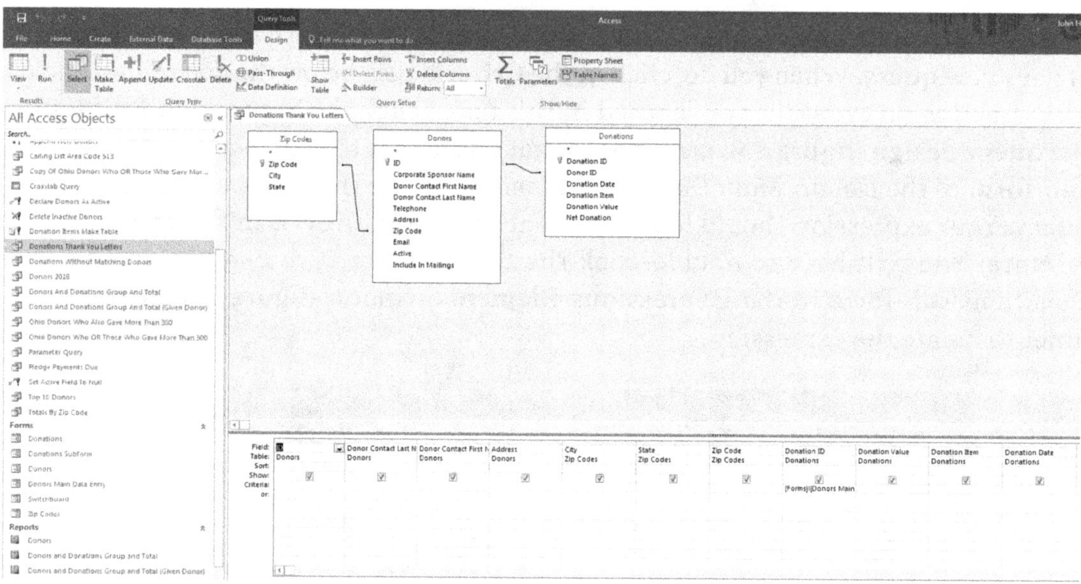

FIGURE 8.4   Creating the document in document management

TO CREATE THE DOCUMENT

Step 1: In the **Create** ribbon, create a new, blank report by clicking **Report Design** (figure 8.5, arrow).

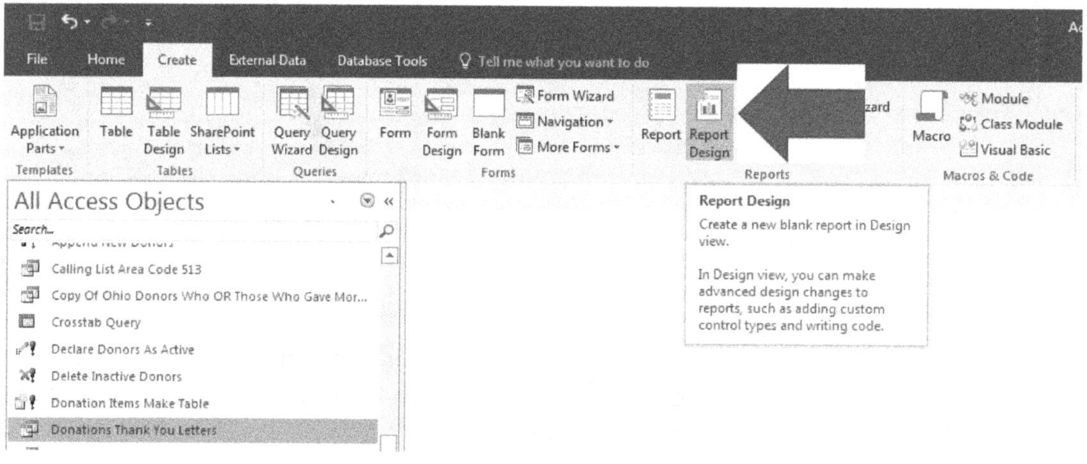

FIGURE 8.5   Creating the document in document management

Step 2: That will take you to what you see in figure 8.6a. Double-click the **Report selection box** in the upper left of the **Design View grid** (figure 8.6a, arrow A). When you do, it will open the **property sheet** for the report.

Step 3: Add a **record source** in the **All** tab of the **property sheet** (figure 8.6a, arrow B) so that the **report** will now read the *Donations Thank You Letters* **query** you have just created.

Step 4: Close the property sheet.

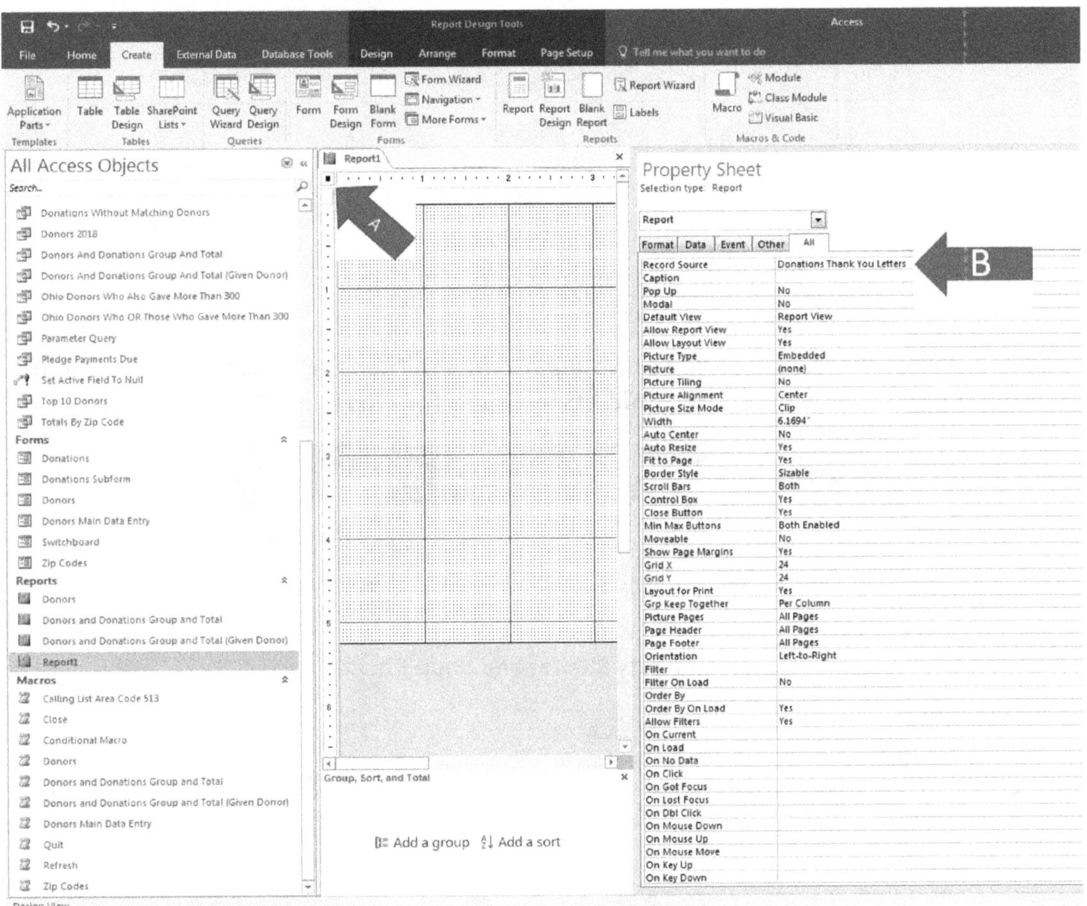

FIGURE 8.6a Creating the document in document management

Step 5: Click once on the **label control** (marked **Aa**, figure 8.6b, arrow) in the **controls toolbox** area of the **design tab** in the **report design tools**, just as you did when you created labels in a previous chapter. Place the **label** at the 3/4 inch mark at the top of the **report** (figure 8.7).

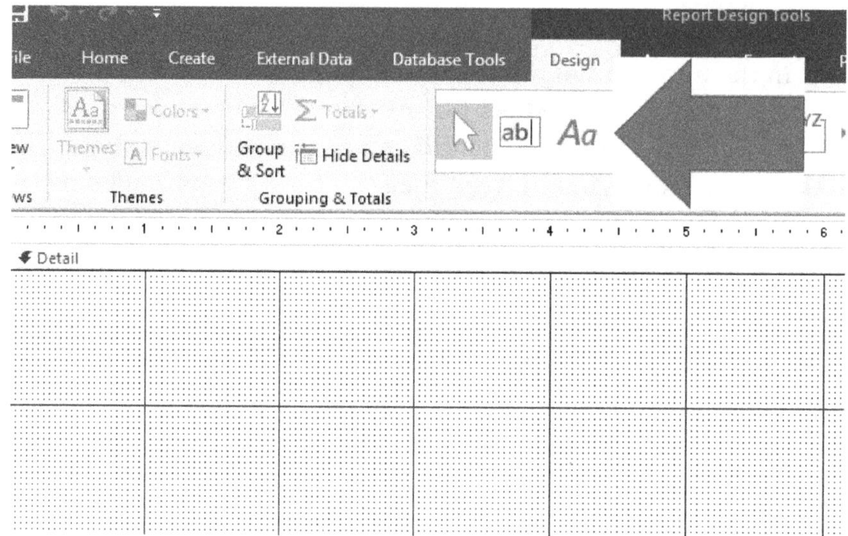

FIGURE 8.6b Creating the document in document management (labels)

Step 6: Type *Community Shelter and Soup Kitchen* in the **label** and format the it with the **font type** of **Calibri (Detail)**, **bold style**, and a **font size** of **22** (figure 8.7).

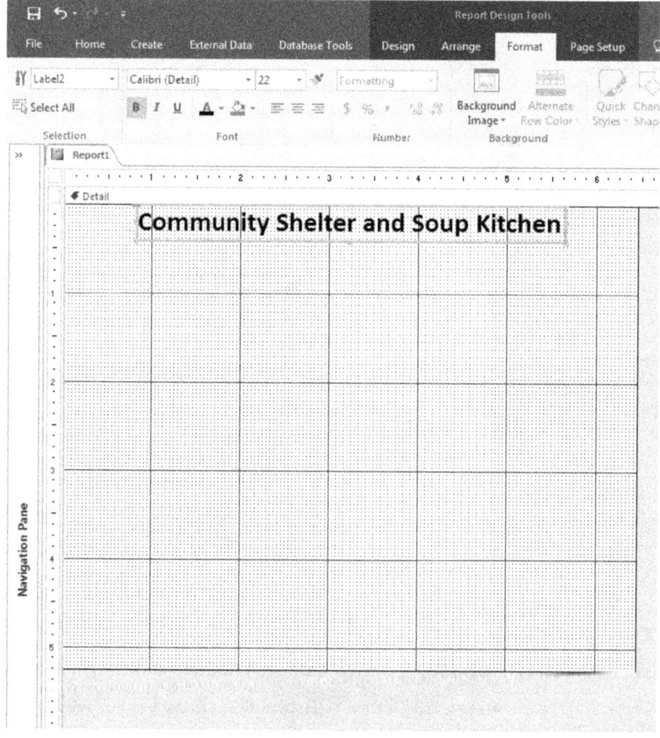

FIGURE 8.7 Creating the document in document management

Step 7: Click once (do not click and drag) on the **Text Box Control** (marked **ab|**) in the **controls** area of the **design tab** in the **report design tools**, just as you did when you created expressions in forms in a different chapter. Place it 1 inch below the top of the **report** and 1/2 inch from the left margin of the **Report Design grid** (figure 8.8).

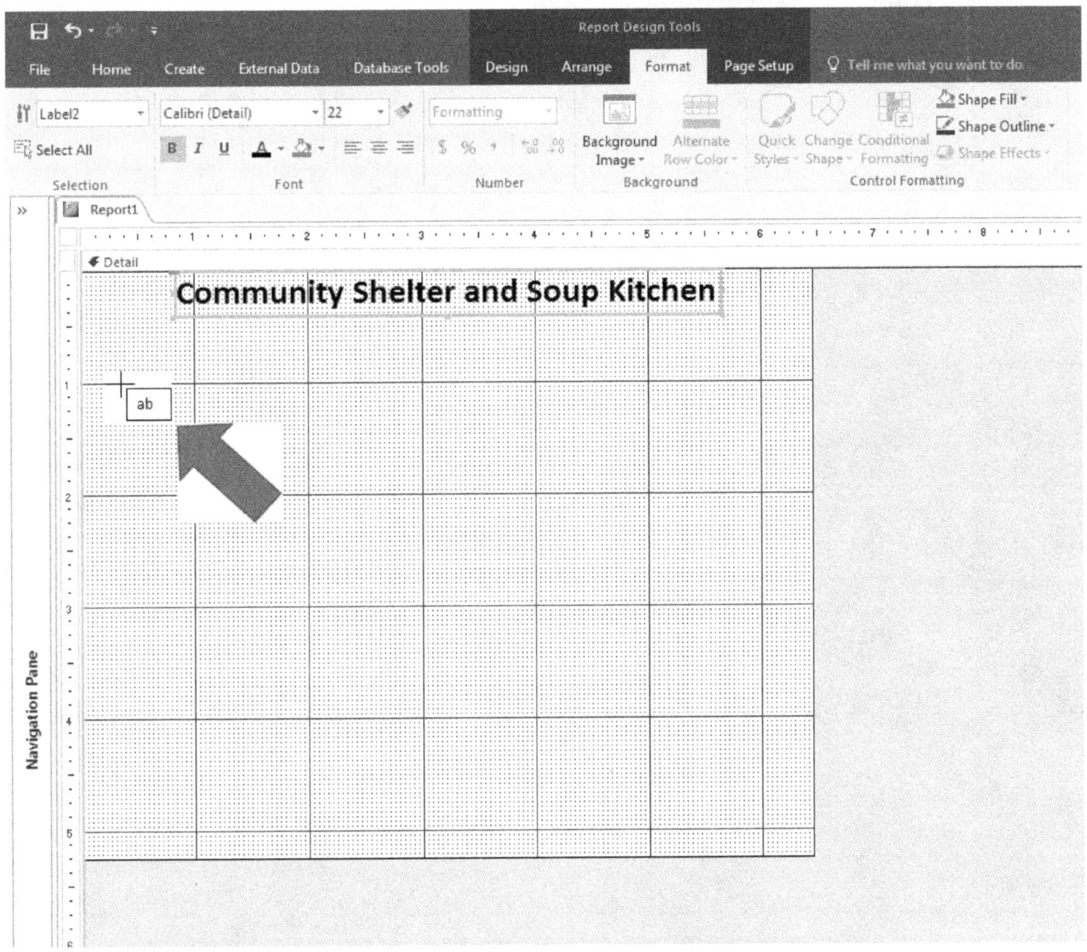

FIGURE 8.8   Creating the document in document management

Step 8: Once you do so, resize the **text box** and make it 6 inches wide by 4 inches tall (figure 8.9).

Step 9: Click on the **label** to the left of the **text box** (in this example the label reads *Text3*, figure 8.9, arrow) and press the **Delete** key on your **keyboard** to remove the **label**.

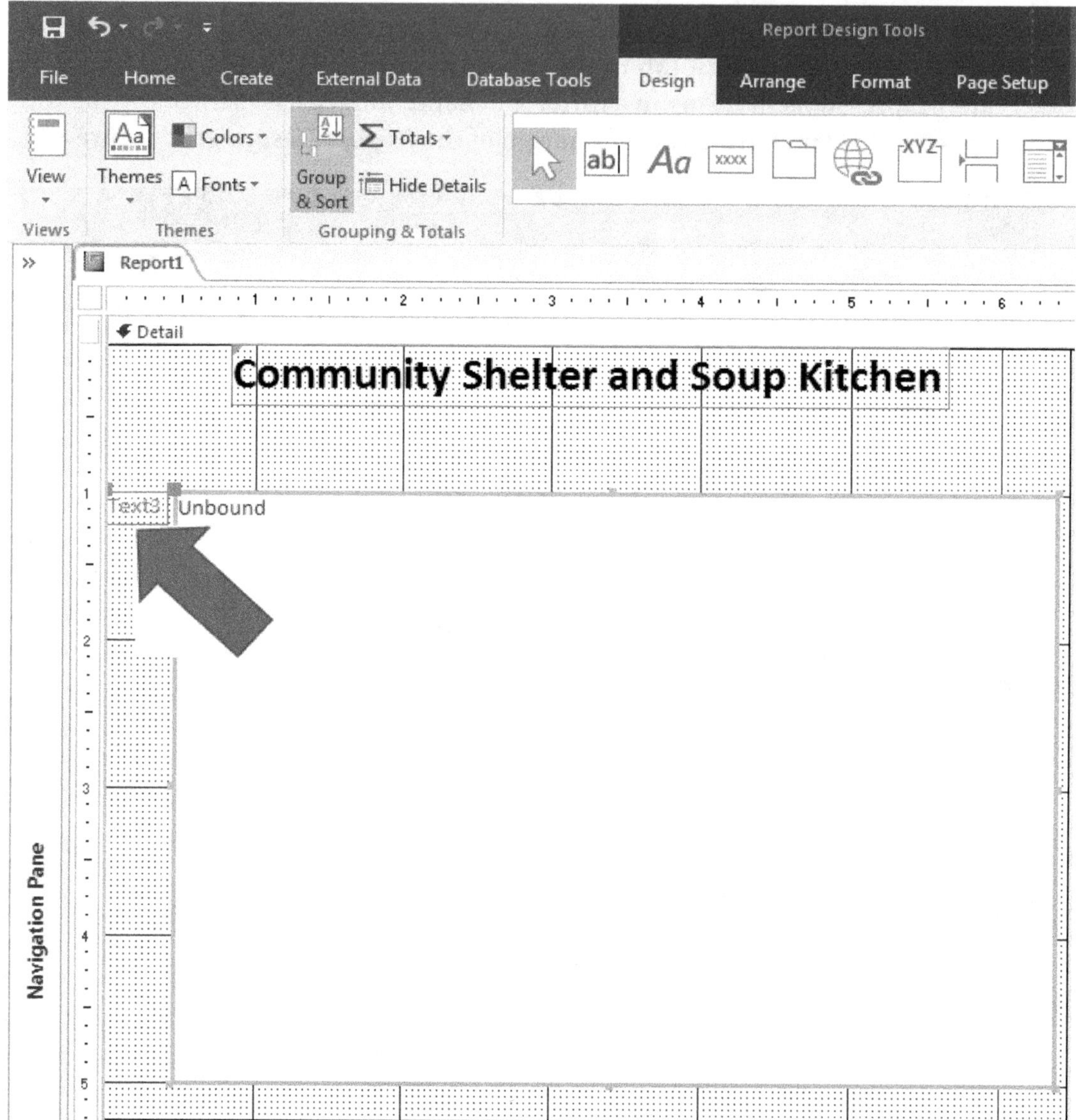

FIGURE 8.9   Creating the document in document management

TO CONCATENATE DATA

Step 10: Type the following expression in the **unbound text box** and then press **Enter** on your keyboard. Every space and character you time must be exactly as you see here. This expression uses the concept of **concatenating** where **field names** are embedded or mingled within text to make a complete sentence, using the data from your database as part of your sentences. At the end of entering this expression, you will need to read some major pieces of information about some of the principles that must be followed when you **concatenate** data.

```
=Format(Date(),"mmmm d"", ""yyyy") & "
   " & [Donor Contact First Name] & " " & [Donor Contact Last Name] & "
   " & [Address] & "
   " & [City] & ", " & [State] & " " & [Zip Code] & "
   Dear " & [Donor Contact First Name] & "
   We at the Community Shelter And Soup Kitchen want to extend to you our
   most heartfelt thanks for your " & [Kind or Very Generous] & " valued at " &
   Format([Donation Value],"Currency") & ". The " & [Donation Item] & " will be
   most useful.
   Please let us know if there is ever anything you would like us to do for you.
   Sincerely,
   YOUR NAME
   Executive Director
   Community Shelter and Soup Kitchen"
```

If you view the **report** in the **print preview**, you will see that the document will slowly begin to look like what is displayed in figure 8.1.

CREATING A SUBROUTINE

In many cases, we will often want a **form** or **report** to display data after looking through some special instruction you may have. For example, if a *donor* is a particularly generous *donor*, you may want to word your document in a way that reflects that. This can be accomplished by what is known as a **subroutine**. To create special instructions to indicate what is defined as a generous *donor*, we will resume our process as follows:

Step 11: Continuing now with this project at step 11, insert an additional **text box** in the **detail** area of the **report** (figure 8.10), below the one you have been using to this point and type the following **IIF function** inside of it:

```
=IIf([Donation Value]>=200,"your very generous contribution","kind contribution")
```

Press **Enter** when finished.

NOTE: You need to make sure you understand the logic of this **IIF function**. It is basically saying that if the *donation* is greater than or equal to *200*, the words *"your very generous contribution"* should be displayed in the body of the letter. Otherwise, if the *donation* is lower than *200*, the words *"kind contribution"* should be displayed. They are to remain in a lower-case format, because they will later be displayed within a sentence.

Step 12: Open the **property sheet** and in the **Format tab**, change the **visible** setting of the **text box** to **No**. This is because we don't want the person who reads this document to see the instructions in the **subroutine**.

Step 13: Give the name of *Kind or Very Generous* to the **text box**.

You will see what is shown in figure 8.10.

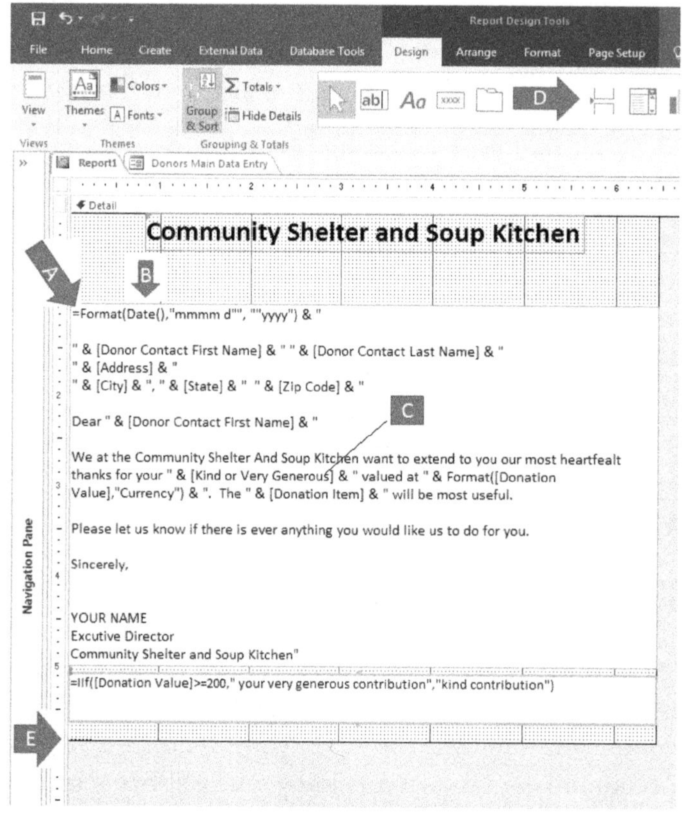

FIGURE 8.10   Creating the document in document management

THERE ARE SEVERAL IMPORTANT THINGS TO UNDERSTAND HERE ABOUT THE ENTIRE LETTER BODY AND THE SUBROUTINE. IT IS IMPORTANT THAT YOU TAKE SOME TIME TO STUDY THESE EXPRESSIONS IN DETAIL!

1. As mentioned before, you must start an expression (as you did with calculated expressions in **forms**) with an **equal sign** (=) (figure 8.10, arrow A).
2. The **Date() function** (figure 8.10, arrow B) is the same as the **Today() function** that is used in Microsoft Excel, and it will display the current date.

   a. To make a date function appear with the **month day, and year** format (such as January 1, 2019), as you did in the expression, you can surround the **Date() function** with the **Format function** as follows (and is also shown at figure 8.10, arrow B):

Format(Date(),"mmmm d""," ""yyyy")

    As a result, that would make the **Date() function** an **embedded function** (one **function** placed inside another **function**).

    NOTE: THERE IS A SPACE AFTER THE COMMA IN THIS FORMATTING FUNCTION. Once you type in the function, it will put additional quotation marks in their proper place. It is always best to allow Access to put those quotation marks there rather than typing them in manually. Typing them in manually can cause the formatting to function to fail, which would give you an error.

3. When you **concatenate** in an expression, as we have done here, you must use the following principles:
   a. All text (words and sentences) must be preceded and followed by **quotation marks** (").
   b. The **ampersand (&)** must **ALWAYS** precede and follow **fields, functions,** or **embedded functions** with the following exceptions:
      i. If you **START** an expression with a **field, function,** or **embedded function**, the **ampersand (&)** must not precede it, but it must still follow it (figure 8.10, arrow A).
      ii. If you **END** an expression with a **field, function,** or **embedded function**, the **ampersand (&)** must precede it, but must not follow it.
   c. Put spaces and punctuation in text as you would if you were writing text in any word processing document.
   d. Never put &" at the end of an expression.
   e. To start a new line when you **concatenate**, you can only do so between **quotation marks** (").
      i. It is done by pressing and holding the **Ctrl** key and then pressing **Enter**.
4. The expression will reference the **text box** that you named *Kind or Very Generous* (figure 8.10, arrow C) by treating it as it would a field name. That is why the reference to that **text box** name is surrounded by **square brackets** [].
   a. Some programmers would call this a **subroutine**, because it goes below your coding to get information before it continues to run your **report**.
5. It is important to put a **page break** in a report if each page (or series of pages) is to be sent to different recipients. You can do so by clicking **Page Break** once (figure 8.10, arrow D) in the **controls** area of the **design tab** of the **report design tools**. After you do, you need only click once at the very bottom of the

Report Design grid and a **page break** will be inserted for you there (figure 8.10, arrow E).

Step 14: Click on the **text box** and then click the **Format** ribbon (figure 8.11, arrow A).

Step 15: Select the **Shape Outline option** (figure 8.11, arrow B).

Step 16: Click **Transparent** (figure 8.11, arrow C). This will remove the **border** to the **text box** holding the expression. **Borders** surrounding expressions such as these in the middle of letters and/or legal documents make a very bad impression; hence, we have just removed them.

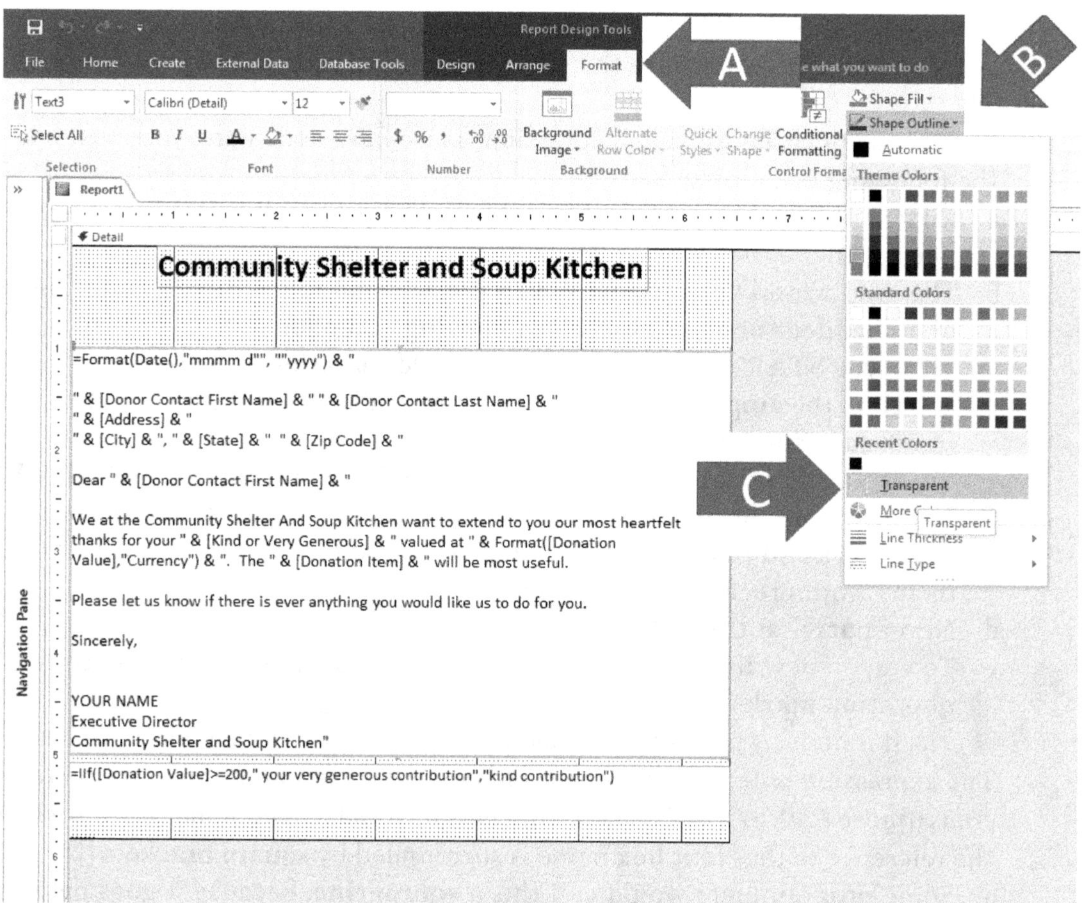

FIGURE 8.11 Creating the document in document management

Step 17: Making sure the *Donors Main Data Entry* is still displaying a given donation, run the report in **print preview**. It will appear as what you see in figure 8.12.

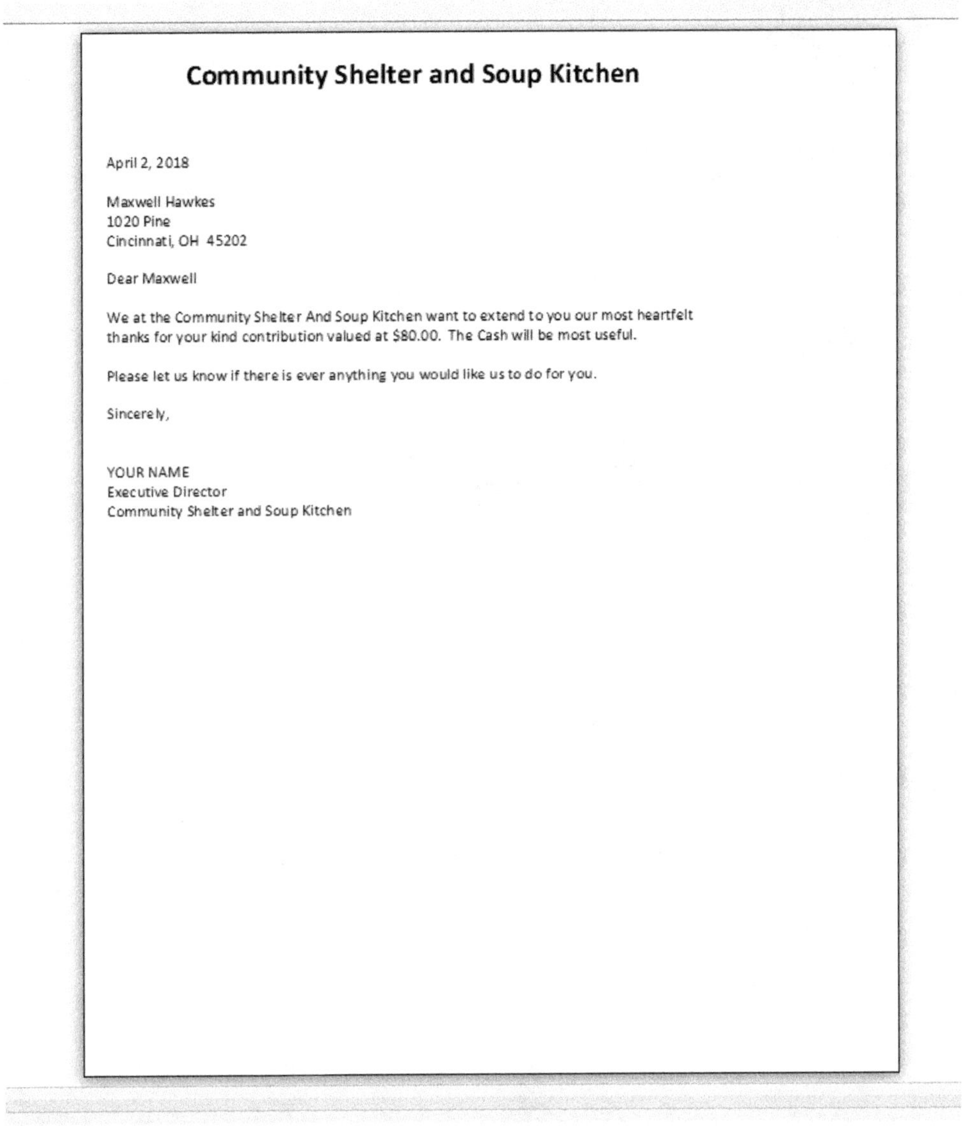

Community Shelter and Soup Kitchen

April 2, 2018

Maxwell Hawkes
1020 Pine
Cincinnati, OH 45202

Dear Maxwell

We at the Community Shelter And Soup Kitchen want to extend to you our most heartfelt thanks for your kind contribution valued at $80.00. The Cash will be most useful.

Please let us know if there is ever anything you would like us to do for you.

Sincerely,

YOUR NAME
Executive Director
Community Shelter and Soup Kitchen

FIGURE 8.12   Creating the document in document management result

Step 18: Using the concepts you have already learned in this textbook, place a **command button** in the **sub-form** of the *Donors Main Data Entry* that will do the following:

a. Use a **macro** that will **refresh** the record and will open the *Donations Thank You Letter* in the **print preview**.
b. Have a caption of *Thank You Letter* (figure 8.13, arrow).

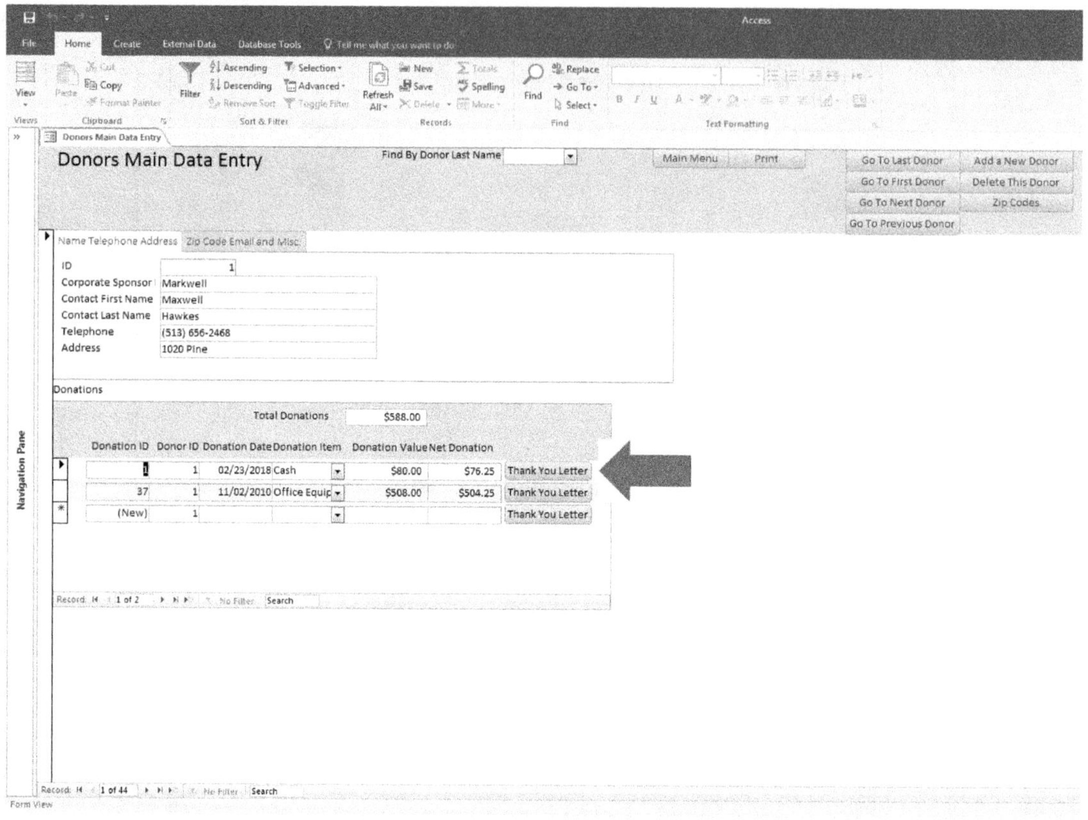

FIGURE 8.13 Creating a button to run the document management button

Chapter 8: Assignment (Summary of Tasks)

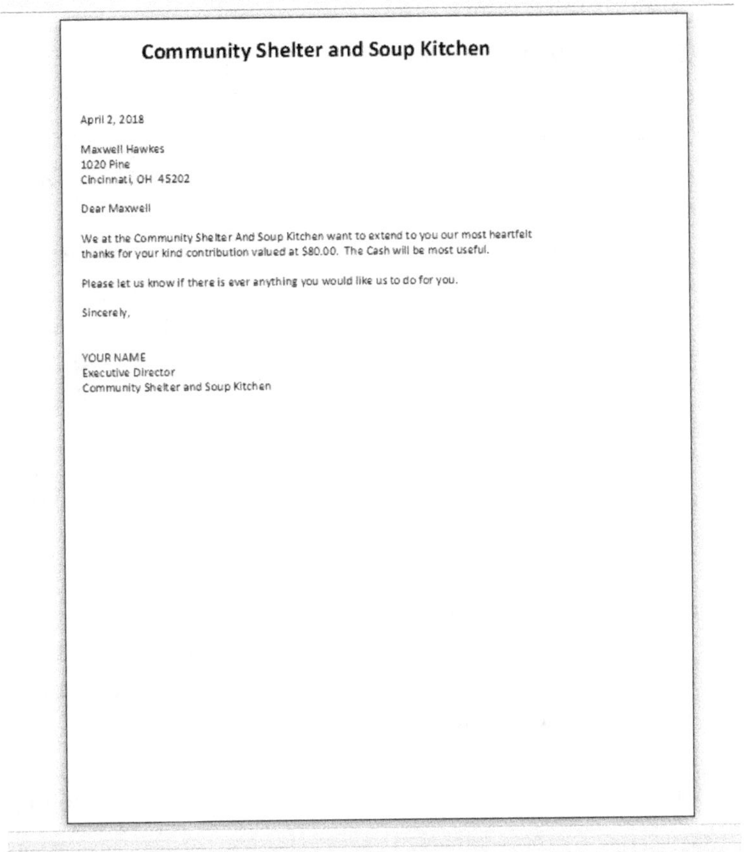

Community Shelter and Soup Kitchen

April 2, 2018

Maxwell Hawkes
1020 Pine
Cincinnati, OH 45202

Dear Maxwell

We at the Community Shelter And Soup Kitchen want to extend to you our most heartfelt thanks for your kind contribution valued at $80.00. The Cash will be most useful.

Please let us know if there is ever anything you would like us to do for you.

Sincerely,

YOUR NAME
Executive Director
Community Shelter and Soup Kitchen

FIGURE 8.14 Finished document management assignment sample

1. Open the Community Shelter and Soup Kitchen System database and open the *Donors Main Data Entry* form.
   a. Make sure a *donor* is displayed.
2. Create a new **query** using the *donors*, *donations*, and *zip code* tables.
   a. Use the ID, donor contact first name, donor contact last name, address, and zip code fields from the donors **table**.
   b. Use the *city* and *state* fields from the *zip code* **table**.
   c. Use the donation ID, donation value, donation item, and donation date fields from the donations **table**.

3. Use the **expression builder** in the **query design** to make sure that the currently displayed *donation* in the **sub-form** of the *Donors Main Data Entry* is the only one that will be displayed when this query is run.
4. Name the query *Donations Thank You Letters*.
5. Create a **NEW report** in the **design view** that will read (or have as its **record source**) the *Donations Thank You Letters* query you just created.
6. Put a label at the top and middle of the report that reads *Community Shelter* and *Soup Kitchen*, using the **font type** of **Calibri (Detail)**, **bold style**, and a **font size** of **22**.
7. Insert a **text box** control, 1 inch from the top of the **report** that is approximately 6 inches wide and 4 inches tall and remove its label.
8. Click on the **text box** and select the **shape outline** option setting as **transparent**.
9. Enter the following complex, extensive, concatenated expression in the **text box** control with spacing exactly as what is shown next:

=Format(Date(),"mmmm d"", ""yyyy") & "
" & [Donor Contact First Name] & " " & [Donor Contact Last Name] & "
" & [Address] & "
" & [City] & ", " & [State] & " " & [Zip Code] & "
Dear " & [Donor Contact First Name] & "
We at the Community Shelter And Soup Kitchen want to extend to you our most heartfelt thanks for your " & [Kind or Very Generous] & " valued at " & Format([Donation Value],"Currency") & ". The " & [Donation Item] & " will be most useful.
Please let us know if there is ever anything you would like us to do for you.
Sincerely,
YOUR NAME
Executive Director
Community Shelter and Soup Kitchen"

10. Insert an additional **text box** below the one you have been using to this point and type the following **IIF function**:

=IIF([Donation Value]>=200,"your very generous contribution","kind contribution")

11. Open the **property sheet** and in the **Format tab**, change the **visible** setting of this **text box** to **No**.
12. Give the name of *Kind or Very Generous* to the **text box**.
13. Figure 8.14 shows what the finished product should look like. Close the report and save it as *Donations Thank You Letters*.

14. Place a command button in the sub-form of the donors main data entry that will do the following:
    a. **Refresh** the record.
    b. Open the *Donations Thank You Letter* in the **print preview**.
    c. Have a caption of *Thank You Letter*.

Chapter 8: Assignment (Alternate) (Summary of Tasks)

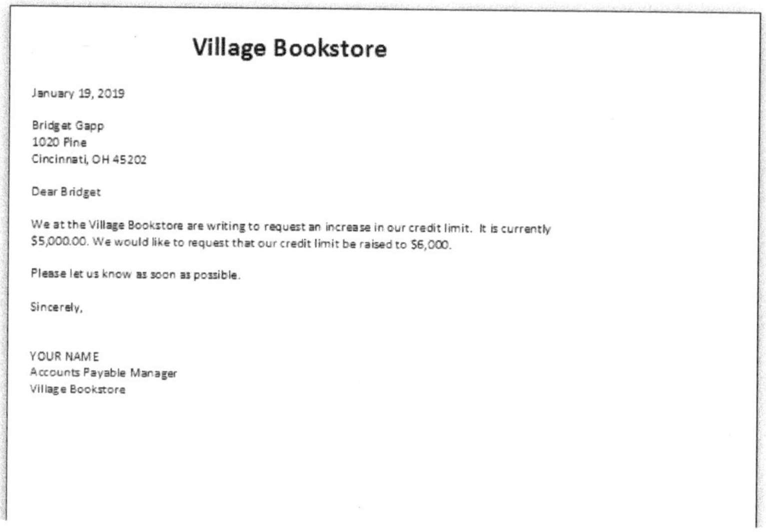

Village Bookstore

January 19, 2019

Bridget Gapp
1020 Pine
Cincinnati, OH 45202

Dear Bridget

We at the Village Bookstore are writing to request an increase in our credit limit. It is currently $5,000.00. We would like to request that our credit limit be raised to $6,000.

Please let us know as soon as possible.

Sincerely,

YOUR NAME
Accounts Payable Manager
Village Bookstore

FIGURE 8.15   Finished document management alternate assignment sample

1. Open the Village Bookstore Database System and open the Vendors and Inventory Main Data Entry form.
   a. Make sure a *vendor* is displayed.
2. Create a new **query** using the *vendors* and *zip code* tables.
   a. Use the ID, vendor contact first name, vendor contact last name, vendor address, credit limit, and zip code fields from the vendors **table**.
   b. Use the *city* and *state* fields from the *zip code* **table**.
3. Use the **expression builder** in the **query design** to make sure that the currently displayed *vendor* in the **sub-form** of the *vendors and inventory main data entry* is the only one that will be displayed when this query is run.
4. Name the query *Credit Limit Increase Requests*.
5. Create a **NEW report** in the **design view** that will read (or have as its **record source**) the *Credit Limit Increase Requests* query you just created.
6. Put a label at the top and middle of the report that reads *Village Bookstore*, using the **font type** of **Calibri (Detail)**, **bold style**, and a **font size** of **22**.
7. Insert a **text box** control, 1 inch from the top of the **report** that is approximately 6 inches wide and 4 inches tall and remove its label.
8. Click on the **text box** and select the **shape outline** option setting as **Transparent**.

9. Enter the following complex, extensive, concatenated expression in the **text box** control with spacing exactly as what is shown below:

=Format(Date(),"mmmm d"", ""yyyy") & "
 " & [Vendor Contact First Name] & " " & [Vendor Contact Last Name] & "
 " & [Vendor Address] & "
 " & [City] & ", " & [State] & " " & [Zip Code] & "
 Dear " & [Vendor Contact First Name] & "
 We at the Village Bookstore are writing to request an increase in our credit limit.
It is currently " & Format([Vendor Credit Limit], "Currency") & ". We would like
to request that our credit limit be raised to "&[Credit Limit Request Amount]&".
 Please let us know as soon as possible.
 Sincerely,
 YOUR NAME
 Accounts Payable Manager
 Village Bookstore"

10. Insert an additional **text box** below the one you have been using to this point and type the following **IIF function**:

=IIF([Vendor Credit Limit]>=5000, "$6,000","$4,000")

NOTE: This letter is aimed toward asking the *Village Bookstore Vendors* for an increase to *$4,000* if their current limit is below *$5,000*. The letter is also aimed toward asking the *Village Bookstore Vendors* for an increase to *$6,000* if their current limit is at or above *$5,000*.

11. Open the **property sheet** and in the **Format tab**, change the **visible** setting of this **text box** to **No**.
12. Give the name of *Credit Limit Request Amount* to the **text box**.
13. Figure 8.15 shows what the finished product should look like. Close the report and save it as *Credit Limit Increase Requests*.
14. Place a **command button** in the **main form** of the *Vendors and Inventory Main Data Entry* form that will do the following:
    a. Run a **macro** that will **refresh** the record and open the *Credit Limit Increase Requests* **report** in the **print preview**.
    b. Have a caption of *Credit Limit Increase Requests*.

Epilogue

You are now a programmer in a **high-level programming language**. With what you have learned here, you can build just about any program that any organization could need. This scenario and the concepts used can be applied to thousands of different organizations and to the data they wish to capture.

Index

Quick Access Toolbar, 20, 60, 63, 90, 328

QuitAccess, 276

R

record, 5, 19, 24, 25, 34, 35, 69, 75, 126, 136, 152, 198, 200, 203, 229, 272, 273, 274, 293, 294, 298, 300, 315, 316, 364, 366

referential integrity, 2, 70, 71, 79, 80, 118, 121, 322

Referential Integrity, 69, 70, 71, 72, 73, 74, 75, 76, 82, 118, 119, 121, 122

Refresh, 329, 330, 361, 365, 367

relational, 5, 7, 8

relational database, 5

Relationships Menu, 81, 83

Relationships Screen, 77–78

Report Design, 186, 188, 189, 191, 192, 194, 196, 248, 250, 352, 353, 355, 359

reports, 1, 4, 6, 11, 15, 16, 162, 254, 257, 268, 279, 318

Report Wizard, 63, 163

Return, 100, 103, 152, 202, 223, 228, 234, 241, 278, 279, 323, 337, 340, 341

Ruler Bar, 224, 301

Run, 89, 90, 94, 97, 101, 102, 104, 108, 140, 151, 152, 154, 158, 161, 178, 180, 296, 298, 300, 315, 316, 320, 324, 360

Run button, 89, 90, 97, 102, 104, 151, 152, 154, 158, 178, 180

RunMenuCommand, 330

S

Save, 20, 21, 34, 60, 63, 66, 68, 90, 94, 95, 96, 103, 108, 119, 122, 123, 130, 138, 152, 154, 161, 177, 178, 179, 180, 229, 263, 273, 280, 297, 307, 322, 326, 328

Selected Fields, 164, 183, 200, 201, 219, 235

Shape Outline Option, 364, 366

shore up, 197, 327

Short-Cut Menu, 224

Short Text, 19, 24, 65, 67

Show All Records, 319

Show Row, 103

SINGLE ARROW, 200

Size/Space Menu, 189

Sort Box, 87

Standard Deviation, 129

static, 6, 7

Subform, 181, 182, 184, 185, 186, 196, 223, 224, 229, 230, 232, 233, 247, 248, 249, 250, 251, 252, 307

Subroutine, 359

SUM, 107, 129, 167, 177, 178, 179, 180, 227

Summary Options, 167, 178, 180

Switchboard, 253, 254, 255, 256, 257, 263, 264, 266, 278, 279, 281, 283, 285, 289, 290, 291, 292, 333, 336, 337, 340, 341, 346, 348

T

Tab, 107, 235, 236, 237, 249, 252, 281, 298, 301, 303, 305, 306, 307, 310, 315, 316, 327, 353, 355, 358, 359, 364, 367

Tab Index, 236

Table, 16, 30, 36, 46, 53, 60, 62, 70, 76, 78, 81, 82, 83, 84, 91, 93, 107, 118, 119, 120, 121, 122, 123, 125, 126, 127, 129, 130, 150, 152, 153, 154, 159, 160, 161, 163, 178, 180, 183, 233, 239, 263, 317, 327, 345, 347

Tab Order, 235, 236, 237, 249, 252

Template databases, 12

Text Box, 131, 133, 134, 177, 179, 355, 364, 366

Text File, 46, 48

Top 10, 101, 119, 123

Total Row, 34, 129, 139, 140, 177, 179

Transparent Setting, 360

U

Unbound, 133, 200, 227, 228, 356

Unique Value, 162

Update Query, 150, 152, 153, 177, 179

Update To, 150, 152

user friendly, 5, 26, 186, 306, 310

user-friendly, 1, 262, 283

V

Validation Rule, 27

Variance, 129

View, 30, 32, 59, 96, 126, 128, 130, 135, 165, 186, 202, 214, 223, 225, 228, 234, 241, 273, 279, 307, 311, 327, 332, 333, 335, 336, 337, 338, 339, 341, 346, 348

View Button, 59

Z

Zoom, 133, 136, 137, 323

CPSIA information can be obtained
at www.ICGtesting.com
Printed in the USA
FSHW011811200820
73156FS